Computational Psychiatry

New Perspectives on Mental Illness

Strüngmann Forum Reports

Julia Lupp, series editor

The Ernst Strüngmann Forum is made possible through the generous support of the Ernst Strüngmann Foundation, inaugurated by Dr. Andreas and Dr. Thomas Strüngmann.

This Forum was supported by the
Deutsche Forschungsgemeinschaft
(The German Research Foundation)

Computational Psychiatry

New Perspectives on Mental Illness

Edited by

A. David Redish and Joshua A. Gordon

Program Advisory Committee:

Huda Akil, Joshua A. Gordon, Julia Lupp, John P. O'Doherty,
Daniel S. Pine, A. David Redish, and Klaas Enno Stephan

The MIT Press

Cambridge, Massachusetts
London, England

Series Editor: J. Lupp
Assistant Editor: M. Turner
Photographs: N. Miguletz
Lektorat: BerlinScienceWorks

The book was set in TimesNewRoman and Arial.
Printed and bound in the United States of America.

Library of Congress Cataloging-in-Publication Data

Names: Redish, A. David, editor. | Gordon, Joshua A., editor.
Title: Computational psychiatry : new perspectives on mental illness / edited
  by A. David Redish and Joshua A. Gordon.
Other titles: Strüngmann Forum reports.
Description: Cambridge, MA : The MIT Press, [2016] | Series: Strüngmann
  Forum reports | Includes bibliographical references and index.
Identifiers: LCCN 2016033927 | ISBN 9780262035422 (hardcover : alk.
  paper)
Subjects: | MESH: Mental Disorders | Computational Biology |
  Neurosciences--methods
Classification: LCC RC455.2.D38 | NLM WM 26.5 | DDC 616.89/140285--
dc23 LC record available at https://lccn.loc.gov/2016033927

10 9 8 7 6 5 4 3 2 1

Contents

The Ernst Strüngmann Forum

Science is a highly specialized enterprise—one that enables areas of enquiry to be minutely pursued, establishes working paradigms and normative standards, and supports rigor in experimental research. Some issues do not fall neatly into the purview of a single disciplinary field, and for these, specialization can hinder conceptualization and limit the generation of potential problem-solving approaches. The Ernst Strüngmann Forum was created to explore these types of problems.

Founded on the tenets of scientific independence and the inquisitive nature of the human mind, the Ernst Strüngmann Forum is dedicated to the continual expansion of knowledge. Its activities promote interdisciplinary communication on high-priority issues encountered in basic science. Through its innovative communication process, the Ernst Strüngmann Forum provides a creative environment within which experts scrutinize high-priority issues from multiple vantage points.

This process begins with the identification of themes. By nature, a theme constitutes a problem area that transcends classic disciplinary boundaries, is of high-priority interest, and requires concentrated, multidisciplinary input to address the issues. Proposals are received from leading scientists active in their field and reviewed by an independent Scientific Advisory Board. Once approved, the Ernst Strüngmann Forum convenes an advisory committee to refine the scientific parameters of the proposal and select participants. Approximately one year later, a central gathering, or Forum, is held to which circa forty experts are invited. Expansive discourse is employed to approach the problem. Often, this necessitates reexamining long-established ideas and relinquishing conventional perspectives, yet when accomplished, new insights begin to emerge. As a final step, the resultant ideas and newly gained perspectives from the entire process are communicated to the scientific community for further consideration and implementation.

Preliminary discussion for this topic began in 2012 (at the Ernst Strüngmann Forum reception during the Society for Neuroscience Meeting in New Orleans), when I was approached by David Redish and Josh Gordon with the following idea: if we could take what is known about how the brain works (particularly the new mechanistic models of memory and decision making), couldn't we look for failure modes within that mechanistic model to derive a better taxonomy? The idea eventually took on a concrete form and was approved. From May 1–3, 2014, the Program Advisory Committee (Huda Akil, Joshua A. Gordon, Julia Lupp, John P. O'Doherty, Daniel S. Pine, A. David Redish, and Klaas Enno Stephan) met to refine the scientific framework for the Forum and select participants, who were placed into working groups. From June 28–July 3, 2015, this invited group of computational neuroscientists, clinical psychiatrists, and theoretical neurobiologists met in Frankfurt am Main to "discuss how new computational perspectives might be used to broaden our mechanistic understanding of psychiatric dysfunction and improve identification and treatment of psychiatric disorders."

The extended discourse is captured in this volume, which is comprised of two types of contributions. Background information is provided on key aspects of the overall theme. These chapters, drafted before the Forum, were subsequently revised based on formal reviews as well as input received at the Forum. Chapters 3, 5, 10, and 12 summarize the extensive discussions of the working groups. These chapters are not consensus documents nor are they proceedings; they transfer the essence of this multifaceted discourse, expose areas where opinions diverge, and highlight topics in need of future enquiry.

An endeavor of this kind creates unique group dynamics and puts demands on everyone who participates. Each invitee played an active role, and for their efforts, I am grateful to all. A special word of thanks goes to the Program Advisory Committee, to the authors and reviewers of the background papers, as well as to the moderators of the individual working groups (John O'Doherty, Huda Akil, Danny Pine, and Klaas Stephan). The rapporteurs of the working groups (Zeb Kurth-Nelson, Nelson Totah, Shelly Flagel, and Rosalyn Moran) deserve special recognition, for to draft a report during the Forum and finalize it in the months thereafter is never a simple matter. Finally, I extend my sincere appreciation to David Redish and Josh Gordon: as the scientific chairpersons of this 20th Ernst Strüngmann Forum, their commitment and cooperation ensured a most vibrant intellectual gathering.

A communication process of this nature relies on institutional stability and an environment that encourages free thought. The generous support of the Ernst Strüngmann Foundation, established by Dr. Andreas and Dr. Thomas Strüngmann in honor of their father, enables the Ernst Strüngmann Forum to pursue its work in the service of science. In addition, the following valuable partnerships are gratefully acknowledged: the Scientific Advisory Board, which ensures the scientific independence of the Forum; the German Science Foundation, for its supplemental financial support; and the Frankfurt Institute for Advanced Studies, which shares its intellectual setting with the Forum.

Long-held views are never easy to put aside. Yet when this is achieved, when the edges of the unknown begin to appear and the resulting gaps in knowledge are able to be identified, the act of formulating strategies to fill such gaps becomes a most invigorating activity. On behalf of everyone involved, I hope that this volume will convey a sense of this lively exercise and promote further enquiry and cross-field interactions.

Julia Lupp, Director Ernst Strüngmann Forum
Frankfurt Institute for Advanced Studies (FIAS)
Ruth-Moufang-Str. 1, 60438 Frankfurt am Main, Germany
https://esforum.de/

List of Contributors

Ahmari, Susanne E.   Department of Psychiatry, University of Pittsburgh, PA 15219, U.S.A.

Akil, Huda   The Molecular and Behavioral Neuroscience Institute, University of Michigan, Ann Arbor, MI 48109-5720, U.S.A.

Anticevic, Alan   Department of Psychiatry, and Department of Neurobiology, Yale University School of Medicine, New Haven, CT 06511, and Abraham Ribicoff Research Facilities, Connecticut Mental Health Center, New Haven, CT 06511, U.S.A.

Barch, Deanna M.   Department of Psychological and Brain Sciences, Psychiatry, and Radiology, Washington University, St. Louis, MO 63130, U.S.A.

Botvinick, Matthew   Department of Psychology, Princeton Neuroscience Institute, Princeton University, Princeton, NJ 08540, U.S.A.

Breakspear, Michael   Queensland Institute of Medical Research, Royal Brisbane Hospital, Brisbane, QLD 4029, Australia

Carter, Cameron S.   Center for Neuroscience and the Imaging Research Center, University of California at Davis, Sacramento, CA 95817, U.S.A.

Chafee, Matthew V.   Department of Neuroscience, and Veterans Affairs Medical Center, Minneapolis, MN 55455, U.S.A.

Denève, Sophie   Ecole Normale Supérieure, Département d'Études Cognitives 29, 75005 Paris, France

Driesen, Naomi   Department of Psychiatry, Yale University School of Medicine, New Haven, CT 06511, and Clinical Neuroscience Division, VA National Center for PTSD, VA Connecticut Healthcare System, West Haven, CT 06516, U.S.A.

Durstewitz, Daniel   Department Theoretical Neuroscience, Mannheim/ Heidelberg University, Germany, and Faculty of Science and Environment, Plymouth University, U.K.

First, Michael B.   Department of Psychiatry, Columbia University and Department of Clinical Phenomenology, New York State Psychiatric Institute, New York, NY 10032, U.S.A.

Flagel, Shelly B.   Department of Psychiatry, Molecular and Behavioral, Neuroscience Institute, University of Michigan, Ann Arbor, MI 48109, U.S.A.

Frank, Michael J.   Cognitive, Linguistic and Psychological Sciences, Psychiatry and Human Behavior, Brown Institute for Brain Science, Brown University, Providence RI 02912-1821, U.S.A.

Friston, Karl J. Wellcome Trust Centre for Neuroimaging, London WC1N 3BG, U.K.

Glahn, David Department of Psychiatry, Yale University School of Medicine, New Haven, CT 06511, and Olin Center for Neuropsychiatry Research, The Institute of Living, Hartford Hospital, Hartford, CT 06106, U.S.A.

Gordon, Joshua A. Department of Psychiatry, Columbia University and Division of Integrative Neuroscience, New York State Psychiatric Institute, New York, NY 10032, U.S.A.

Harlé, Katia M. Laboratory of Biological Dynamics and Theoretical Medicine, Department of Psychiatry, University of California San Diego, La Jolla, CA 92037-0985, U.S.A.

Huang, Crane Laureate Institute for Brain Research, Tulsa, OK 74136-3326, U.S.A.

Huys, Quentin J. M. Translational Neuromodeling Unit (TUN), Institute for Biomedical Engineering, University of Zurich and Swiss Federal Institute of Technology (ETH), and Department of Psychiatry, Psychotherapy and Psychosomatics, Hospital of Psychiatry, University of Zurich, 8032 Zurich, Switzerland

Kalivas, Peter W. Department of Neurosciences, Medical University of South Carolina, Charleston, SC 29425-6160, U.S.A.

Krystal, John H. Yale University School of Medicine, Yale-New Haven Hospital, New Haven, CT 06511, U.S.A.

Kurth-Nelson, Zeb Wellcome Trust Centre for Neuroimaging and MPS-UCL Centre for Computational Psychiatry and Ageing, University College London, London, U.K.

MacDonald III, Angus W. Department of Psychology, University of Minnesota, Minneapolis, MN 55113, U.S.A.

Maia, Tiago V. School of Medicine, University of Lisbon, 1649-028 Lisbon, Portugal, and Department of Psychiatry, Columbia University, New York, NY 10032, U.S.A.

Malenka, Robert C. Department of Psychiatry and Behavioral Sciences, Stanford Neurosciences Institute, Lorry Lokey Stem Cell Research Building, Stanford, CA 94305, U.S.A.

Mathew, Sanjay J. Mental Health Care Line, Michael E. Debakey VA Medical Center and Menninger Department of Psychiatry and Behavioral Sciences, Baylor College of Medicine, Houston, TX 77030, U.S.A.

Mathys, Christoph Max Planck UCL Centre for Computational Psychiatry and Ageing Research, London, U.K.; Wellcome Trust Centre for Neuroimaging, Institute of Neurology, UCL, London WC1N 3BG, U.K.; and Translational Neuromodeling Unit (TNU), Institute for Biomedical Engineering, University of Zurich and ETH Zurich, Switzerland

Montague, P. Read   Computational Psychiatry Unit, and Department of Physics, Virginia Tech Carilion Research Institute, Roanoke, VA 24016, U.S.A., and Wellcome Trust Centre for Neuroimaging, University College London, London WC1N 3BG, U.K.

Moran, Rosalyn   Virginia Tech Carilion Research Institute, Roanoke, VA 24016, U.S.A.

Murray, John D.   Department of Psychiatry, Yale University School of Medicine, New Haven, CT 06511, and Center for Neural Science, New York University, 4 Washington Place, New York, NY, U.S.A.

Netoff, Theoden I.   Department of Biomedical Engineering, Minneapolis, MN 55455, U.S.A.

Niv, Yael   Department of Psychology and Neuroscience Institute, Princeton University, PNI, Princeton, NJ 08540, U.S.A.

O'Doherty, John P.   Division of Humanities and Social Sciences and Computation and Neural Systems Program, California Institute of Technology, Pasadena, CA 91125, U.S.A.

Pauli, Wolfgang M.   Caltech Brain Imaging Center, California Institute of Technology, Pasadena, CA 91125, U.S.A.

Paulus, Martin P.   Laureate Institute for Brain Research, Tulsa, OK 74136-3326, and Laboratory of Biological Dynamics and Theoretical Medicine, Department of Psychiatry, University of California San Diego, La Jolla, CA, 92037-0985, U.S.A.

Petzschner, Frederike   Translational Neuromodeling Unit (TNU), University of Zurich and ETH Zurich, 8032 Zurich, Switzerland

Pine, Daniel S.   National Institute of Mental Health (NIMH), Bethesda, MD 20892-2670, U.S.A.

Redish, A. David   Department of Neuroscience, University of Minnesota, Minneapolis, MN 55455, U.S.A.

Ressler, Kerry   Department of Psychiatry and Behavioral Sciences, Yerkes Research Center, Emory University, Atlanta, GA 30329, U.S.A.

Schmack, Katharina   Klinik für Psychiatrie und Psychotherapie, Charité-Universitätsmedizin Berlin, 10117 Berlin, Germany

Smoller, Jordan W.   Psychiatric and Neurodevelopmental Genetics Unit (PNGU), Department of Psychiatry and Center for Human Genetic Research, Massachusetts General Hospital, Simches Research Building, Boston, MA 02114, U.S.A.

Stephan, Klaas Enno   Institute for Biomedical Engineering, University of Zurich and ETH Zurich, 8032 Zurich, Switzerland

Thapar, Anita   Institute of Psychological Medicine and Clinical Neurosciences, MRC Centre for Neuropsychiatric Genetics andGenomics, Cardiff University School of Medicine, Cathays, Cardiff, CF24 4HQ, U.K.

Tost, Heike   Central Institute of Mental Health, J 5, 68159 Mannheim, Germany

Totah, Nelson   Department of Physiology of Cognitive Processes, Max Planck Institute for Biological Cybernetics, 72076 Tübingen, Germany

Wang, Xiao-Jing   Center for Neural Science, New York University, New York, NY, U.S.A., and NYU-ECNU Joint Institute of Brain and Cognitive Science, NYU Shanghai, Shanghai, China

Yang, Genevieve   Department of Psychiatry, and Department of Neurobiology, Yale University School of Medicine, New Haven, CT 06511, U.S.A.

Zick, Jennifer L.   Department of Neuroscience, Netoff Epilepsy and Neuroengineering Laboratory, SE, Minneapolis, MN 55455, U.S.A.

Introduction

1

On the Cusp

Current Challenges and Promises in Psychiatry

Joshua A. Gordon and A. David Redish

Abstract

Modern psychiatry seeks to treat disorders of the brain, the most complex and least understood organ in the human body. This complexity poses a set of challenges that make progress in psychiatric research particularly difficult, despite the development of several promising novel avenues of research. New tools that explore the neural basis of behavior have accelerated the discovery in neuroscience, yet discovery into better psychiatric treatments has not kept pace. This chapter focuses on this disconnect between the challenges and promises of psychiatric neuroscience. It highlights the need for diagnostic nosology, biomarkers, and better treatments in psychiatry, and discusses three promising conceptual advances in psychiatric neuroscience. It holds that rigorous theory is needed to address the challenges faced by psychiatrists.

Introduction

Psychiatry attempts to treat mental disorders. Modern psychiatry recognizes that since all mental phenomena are products of the brain, it follows that psychiatry attempts to treat disorders of the brain, the most complex and least understood organ in the human body. In their everyday practice, psychiatrists face a set of challenges that are fundamental to how they care for patients. Nearly every element of the clinical endeavor—from diagnosis, to treatment selection, to monitoring efficacy, to maximizing stability—is fraught with uncertainty. Diagnoses rarely represent specific entities; treatment selection is more like educated guesswork than evidence-based decision making; treatments are at best partially effective; and there are neither objective measures of treatment response nor clear paths to a cure. Although many patients benefit tremendously from the various treatments psychiatrists have at their disposal, many others are left with intractable symptoms that cause considerable morbidity and mortality.

These challenges, while not unique to psychiatric disorders, are driven by the complexity of the brain. This complexity is manifest at multiple levels. At the genetic level, the incredible diversity of cell types, gene expression profiles, and developmental progressions prevents simplistic genotype-phenotype correlations. At the physiological level, linking specific brain processes or physiological states to specific symptoms has been confounded by multiple factors, including a lack of understanding of how circuits perform or how these circuits are wired together. At the level of symptoms, the complexity of human mental phenomena and a relatively poorly developed tool set leaves us without clear ideas of how to characterize and study patients' experiences. Finally, there is the added complexity of how these various levels are bridged—there are multiple potential routes from genes to circuits, and from circuits to behavior.

Despite these challenges, there has been recent cause for optimism. New tools for exploring the neural basis of behavior have revolutionized neuroscience. Novel functional and structural neuroimaging technologies allow us to peer into the brains of patients even as they experience symptoms. Large-scale genetic studies identify literally hundreds of susceptibility genes that correlate with risk for psychiatric disease. Increasingly powerful tools permit the observation and manipulation of neurons and neural circuits in model organisms with exquisite specificity. These tools have facilitated an accelerating pace of discovery in neuroscience over the past two decades.

Yet the pace of discovery in psychiatric treatments has not accelerated. Rather, it has stagnated. Virtually no mechanistically novel treatments have emerged from this explosion in neuroscience. Many drug companies have completely abandoned their psychiatric drug development pipelines, while others have so radically restructured their endeavors that they appear to be starting from scratch.

In this introductory chapter, we focus on the disconnect between the challenges and promises of psychiatric neuroscience. We explore three specific challenges that psychiatry faces. First, psychiatry needs a more accurate, more neurobiologically based, *diagnostic nosology*: before one can treat a disorder, one must know what the disorder is. Second, an informed clinical practice requires *biomarkers*, measurable indicators that are associated with disorders and/or track treatment response. Third, psychiatry needs *better treatments* with enhanced efficacy and reduced side effects. Addressing these challenges would dramatically improve the practice of psychiatry.

After considering these challenges, we discuss three promising conceptual advances in psychiatric neuroscience. First is the notion of *genetics as destiny*. The veritable explosion in genetic information, facilitated by large collaborations and even larger data sets, is clarifying the role of genetic risk in the etiology of many psychiatric disorders. Second, modern neuroscience techniques have led to the clear demonstration that *circuits drive behavior*, inspiring efforts to characterize circuit-level disturbances in patients and in animal models of psychiatric disease. Finally, we will consider *personalized medicine*, which presupposes that quantifiable factors can guide treatment selection and predict treatment response on an individual basis. These three areas are currently the focus of intense effort; the extent to which these efforts can impact on the three challenges above may determine whether psychiatry can overcome the complexity of the brain.

This discussion leads directly into the second chapter (Redish and Gordon, this volume), which will open up discussion on how computational neuroscience might contribute to psychiatry. The premise behind these introductory chapters is that rigorous theory can help fulfill the promises of modern psychiatry in addressing the challenges psychiatrists face. The remainder of the book offers a more detailed, and hopefully compelling, consideration of this premise.

Challenges

Challenge 1: Diagnostic Nosology

A proper diagnosis serves as a crucial starting point in the patient–physician relationship. It determines how the physician approaches the patient, predicts the course and outcome of the illness, and guides treatment planning. Ideally, diagnoses are defined as part of a disease-categorizing system—a nosology—that defines an illness in a manner that is true to its biology. Individuals assigned a given diagnosis should share some common biological feature, or set of features. A diagnosis might imply a specific etiology (e.g., a gene, infectious agent, or dietary deficiency) or a specific pathophysiology (e.g., loss of insulin-secreting cells, elevated blood pressure, or uncontrolled cellular replication). The biological feature serves as a point of attack that allows the physician to understand something about the illness; a biological cornerstone around which the physician can construct a patient-specific care plan. In addition and particularly important for our discussion, the biological feature can serve as an important starting point for research aimed at improving patient care.

Psychiatric nosology lacks these cornerstones. It is not built around biology; rather, it is built upon symptoms. The classification system psychiatrists use today, codified in the fifth edition of the Diagnostic and Statistical Manual (APA 2013), relies on symptom lists. If you have two or more psychotic symptoms (delusions, hallucinations, thought disorder, catatonia, amotivation/flat affect) for at least one month, you have "schizophrenia." If you have five or more depressive symptoms (depressed mood, decreased enjoyment, weight change, sleep change, low energy, feelings of worthlessness, decreased concentration, thoughts of death) for at least two weeks, you have "major depressive disorder." On the face of it, establishing a diagnosis involves checking off symptom lists; for research especially, structured clinical interviews built around these symptom lists are the basis for categorizing individual patient presentations into disorders.

In many ways, this system works well. First and foremost, it provides a framework for classifying patients upon which psychiatrists can agree. It has a high degree of inter-rater reliability, at least as high as many "medical" diagnoses (Pies 2007; Freedman et al. 2013). Second it helps guide treatment and research into new treatments. Diagnostic categories suggest classes of treatments (antipsychotics for schizophrenia; antidepressants for depression), and those treatments can be reasonably effective: antipsychotics reduce psychotic symptoms in 30–70% of patients with schizophrenia (Miyamoto et al. 2002; Lieberman et al. 2005); antidepressants induce remission in 35–40% of patients with major depressive disorder and significant improvement in 50–60% (Rush et al. 2006a). Furthermore, since there is a good inter-rater reliability, research done on a particular diagnostic category can be compared across research groups in a straightforward manner. The results of such studies can often be applied to patients who meet criteria for that category with reasonable expectation of success. Finally, diagnoses are often extremely helpful for patients, allowing them to see that they are not alone in their suffering, and giving them a label to hold on to. The importance of this last point should not be underestimated, especially for psychiatric patients. Naming what they have provides immense comfort to many patients, who otherwise blame themselves for problems and symptoms they often see as integral with their personalities and sense of selves.

In other ways, the current diagnostic nosology does not work nearly so well. Categories have multiple, overlapping symptoms. For example, sleep disturbances are an official diagnostic criterion for major depressive disorder, bipolar disorder, posttraumatic stress disorder, and primary sleep-wake disorders, yet sleep disturbances are also found in other disorders, even if they are not part of the official criteria. The practical impact of this symptom overlap is high rates of comorbidity; some studies estimate that as many as 75% of patients with major depressive disorder are also diagnosed with an anxiety disorder at some point in their lifetime (Lamers et al. 2011). Moreover, some fraction of patients seeking help do not clearly meet criteria for any given disorder, resulting in "catch-all" categories such as "anxiety disorder NOS" (i.e., not otherwise specified). These and other issues decrease the ability of the physician to make reliable predictions as to the course of an illness and the response to treatment, since many patients do not match the "pure" forms of diagnoses typically studied in research protocols.

Perhaps the most troubling aspect of the current diagnostic nosology for psychiatric disorders is the lack of biological relevance. Over and over, attempts to characterize biological correlates of diagnostic categories have by and large failed. Biomarkers based in biology were among the first to be studied; they fail, however, to adhere reliably to disorder boundaries (see below). Modern imaging studies are little different: multiple anxiety, mood, and substance use disorders show enhanced activity in the amygdala (Gilpin et al. 2014); posttraumatic stress disorder, schizophrenia, and depression are all associated with decreased size of the hippocampus (Videbech and Ravnkilde 2004; Smith 2005; Adriano et al. 2012). Similarly, early family studies that demonstrated heritability of psychiatric disease risk showed that genetic risk factors typically predisposed to multiple disorders (Kendler 2006). This is true even for large pedigrees with defined genetic lesions of relatively large effect, such as the Scottish DISC1 translocation pedigree (Brandon and Sawa 2011). Modern molecular genetics confirms what was already known from the family studies at the single gene level: many specific genetic lesions raise risk for multiple diagnoses (Intl. Schizophrenia Consortium 2009; Williams et al. 2011). For example, a calcium channel gene, *CACNA1C*, increases risk for both bipolar disorder and schizophrenia (Green et al. 2010; Curtis et al. 2011); a microdeletion on chromosome 22 raises risk for both autism and schizophrenia, among other psychiatric diagnoses (Schneider et al. 2014). The bulk of the biological evidence makes it very clear that our current diagnostic categories are missing the mark in terms of carving out psychiatric disease at its neurobiological joints.

Challenge 2: Biomarkers

Biomarkers of disease can be the key to accurate diagnosis and optimal treatment. Think of hemoglobin A1c levels in diabetes.[1] This marker, elevated in patients with chronic uncontrolled diabetes, gives an indication of how dysregulated blood glucose levels have been over the recent past. It is key to the diagnosis of diabetes, particularly type II diabetes, the kind with (usually) adult onset and an association with obesity. Like psychiatric disorders, type II diabetes is of complex etiology, with multiple small effect genetic risk factors and a host of possible environmental precipitants. Yet unlike psychiatric disorders, the diagnosis is made more straightforward by testing for hemoglobin A1c levels. Should these levels rise above a threshold, the physician and patient can discuss various interventions. However, the utility of this biomarker does not end with diagnosis. Proper management of blood glucose will result in a gradual decline in hemoglobin A1c levels. By monitoring these levels regularly, the efficacy of the treatment can be tracked over time.

Considerable effort has been expended to try and develop biomarkers for psychiatric disease that might be similarly useful. Among the earliest biomarkers were neuroendocrine markers, such as dysregulation of the glucocorticoid system for major depressive disorder (Plotsky et al. 1998). The dexamethasone

---

[1]   http://www.webmd.com/diabetes/guide/glycated-hemoglobin-test-hba1c (accessed May 5, 2016).

suppression test was one of the first tests to be proposed for any psychiatric disease (Kalin et al. 1981; Hayes and Ettigi 1983). It takes advantage of the negative feedback system instantiated in the hypothalamic-pituitary-adrenal axis; exogenous glucocorticoids (dexamethasone) induce a downregulation of endogenous cortisone. Patients with depression have a relative failure to downregulate cortisone in response to exogenous dexamethasone. However, the dexamethasone suppression test has relatively poor sensitivity (as low as 40–60%) and specificity (as low as 70%) and has not proven useful in clinical prediction (APA Task Force on Laboratory Tests in Psychiatry 1987).

Similar issues have befallen other attempts to develop biomarkers using various modalities. Neurophysiological biomarkers examining, for example, electroencephalographic activity at baseline or in response to various stimulation paradigms have been proposed for schizophrenia (Rosen et al. 2015). Similarly, neurobehavioral tests, such as smooth pursuit eye movements, have been touted as likely possibilities (Calkins and Iacono 2000). Neither has led to usable biomarkers, either due to a failure to reliably distinguish controls and patients when tested broadly, or because of a lack of specificity for schizophrenia. Moreover, several such tests demonstrate schizophrenia-related phenotypes in unaffected relatives. This can be advantageous in that it suggests that these biomarkers reflect the underlying traits that correlate with schizophrenia susceptibility. However, the presence of a biomarker in unaffected relatives suggests limited utility for differentiating individuals with schizophrenia from those with other diagnoses, and for the state of the patient during treatment.

The advent of neuroimaging has led to increasingly sophisticated attempts to utilize patterns of brain structure or activity as biomarkers for psychiatric illness. Here some studies have demonstrated effects of treatment. For example, successful treatment of intractable depression results in the reversal of abnormal patterns of activity in the medial prefrontal cortex, regardless of the type of treatment that was used (Mayberg et al. 1999). Similarly, both psychotherapy and medication treatment of obsessive-compulsive disorder reverses abnormal patterns of activity in the striatum (Baxter et al. 1992). Nonetheless, the general applicability of these findings to clinical situations is unclear. Moreover, as noted above, numerous imaging findings have proven to be nonspecific, with considerable overlap even between seemingly disparate psychiatric disorders.

One possibility for improving upon traditional biomarker studies would be to combine multiple biomarkers and compare across diagnoses. While attempts to do so have not met with success in the past, increasing the power of biomarker studies by using larger data sets, as has been done for genetic studies, may yet meet success (Schwarz and Bahn 2008). Currently, however, it is unclear whether such approaches will yield the kind of useful biomarkers that would aid clinicians in their attempts to diagnose and treat patients.

Challenge 3: Treatments

For psychiatry, treatments are a relative success story. Dopamine 2 receptor-blocking antipsychotics, lithium and other stabilizers, monaminergic boosters, and benzodiazepines have been relatively successful in treating schizophrenia, bipolar disorder, depression, and anxiety, respectively. In addition, there are now many well-studied, tailored psychotherapies for a variety of psychiatric conditions. The best can be at least as efficacious as medication, and for some conditions (e.g., obsessive-compulsive disorder) they can be even better (Foa et al. 2005). Finally, electroconvulsive therapy (ECT) has established itself as a bona fide treatment for mood disorders with proven efficacy (McDonald et al. 2002). Other somatic therapies, from magnetic stimulation to deep brain stimulation, may not be far behind.

Yet for all this success, treatment remains inadequate for many patients, particularly in real-world situations. Most academic studies report treatment response rates of 50–90%. Responses represent significant improvement, which itself is typically defined as some threshold decrease in symptoms, scored on a standardized scale to reduce subjectivity. Remission rates are considerably lower. Remission requires the symptom scores to be lower than would be required to make the diagnosis in the first place. Most clinicians would consider remission to be the goal with their patients, where possible. In most academic studies, remission is achieved only 30–60%, though of course these rates vary by disease and the population studied. Clinical studies that attempt to mimic real-world clinical situations (by allowing for complicating factors such as co-morbid conditions) have even lower rates. For example, in two large U.S. trials, one for depression and one for schizophrenia, remission rates were 30% and 15%, respectively (Sinyor et al. 2010; Levine et al. 2011b). Although some individual treatments for specific disorders can achieve higher rates (response rates to ECT for depression can be as high as 90%; Petrides et al. 2001), on the whole, available therapies leave many psychiatric patients inadequately treated.

Of course, even once remitted, psychiatric disorders can relapse. Relapse rates for most major psychiatric disorders are quite high: 10-year relapse rates for patients successfully treated with antidepressants are as high as 90% if patients stop their medication (Boland and Keller 2002). Even continuing medication is no guarantee of remaining well. Relapse rates on (previously effective) mood stabilizers, for example, can be as high as 50% (Keck and Manji 2002).

Even when current treatments work, tolerability becomes a significant issue. Side effects of psychiatric medications can be considerable. For antipsychotic medications, weight gain, hyperglycemia, and motor symptoms can be significant. For antidepressants, weight gain and sexual dysfunction are often given as reasons why medications are dropped. Anxiolytic medications can be addictive. In addition, tolerability is not just an issue for medication. Many patients have considerable memory loss with ECT, and compliance with

psychotherapies can be difficult. The costs in time and money of psychotherapy are also of concern to some patients. Finally, few psychiatric medications
work quickly; many take weeks or even a few months to exert their effects,
leaving patients suffering considerably even after treatment has been started.

Given the considerable weaknesses of currently available treatments for
psychiatric disorders, one would expect considerable activity aimed at improving upon them. Unfortunately, there has been little truly novel treatment
development for some time. Most current medications were either discovered
to be efficacious by happenstance—like the first antipsychotics, mood stabilizers and antidepressants, which were all under study for other purposes before accidentally being found to be helpful in their respective areas—or are
"me-too" drugs designed to tweak the molecular structure of existing drugs
but relying on the same underlying mechanism. Even research in somatic and
psychological treatments suffers from the me-too problem, with the successful
treatment-du-jour being tried repeatedly for a number of different disorders,
on the premise that if it works for one thing, it just might work for them all.
Meanwhile, as noted above, pharmaceutical companies have little in the development pipeline after multiple, high-cost failures (Insel 2011). Part of the issue
is the difficulty of translating findings from animal models into the human; the
history of innovation in psychiatric pharmacotherapy is littered with examples
of therapeutics that worked wonderfully in rodent models but failed in clinical
trials (Hyman 2014).

Promises

Promise 1: Genetics as Destiny

Psychiatric disorders are overwhelmingly familial, with inheritance rates estimated at 30–70%, depending on the diagnosis (Kendler 2006). Understanding
this inherited risk has been incredibly difficult. For a long time it was not at
all clear where the destiny of psychiatric genetics led: to the holy grail of a
thorough understanding of the neurobiological basis for psychiatric disease, or
the trash heap of promising but eventually discarded technological advances.
Early attempts at identifying psychiatric risk genes failed, with a few notable
exceptions, such as Huntington disease and some Mendelian forms of autism.
Several factors contributed to these failures, including incorrect assumptions
about the form of genetic risk (simple vs. complex genetics), as well as the fact
that genetics crosses diagnostic boundaries, as noted above. More recently,
some success has been made in identifying the genes that contribute to disease
risk, but this has not yet had an impact on psychiatric practice.

Gene identification has benefited from several developments. First, the genomic era has dramatically reduced the price of genotyping while increasing
its speed and accuracy. Second, geneticists have realized that progress requires

enormous sample sizes, which are facilitated by these technical advances. Third, these same geneticists have formed large international collaborations to generate samples of sufficient size to carry out such studies. One recent genome-wide study of schizophrenia used samples of nearly 40,000 cases and over 100,000 controls (Schizophrenia Working Group of the Psychiatric Genomics Consortium 2014).

With this increase in scale comes a greater understanding of the landscape of genetic risk, at least for schizophrenia and autism. Risk genes fall into two categories: common alleles of small effect size and rare alleles of large effect size. The common alleles are frequently present in the general population; any one risk allele raises the risk of disease only slightly (on the order of a few percentage points increased risk). The 108 loci identified in the above study are of this sort, each conferring a very small amount of risk; estimates based on modeling suggest that there may be as many as 2000 risk loci of the common allele, small effect size variety for schizophrenia.

The rare alleles occur very infrequently in the general population, often arising *de novo* in the patient and not inherited from either parent. The presence of one of these rare alleles signifies a considerable increase in risk—as much as 30-fold (3000%!) (Karayiorgou et al. 1995). Many of these rare alleles are copy number variants (large deletions or duplications of many genes). Others are sequence variants identified only with whole exome (all expressed genetic material) sequencing. It is unclear how many of these rare variants exist; for autism, hundreds have already been identified (Sebat et al. 2007; Levy et al. 2011; Iossifov et al. 2014).

Of course, feeding into this complexity is the fact that these genetic studies are being conducted on samples derived from current diagnostic criteria, which as noted above are crude. At least part of the complexity might be reduced by an improved diagnostic system. Indeed, it may be that genetic information could be used to improve diagnostic nosology, given that it is inherently biological. The complexity of the genetic landscape may be already apparent in the numbers of identified genes, which provide an embarrassment of riches to those wishing to use the clues to unravel the neurobiology of psychiatric illness noted above. Yet the sheer number of genes represents only the tip of the iceberg in terms of complexity. The relationship between genotype and phenotype is likely to be complex. A given gene can result in different psychiatric phenotypes in different patients, and nearly identical phenotypes can be caused by remarkably different genotypes. Gene–environment interactions and epigenetic modifiers complicate matters even further.

Methods are needed to address this complexity. The first attempts have been aimed at organizing these large numbers of genes into pathways and networks, so that their effects on biology can be understood. But these attempts are far removed from the behavioral endpoints of psychiatric disease. In the end, translation from genetics to behavior occurs through neural structures, which are fundamentally about the computations that support behavior. Understanding this translation will be fundamental to demonstrating how genetics influence risk for psychiatric disease.

Promise 2: Circuits Drive Behavior

Increasingly, the focus of neuroscientists trying to understand how the brain produces behavior has been drawn to the level of the neural circuit. Neural circuits may be localized within a given brain region or distributed across several areas. Their building blocks are neurons of various, specific types, as well as the neural processes and synapses that connect them. Circuits have the potential to carry out neural computations; that is, to take in information, transform that information into commands, and output those commands appropriately, driving behavior. The ascendency of circuit-based analyses in neuroscience has led to the corresponding circuit hypothesis of psychiatric disease, in which abnormal function of neural circuits leads to psychiatric symptomatology (Ressler and Mayberg 2007; Akil et al. 2010). Methods to change circuit function thus become more than just research tools, but potential therapies.

This focus on circuits has been driven by technological advances in animal models. Anatomical techniques, starting with Golgi and Ramon y Cajal and progressing to engineered viruses and specialized tissue processing and microscopy techniques, enable a description of the fine details of neurons and the connections which make up circuits. Physiology, beginning with electrophysiology but now including fluorescent activity indicators, permits the monitoring of these circuits with exquisite specificity. Genetic manipulation and molecular biology facilitate increasing knowledge regarding the cellular machinery underlying the formation and maintenance of circuitry. More recently, optogenetic and pharmacogenetic technology permits the manipulation of circuit function with cellular, anatomical, and temporal specificity. Using these tools, the precise wiring diagram for a given circuit can be mapped, the activity patterns of each of the elements in the circuit can be monitored during behavior, and these patterns can be mimicked or interrupted to test whether they are necessary and/or sufficient for the behavior.

A considerable amount of these efforts are directed at circuits and behaviors with relevance to psychiatry, albeit in animal models. Studies of depression- and anxiety-like behaviors have implicated amygdala and prefrontal circuits as well as neuromodulatory centers (Tye et al. 2011, 2013). The circuits underlying social behavior have been explored, with attention paid to some of these same brain regions (Yizhar 2012). Cognitive behaviors disrupted in schizophrenia, particularly working memory and executive function, have also been examined (Cho et al. 2015; Spellman and Gordon 2015). The rapidly expanding tool set of the circuit neuroscientist has given traction to efforts to understand the complex neurobiology underlying these phenomena. Importantly, these findings demonstrate that the key building block of behavior is the neural circuit. Perhaps more importantly for psychiatrists, they also show that one can have tremendous impacts on behavioral output with relatively specific manipulations at the circuit level.

Meanwhile, cutting-edge clinical work would suggest that the focus on circuits might translate to patients and be useful to clinicians. Neuroimaging studies clearly point out that these same brain regions, important for specific behaviors in the rodent, are engaged in humans in similar tasks, and often are dysregulated in patients. Moreover, imaging and other experiments in humans can help make the connection between circuit function and subjective experiences that play such an important role in psychiatric disorders and yet can be challenging to study in animal models. Finally, studies of the effects of various therapies—whether pharmacological, psychotherapeutic, or brain stimulation—demonstrate that it is possible to modulate activity at the level of the (macro) circuit in humans (Baxter et al. 1992; Mayberg et al. 1999).

Promise 3: Personalized Medicine

Personalized medicine is a movement throughout the entire field of medicine to tailor therapy to the individual patient. In some ways this is an old idea, expressed in a new way. A physician has many options to offer to treat the hypertensive patient. One guide can be simply taking the patient's pulse. Patients with hypertension and low resting heart rate tend not to respond well to beta blockers, drugs that block the beta-adrenergic receptor. For such a patient, choosing an alternative medicine represents a form of personalized medicine.

The modern advance is to consider not just the specifics of the illness but the specifics of the patient as well. For example, certain patients metabolize certain medications faster, or slower, meaning that doses should be adjusted or medications avoided. Metabolizer types could be identified from the genome or tested biochemically. Beyond metabolism, biomarkers could help stratify patients into those more likely to contract a specific subtype of an illness, and/or respond to a particular treatment.

Currently in psychiatry, treatment selection is not guided by such information. Patients are evaluated and diagnosed, and treatments are applied to symptoms. But the selection of a given treatment has more to do with the avoidance of side effects (e.g., a patient's preference weight gain or insomnia) than with the efficacy of the medication for any particular kind of patient. Better diagnostics would help, if improved diagnostics would lead to improved predictions about treatment responses. But even in the absence of improved diagnostics, it may be that certain biomarkers—genetic or otherwise—would help guide treatment selection in a meaningful way.

There are some examples of this kind of research in psychiatry. For example, early behavioral subtyping of depression led to the demonstration that patients with atypical depression, which is characterized by mood reactivity, will respond better to monoamine oxidase inhibitors than to tricyclic antidepressants (Liebowitz et al. 1988). The true promise of personalized medicine is that buried somewhere in the immense data sets being collected lies the secret to determining which treatment will result in the best response for a particular, individual patient.

Waiting for Godot

The disconnect between the promises and challenges of psychiatric neuroscience begs for a solution, for a savior—some breakthrough that will solve all the problems of psychiatry in one fell swoop. Indeed, the history of psychiatric research is riddled with failed saviors, from psychoanalysis to behaviorism, to pharmacology, to neuroimaging, to molecular genetics, each hailed by its own generation as the miracle that will help us define and understand mental illness.

The truth is that there are no saviors. No all-encompassing breakthrough will lead us down the garden path toward improved understanding and better treatments. The complexity of the brain stands in our way. Psychiatric disorders are the products of hundreds of genes and thousands of cell types and millions of connections. Answers will surely not come from one sole technological advance, and they will most assuredly be as complex as the questions.

Nonetheless, computational neuroscience may be poised to influence the field, to help the promises of psychiatry overcome the challenges. Genetics has generated lists of hundreds of genes that raise risk for schizophrenia, autism, and other psychiatric disorders. How can we organize and understand these genes? Circuits are the fundamental building blocks of behavior. How can we understand which circuits are broken in mental illness? How can we learn enough about these circuits to repair them? Finally, patients may respond better if therapies are tailored to their unique biology. How do we learn enough about any individual patient to guide treatment appropriately?

In this volume, we explore the potential role that computational approaches can play in addressing these questions. We will wonder openly whether and how understanding the biological system of the brain through a set of rigorously constructed and quantitatively tested theoretical constructs will help clarify diagnostic issues by identifying where these biological systems can break down. We will contemplate how such an approach could lead to biologically and pathophysiologically relevant biomarkers. We will attempt to generate ideas about how psychiatry can help models become more focused on issues of importance to patients and physicians. And we will speculate on what a success might look like, what form the first "killer app" of computational psychiatry neuroscience might take. We know that computational psychiatry will not be the next savior, but we hope, at least, that failure might be averted through a careful consideration of how to use computational approaches to address the challenges psychiatry faces.

2

Breakdowns and Failure Modes

An Engineer's View

A. David Redish and Joshua A. Gordon

Abstract

Psychiatry faces a number of challenges due largely to the complexity of the relationship between mind and brain. Starting from the now well-justified assumption that the mind is instantiated in the physical substrate of the brain, understanding this relationship is going to be critical to any understanding of function and dysfunction. Key to that translation from physical substrate to mental function and dysfunction is the computational perspective: it provides a way of translating knowledge and understanding between levels of analysis (Churchland and Sejnowski 1994). Importantly, the computational perspective enables translation to both identify emergent properties (e.g., how a molecular change in a receptor affects behavior) and consequential properties (e.g., how an external sociological trauma can lead to circuit changes in neural processing). Given that psychiatry is about treating harmful dysfunction interacting across many levels (from subcellular to sociological), this chapter argues that the computational perspective is fundamental to understanding the relationship between mind and brain, and thus offers a new perspective on psychiatry.

The Computational Perspective

Fundamentally, the computational perspective is about how information is processed within neural circuits; it uses formal methods to identify how inputs and recurrent processing combine to create outputs. By being formal, the computational perspective enables an explanation and understanding of neuropsychology in its elemental basis so that we can identify measurable changes and determine where breakdowns occur. This perspective allows us to define *computational psychiatry* as a methodology using formal computational perspectives to address psychiatric dysfunction. It is only with computational explanations of function that we can begin to identify how physiological, sociological, and other changes can lead to *dys*function.

The computational perspective hypothesizes that the role of brains is to perform computations to improve behavior. For instance, maintaining thermodynamic equilibrium is a complex computational process (Goldstein and McEwen 2002), as is sensory recognition for improved motor control (Llinas 2001), escape from predators or the identification of prey (Eaton 1984), and social prediction for interaction with conspecifics (Cheney and Seyfarth 1990). To understand how the brain computes these mental processes, we need to understand the mechanism of the computational process. While the definition may sound tautological, the key to the computational perspective, whether in computational neuroscience or computational psychiatry, is that it requires explanations to be specified in a formalism that forces a more complete story and often reveals obscure consequences. The process by which pathologies in physiological processes engender pathologies in psychological processes is often not obvious: complex consequences can only be derived from computational models and formal analyses.

When talking about a computational perspective, it is important to make clear what it is not. Although computer models can play important roles in computational psychiatry, it is possible to construct computational formalisms that provide explanations without an explicit computer model of pathopsychology. For instance, Kurth-Nelson et al. (this volume) describe examples of formalisms of families of models that are all susceptible to specific pathophysiologies. Similarly, computational psychiatry is more than applying computer algorithms to large data sets, such as clustering genetics or behavioral distributions or what is known as "big data" (Pevsner 2005; Schadt et al. 2010; Mayer-Schönberger and Cukier 2013). Neither computer models nor algorithmic clustering of large data sets provides understanding (although both can guide and test development of theories). For instance, Flagel et al. (this volume) provide a novel formalism for nosology which implies a clear use for big data in deriving consequences of diagnosis and treatment, but emphasizes the importance of the computational perspective to provide the underlying latent constructs.

We can define a theory as an explanation of how an observation arises from a lower-level phenomenon, such as how Parkinsonian movement disorders could arise from circuit changes derived from changes in dopaminergic tone (Albin et al. 1989). Models can then be used to test these theoretical hypotheses. For instance, the theory that Parkinsonian movement disorders arise from depletion of dopaminergic tone implies that one can create an animal model of Parkinson disease by depleting dopamine from an otherwise normal animal subject (Langston et al. 1984; Deumens et al. 2002). This animal model is really creating an analogous situation to that of Parkinson disease, which can then be explored. Similarly, a computational model can be used as an analogous situation which can then be explored. For instance, in a recent paper, Schroll et al. (2015) examined a detailed computational model of three theories of the progress of Huntington disease and found that only a progressive degeneration of medium spiny neurons in the direct and indirect pathways provides compatible behavioral deficits to those seen in real patients. However, just as the animal model embodies a theory of Parkinsonian mechanism, but is not the theory of Parkinson disease, neither does a computational model deliver a theory of Huntington disease. Rather, these models are only interpretable when taken from a theoretical (computational) perspective about the computations being performed by the direct and indirect pathways of the basal ganglia (Albin et al. 1989; Kravitz et al. 2012).

Importantly, it is not necessary to provide an explanation down to the cellular, subcellular, or molecular level; the explanation has to be focused at the appropriate explanatory level. For instance, it is well established that hippocampal cells encode information about the spatial location of an animal ("place cells"; O'Keefe and Dostrovsky 1971; O'Keefe 2015). What mechanisms make these cells fire in their given locations is an interesting scientific question. However, if we want to know how a complex firing pattern in hippocampal place cells drives behavior, then it is not necessary to know why the place cells fire in that complex pattern, only that they do. An appropriate theory would start from the statement that hippocampal place cells show a complex spatially related firing pattern and use that to explain how changing those firing patterns changes navigational processes.

The usefulness of computational models should not be underestimated. Computational models force researchers to develop precise, explicitly specified, falsifiable hypotheses. Although it is the theoretical statements (neural mechanism X implies behavioral change Y, behavioral incident Z creates neural mechanism X) that will necessarily drive understanding and treatment, these derivations are rarely straightforward and rarely simple. If you think that neural mechanism X can drive behavior Y, then it should be possible to build a model of X that can perform behavior Y. There are many cases in neuroscience where people have thought that two effects were incompatible, only to find them compatible after a computational model was built (e.g., space and memory in hippocampus; Redish 1999). Biology is complex and neuroscience particularly so. It is dangerous to simply infer backward from symptoms to dysfunction. Flagel et al. (this volume) and Moran et al. (this volume) suggest a new computational formalism to provide a more nuanced inference process through computational mechanisms to connect symptoms, dysfunction, diagnosis, and treatment.

What Does the Computational Perspective Provide?

Computational perspectives bring two things to the table that we believe fundamentally change psychiatry. First, a computational perspective promises to ask different questions about patients than traditional clinical perspectives. These novel questions can guide diagnosis and treatment by getting at fundamental psychological and neural processes which cut across symptomatic and diagnostic boundaries. Computational perspectives suggest that the fundamental question we should be asking is: *What is different about how this patient processes information about the world (including the patient's self)?*

Second, computation is a way of addressing how mechanisms translate between levels, such as how a change in neural structure can lead to a change in behavior, or how an external incident can lead to changes in neural processing. For instance, a deficit in a certain ion channel in a specific neural structure changes the excitatory-inhibitory balance within that neural structure, and can produce behavioral changes observable at a macro level as epilepsy (Soltesz and Staley 2008). Similarly, externally induced (sociological) stress produces changes in hormone levels, neural circuits, and thus the computations a patient makes about interactions with the world (Payne et al. 2007).

These computational perspectives can help psychiatry incorporate increasingly more biological mechanisms into its categorization and treatment. However, as argued in this book, the computational perspective goes beyond connecting biology to psychiatry; psychiatric disorders themselves should be couched in terms of disorders of information processing and computations.

The Failure Mode Hypothesis

Psychiatry starts from the concept of *harmful dysfunction*; that is, there is some underlying dysfunction in the system that is serious enough to warrant intervention and treatment (Wakefield 1992a, 2007; Flagel et al., this volume). Defining psychiatry from dysfunction implies that one must first understand *function* before one can see how it has become disrupted.

We will not assume that the starting dysfunction always proceeds from brain to behavior. For instance, many theories of gambling and addiction suggest that dysfunction arises from interactions between functional neural processes and dysfunctional external situations for which humans have not evolved to accommodate (Wagenaar 1988; Redish et al. 2007; Schüll 2012). While altering the brain changes the mind, the physical nature of the mind implies that altering the mind also changes the brain. Thus, translating across levels (whether from brain to mind or mind to brain) requires an understanding of the computational mechanics of that system, and how those computational mechanics change when underlying structures change. This computational perspective suggests that we can take the engineer's view and ask new questions about how the system can break down.

Applying an engineer's analysis to a specific dysfunctional or misbehaving system means trying to discover what the changes are in the system that have created the problematic behavior. These potential ways that a system can break down are known as "failure modes." Colloquially, we can think of this as: *Where are the weak links? Where and how does this system typically break?*

Importantly, errors (failure modes) can exist at many levels. Huntington disease is a genetic abnormality, a CAG repeat in the Huntingtin gene (Kremer et al. 1994). Parkinsonianism occurs from loss of dopaminergic function, which can arise from genetic dysfunction (Gasser 2009; Shulman et al. 2011) or an external toxin, such as MPTP (Langston et al. 1983; Langston and Palfreman 2013). Physical trauma (e.g., traumatic brain injury) creates an abnormality in the physical structure of the network. Mental trauma (e.g., from prolonged solitary confinement) creates changes in underlying structure leading to a change in the function of neural circuits (Grassian 1983). Drug addiction is an interaction between external causes (the drugs) and internal effects (neural susceptibility). All of these effects, however, fundamentally affect behavior by altering the brain's computation. To understand how these effects occur, we have to understand that computation and how it becomes altered.

From a treatment perspective, the engineering analysis attempts to find the levers of control: *Where are the points that can provide the optimal means of changing the system back into function?* Just as there can be errors (failure modes) at many levels, treatment can occur at many levels: pharmacological manipulation (Schatzberg and Nemeroff 1995), circuitry manipulations (Obeso and Guridi 2001; Mayberg et al. 2005), physical or mental training (Bickel et al. 2011), or even reinterpretations engendered by cognitive reprocessing (Ainslie 1992, 2001; Kurth-Nelson and Redish 2012) or changes in social interactions (Heyman 2009; Petry 2012). All treatments change the underlying physical and mental structure and thus the computational processing. Understanding how the computational process changes in the face of treatment is an important step in understanding when and how treatment should be used.

Computational Perspectives throughout the Components of Processing

We can apply computational perspectives to all aspects of a person's interaction with the world. New computational perspectives have changed how we understand many processes, including decision making, memory, perception, and action.

For instance, computational and theoretical neuroscientists working on decision making have now garnered considerable evidence that there are several different action-selection systems, each of which processes information about the world differently (Redish 2013). Errors can occur within each of these different systems as well as in the interaction between these systems. Failure modes of the decision-making system will arise with drug addiction (Redish et al. 2008), emotional disorders, such as anxiety and depression (Huys 2007; Rangel et al. 2008) as well as motivational disorders, such as obsessive-compulsive disorder (Pitman 1987; Maia and McClelland 2012). Different treatments will be needed for different failure modes (Rangel et al. 2008; Redish et al. 2008; Redish 2013). On the other hand, it may be possible to use an intact action-selection system to counter a failure mode within one of the other systems, either through training (Bickel et al. 2011, 2015) or through changing the situations the subject is experiencing (Heyman 2009; Kurth-Nelson and Redish 2012; Petry 2012).

Although there has been a tremendous amount of work in the relationship between decision making and psychiatry (Huys 2007; Rangel et al. 2008; Redish et al. 2008; Maia and Frank 2011; Montague et al. 2012; van der Meer et al. 2012b; Redish 2013), a similar computational perspective can be applied to other neural components. For instance, one can derive computational explanations for disordered thinking in schizophrenia (Seamans and Yang 2004; Tanaka 2006; Durstewitz and Seamans 2008). Computational perspectives on perception as signal detection (Tougaard 2002), information derivation (Poggio and Bizzi 2004; Serre et al. 2007), or situation recognition and categorization (Redish et al. 2007; Gershman and Niv 2010) can be used to explain disorders of perception, such as in hallucinations (Bressloff et al. 2002), migraine auras (Reggia and Montgomery 1996; Dahlem and Chronicle 2004), or an inability to recognize social cues (Dapretto et al. 2006; Singer 2008).

What Can Computational Perspectives Provide to Psychiatry?

The companion introductory paper (Gordon and Redish, this volume) describes psychiatry as facing three current challenges and notes three current promises being incorporated into psychiatry. We believe that the computational viewpoint can provide help with these challenges and provides novel perspectives on these promises.

Challenge 1: Nosology

Psychiatry has long noted the difference between reliable and valid categories (McHugh and Slavney 1998). A reliable category is one where membership can be reliably assigned. In contrast, a valid category is one that reflects an underlying similarity in process or outcome. As noted in the companion paper (Gordon and Redish, this volume), psychiatry has a high degree of inter-rater reliability that is similar to many medical disorders, but the efficacy of psychiatric treatments lags that of many other medical disorders. This seems to be because the categories (while reliable) do not conform to either biological or treatment boundaries.

In part, this lack of efficacy is because psychiatry has long found itself trapped between the necessity of using categorical constructs for diagnosis (e.g., ICD-10 or DSM-IV-TR) and more parsimonious dimensional constructs to explain behavior (Krueger 1999; Insel et al. 2010; see also chapters by First, MacDonald et al., and Friston, this volume). The computational perspective can provide a novel solution to this category/dimension complexity by integrating both together.

Nosology has a number of goals ranging from communication (between clinicians, as well as to patients and their families) to guiding diagnosis and treatment (see chapters by First as well as Flagel et al., this volume). In part, categories have arisen because of their enhanced simplicity in identifying diagnosis (for insurance reimbursement purposes) and treatment. Given that treatment is in the end an action taken, treatment is a categorical decision (either you treat or you don't). However, the computational perspective provides a more nuanced perspective on this categorical decision.

Several chapters in this book lay out a new computational nosology based on Bayesian inference linking underlying dimensional constructs with categorical diagnoses and actions (Flagel et al., Moran et al., and Friston, this volume). In short, an underlying set of dimensional constructs (computational constructs) predicts observations through a well-defined (formal) mathematical definition known as Bayesian inference (for mathematical details, see chapters by Mathys and Friston, this volume.) This inference process allows inferences to proceed both from observations to constructs and from constructs to observations. In this formulation, observations can be measurements (such as from a biological or behavioral test), psychological instruments (such as answers on a questionnaire), or diagnoses.

Importantly, in this formulation, diagnoses are seen not as a direct reflection of the fundamental dysfunction, but rather as clinical observations that arise from underlying (computational) dysfunctions. For instance, the hypothesis that repeated drug use can arise from many potential failure modes in different neural systems suggests that the clinical identification of dependence is only a partial predictor of the potential underlying dysfunctions (Redish et al. 2008). Similarly, treatment is seen as changing the trajectory of the dimensional (computational) constructs over time and thus as changing the future observations (symptoms).

This new *computational nosology* relies on two aspects of the computational perspective: the multifarious nature of the relationship between the underlying dysfunctions and diagnoses, and the computational nature of the underlying constructs.

Challenge 2: Biomarkers

The computational nosology hinted at in the preceding section and developed fully by Flagel et al., Moran et al., and Friston (this volume), imply a new perspective on biomarkers. The use of computation to drive nosology suggests that computational perspectives should provide new opportunities to identify differences along construct dimensions and thus to measure differences between categories. A classic example is the way that an EKG measures the dynamics of the heart electrophysiology, allowing separation of chest pain.

Heart attack and indigestion both create chest pain, but require different treatments—having an EKG to measure heart function can be critical to identifying the appropriate treatment. One of the main advantages of the new nosology proposed in several chapters in this volume (see MacDonald et al., Flagel et al., Moran et al., and Friston) is that biomarkers don't necessarily have to define categorical diagnoses, as long as they can be used to drive predictions about prognosis and/or treatment response.

It is likely that computation itself will provide new (bio)markers for use throughout psychiatry. In fact, those markers do not have to be biological per se, but might instead be detectable through behavioral tests or even instrumental questionnaires. As noted above, the computational perspective changes the question of psychiatric function and dysfunction to *"What is different about how this patient processes information about itself or about the world?"* This suggests that measures of computational processing can be used to differentiate patients and treatment. For instance, smokers deal with counterfactual (could-have-been) rewards differently than nonsmokers (Chiu et al. 2008). Huntington patients are less able to compensate for force changes applied to motor activities (Smith et al. 2000). And drug-dependent users (on average) discount future outcomes at much faster rates than non-users (Kirby et al. 1999; Odum et al. 2002). However, within any drug-dependent population, some users do show normal discounting rates. Recently, an analysis of drug-addiction treatments found that the more successful treatments normalized the discounting rates of the subset of users who were discounting overly fast (Bickel et al. 2014). This tells us two things—first, that these treatment processes are not just selecting for the subset of users with normal discounting rates (so we can send both fast and slow discounting users to treatment), and second, that there is a relationship between treatment success and changes in discounting rate (which imply a potential marker for treatment success, at least in a subset of patients).

Computational perspectives can also be used to identify how biomarkers produce their effects and what effects they are likely to produce. For instance, changes in genetic underpinnings of dopamine receptor efficacy (in D1, D2, and the COMT variation) produce differences in the efficacy of learning strategies—genetic changes in D1 drive learning from positive rewards, whereas genetic changes in D2 drive learning from punishment signals, and the COMT variation affects ability to reverse responses (Frank et al. 2007a). From a computational understanding of dopamine's role in driving learning, the effects of Parkinson disease (decreasing dopaminergic tone, leading to lower signal-to-noise ratios between phasic bursts of dopamine and baseline levels, making it hard for neurons to recognize phasic increases in dopamine) and levodopa treatment (increasing dopaminergic tone, but reducing the depth of the drop in dopamine that occurs with undelivered reward, thus making it harder for neurons to recognize decreases in dopamine signals) can be predicted (Frank 2011; Moustafa and Gluck 2011).

These predictions have been confirmed experimentally (Frank et al. 2004; Shohamy et al. 2006; Kéri et al. 2010). Whether they can be used to guide treatment remains an open question.

In contrast, the evidence that hippocampal size is a biomarker for a vulnerability to posttraumatic stress disorder (PTSD) in the face of trauma (Gilbertson et al. 2002) does not explain why decreased hippocampal size predicts that vulnerability. Computational explanations of hippocampal dysfunction, such as inability to recognize context (Nadel and Jacobs 1996; Jacobs and Nadel 1998) or an inability to consolidate memories (Redish 2013), may provide some clues. Computational analyses of hippocampally dependent behaviors would likely predict that vulnerability as well, and may be easier to test potential patients behaviorally than with structural MRI scans. (Imagine, e.g., if every soldier was tested for a vulnerability to PTSD by a behavioral measure of hippocampal abilities before being sent into combat).

It will be important to determine which of these computational changes are trait effects (preexisting within the individual) and which are state effects (thus temporary effects due to the physical, mental, and pharmacological situation in which the individual finds itself). While it has long been hypothesized that discounting impulsivities could drive addiction (Ainslie 1992; Belin et al. 2008; Odum and Baumann. 2010), the fact that discounting changes normalize after treatment (Odum et al. 2002; Bickel et al. 2014) suggests that discounting differences with addicts may be more state effects than trait effects. Both state effects and trait effects can be useful biomarkers. At this point, most computational biomarkers have not been as thoroughly examined as the discounting of future options, but computational biomarkers have been proposed for other dysfunctions as well, such as personality disorders and sociopathy differences that can be tested through economic games such as the ultimatum, trustee, or dictator games (Kishida et al. 2010).

Challenge 3: Treatment

Computation provides new perspectives not only on dysfunction (failure modes) but also on how the system can change (where the levers are). This means that computational analyses of treatment paradigms could provide better explanations for how those treatments are working, which can suggest new ways to improve them. Computational analyses should also provide better explanations for which dysfunctions will be ameliorated by treatments, which can suggest better assignments of patients to treatment.

As nosologies and biomarkers are improved, it should become easier to assign specific treatments to specific patients. In addition, we suspect that new treatments will be developed as the underlying computational dysfunctions are identified and as the available manipulations are identified. Computational analyses of treatment will suggest which subsets of patients (i.e., which failure modes) will be best ameliorated by a given treatment. At the extreme, this leads to personalized medicine—identifying a specific battery of treatments optimized for the specific computational processes underlying a given person's mind and brain.

At this point, the computational implications of treatment are only just starting to be explored. For instance, Regier and Redish (2015) suggest that the treatment of contingency management, in which rewards are provided to addicts for staying clean of drugs, is unlikely to be working primarily through the basic economics of making drugs more expensive (taking drugs now loses the addict the alternate reward as well as the usual cost of the drugs), because a computational analysis showed that the alternate rewards are too small. Instead, they suggest that contingency management creates an explicit choice between two concrete rewards (small as one of them may be). Explicit choice ("Take drugs or get that gold star") tends to activate different decision-making systems that are dependent on different neural circuits from Go/No-Go decisions ("take drugs or don't"). While it is not known whether this computational explanation for contingency management is correct or not, this hypothesis suggests that contingency management success would depend on the neural circuits that drive explicit choice behaviors (such as prefrontal–hippocampal interactions). These circuits could be examined (e.g., through structural or functional imaging), and if impaired, could be improved, either through training or through pharmacological means. There is some evidence that prefrontal integrity protects against relapse after treatment (Camchong et al. 2014), and animal models have suggested that pharmacological interventions can change prefrontal integrity (Dalley et al. 2004). Because of the computational understanding of the processes that underlie explicit choice decision making, it is also possible to suggest ways to change the contingency management treatment itself. For instance, making delayed options more explicitly concrete activates the prefrontal–hippocampal interaction and makes people more willing to wait for delayed rewards (Peters and Büchel 2010). This suggests that very concrete options would be particularly effective in contingency management.

Promise 1: Genetics

As noted in the companion paper (Gordon and Redish, this volume), there is a strong genetic component to psychiatric disorders. However, the connection from genetics to behavior (and worse, to dysfunctional behavior) is long and arduous and depends on interactions with many other environmental and social components. It is rare that a single genetic abnormality translates to a behavioral disorder (such as Huntington disease), but even in those cases, the behavioral consequences can be quite complex and varied. In other disorders, hundreds of genetic markers have been found, which suggests that these markers are being transformed through some intervening substrate to generate the behavior.

The computational perspective provides an access point to this intervening substrate: genetics change the physical nature of neural circuits, which changes how they compute. Thus, if we want to understand the role of genetics in psychiatric disorders, we need to use computation to connect genetics with neural circuits and neural circuits with behavior. Some computational progress has been made (e.g., in epilepsy) where many genetic abnormalities have the common effect of changing the balance between excitation and inhibition, which creates a mathematical instability and the potential for a sudden shift from one state (balance) to another (seizure) (Soltesz and Staley 2008).

Promise 2: Circuits

The translation through neural circuits suggests the possibility of identifying psychiatric disorders at the level of those circuits themselves. But just as one needs computation to translate from genetic changes to functional changes in neural circuits, one needs computation to translate from physical changes in neural circuits to behavioral changes. Computational neuroscience has been quite successful over the last several decades identifying how functional circuits create behavior, particularly in the context of normal function and constructed dysfunction.

For instance, it was computational perspectives that suggested a role for the hippocampus in spatial navigation (O'Keefe and Nadel 1978) and led to the development of the water maze (Morris 1981), including the subtle distinctions between variants of it (Eichenbaum et al. 1990; Day et al. 1999). Similarly, computational perspectives have been critical for an understanding of memory transformations and the development of schema and the transformation from episodic to semantic memory (O'Reilly and McClelland 1994; Redish and Touretzky 1997). Computational perspectives on multiple decision-making systems successfully predicted how different neural circuits process information about cues, actions, and rewards (van der Meer et al. 2010). It has long been possible to manipulate circuits, but new techniques can provide exquisite control of neural circuits at unprecedented cellular, connectivity, and temporal scales (Tye and Deisseroth 2012). With advances in both our computational understanding of neural circuits and newly available techniques to manipulate them neurophysiologically, it is now becoming possible to create explicit deficits hypothesized by computational theories and to test whether they produce the expected behavioral consequences. Whether that success can be translated into clinical practice, however, remains an open question.

Promise 3: Personalized Medicine

To tailor therapy to an individual patient, we need to understand both dysfunction and treatment at a deep enough level so that we can match treatment to dysfunction. We argue that computation is the path to that understanding.

By understanding how physical dysfunctions create psychiatric disorders, we can identify the most appropriate probes to identify which dysfunctions exist within a given patient. By understanding how treatment changes the brain (and thus the mind), we can identify the most appropriate treatments (and the most appropriate variations on a given treatment) for a given patient. This is the promise of the computational perspective. But it raises the obvious question: *Are we there yet?*

Open Questions

The first, and most important question, is whether the computational perspective is ready to take to the clinic yet. There are obviously large gaps in our knowledge of the computations being performed by neural systems as well as large gaps in our knowledge of how neural circuits (and subcircuit dynamics) create those computations. However, as we have seen throughout this chapter (and shall see further throughout this volume), computational perspectives are a means of connecting between levels, and it is not necessary to model the whole brain in order to make contributions that can be useful clinically (see, e.g., Kurth-Nelson et al. and Totah et al., this volume).

How Can the Computational Perspective Handle the Heterogeneity of Real Patients?

Clinical presentations are highly heterogeneous, presenting variability both across patients and within individual patients across time. Totah et al. (this volume) directly raises this issue and discusses it in the relation to individuals. Flagel et al. (this volume) addresses the temporal aspects of this by incorporating trajectories through time directly into the proposed nosology.

Psychiatric disorders tend to have a highly complex temporal trajectory. Although current computational perspectives are capable of integrating temporal trajectories (both recurring and progressing) into their constructs, few current computational models have directly addressed this temporal complexity. How will models be able to capture that temporal trajectory in a way that is informative but not constraining? In a sense, treatment is about changing the trajectory of future symptoms. Can computational models that capture the trajectory of a psychiatric dysfunction be used to guide treatment by making predictions about how those trajectories will change?

Patients have extensive social and psychological lives; the dysfunctions in their computational systems will bleed over into these areas, complicating the picture. It is unlikely that a given patient will have an isolated dysfunction that can be identified as a single failure mode that can be treated by a single paradigm. Is it possible to model the complexity of a given patient? On the other hand, it is an open question whether we need to model the complete patient in order to treat a given dysfunction. Even small improvements in the three challenges (nosology, biomarkers, treatment) and three promises (genetics, circuits, personalized medicine) would be a useful contribution.

What Is the Appropriate Level of Analysis?

Another important open question is to ask at what level of analysis the computational perspective should be applied: Do we need to understand the circuit level of a dysfunction? Do we need to understand how genetic variation affects ion channels thus affecting subcircuit interactions leading to changes in computation? Or do we need to understand the computations of whole neural circuits? Humans are fundamentally social animals. Do we need to understand that social computation?

The answer is likely to be that it will depend on the specific psychiatric dysfunction. In general, as noted above, the computational perspective is particularly useful for connecting different levels of understanding. For instance, to understand how a genetic variant can lead to differences in neural responsiveness for an individual neuron type, which can lead to a change in neural circuit dynamics, which can lead to a susceptibility to social or physical stress, one needs to understand the computation being performed at each level. However, if one knows that social interactions depend on decision-making systems, then it may be enough to start from behavioral tests of those decision-making systems.

What Are the Dysfunctions?

As noted above, psychiatry has developed categorizations of dysfunction that are reliable but unlikely to be valid. An important open question is whether those categorizations contain enough validity to start from or whether we need to start over with a new nosology. It would obviously be easier to use the current taxonomy of psychiatric dysfunction to bootstrap a (hopefully more valid) computationally and mechanistically justified taxonomy. It is likely that it will be very difficult to completely throw out the current nosology of psychiatric dysfunction (e.g., DSM-IV-TR or DSM-V), but how much it needs to be modified is going to be a critical open question.

An important aspect to this open question is that the answer may be very different when applied to disorders presenting with a limited number of syndromes (e.g., obsessive-compulsive disorder) and when applied to general broad spectrum disorders such as anxiety or depression. It is likely that many disorders with a wide spectrum of behavioral manifestations actually consist of several different disorders that have been categorized together. Schizophrenia, for example, may be an example of a super-category, where many failure modes drive many behavioral outcomes (Silverstein et al. 2013). Additionally, some disorders may be symptoms, whereby many underlying failure modes can lead to the same general outcome (like heart attack and indigestion both causing chest pain). Addiction has been suggested to be such a disorder, where multiple different failure modes can all lead to continued drug use (Redish et al. 2008). Whether computational biomarkers can pull the underlying disorders out or whether there is actually a single underlying dysfunction remains unresolved.

There has been a recent attempt by the U.S. National Institute for Mental Health (NIMH) to create a new categorization of topics for research and analysis called Research Domain Criteria (RDoC) based on psychological constructs, such as attention, cognition, reward systems, etc. (Insel et al. 2010). The RDoC process is still under development, but it is not clear how much of a role computational perspectives have played, or will play, in RDoC's development. The new nosology proposed in this volume offers a novel, nuanced perspective on this difficulty, providing a way of integrating existing taxonomies with dimensional and computational constructs such as those proposed by RDoC.

What Is the First Exemplar?

At this point, most computational models and explanations for psychiatric dysfunction and treatment have been based on small-scale problems that occur within limited experimental domains, with limited and abstract cue-sets and simple decision components. Although these models have been built at very abstract levels and applied to small-scale (toy) problems, they do capture key factors that have been theorized to drive aspects of psychiatric dysfunction. However, for these models to make an impact on psychiatry, there needs to be a path from computation and theory to clinical practice.

Can we derive a new pathway to understanding psychiatric dysfunction? Is there a general paradigm to apply this computational perspective to psychiatry? A new field, called "computational psychiatry," has begun to emerge (Rangel et al. 2008; Maia and Frank 2011; Montague et al. 2012; van der Meer et al. 2012b; Redish 2013), but it is unclear what that pathway is. The chapters in this book propose first steps toward this new pathway starting from this new computational understanding of psychiatry.

The Strüngmann Forum

A group of scientists, split evenly between computational neuroscientists and clinical psychiatrists met in June 2015 in Frankfurt under the rubric of the Strüngmann Forum to discuss these issues. The goal of this forum and this book that has arisen from it was to bring together leaders in the fields of psychiatry and computational neuroscience to see if we can make progress on these open questions.

To take on these questions, we divided into four working groups, each of which addressed a key topic relating to these issues:

1. *Mechanisms*, examining the way that computational perspectives can provide ways of connecting mechanistic differences (genetic variation, differences in social experiences, their interaction) with psychiatric behavioral outcomes

2. *Modeling realistic psychiatry*, examining how computation can address realistic psychiatric patients, who often show comorbidities, who often shift from diagnosis to diagnosis, and who often express complex compensation mechanisms in response to treatments

3. *Nosology*, examining how the computational perspective changes the taxonomy of diagnoses, addressing how the dimensionality of constructs, particularly computational constructs, can be integrated with the clinical practice

4. *A first example*, looking at what it would take to find specific examples to determine whether we have enough at this point to measurably improve treatment for a given dysfunction

The goal of this book is thus to begin to get at the crux of a new question: *How does the computational perspective change psychiatry?* Our hope is that this forum and book can form a concrete framework for future studies and serve as an opening to jump-start this potentially very important cross-field interdisciplinary interaction.

Open Issues in Psychiatry

3

Complexity and Heterogeneity in Psychiatric Disorders

Opportunities for Computational Psychiatry

Nelson Totah, Huda Akil, Quentin J. M. Huys, John H. Krystal,
Angus W. MacDonald III, Tiago V. Maia,
Robert C. Malenka, and Wolfgang M. Pauli

Abstract

Psychiatry faces a number of challenges, among them are the reconceptualization of symptoms and diagnoses, disease prevention, treatment development and monitoring of its effects, and the provision of individualized, precision medicine. Achieving these goals will require an increase in the biological, quantitative, and theoretical grounding of psychiatry. To address these challenges, psychiatry must confront the complexity and heterogeneity intrinsic to the nature of brain disorders. This chapter seeks to identify the sources of complexity and heterogeneity as a means of confronting the challenges facing the field. These sources include the interplay between genetic and epigenetic factors with the environment and their impact on neural circuits. Moreover, these interactions are expressed dynamically over the course of development and continue to play out during the disease process and treatment.

We propose that computational approaches provide a framework for addressing the complexity and heterogeneity that underlie the challenges facing psychiatry. Central to our argument is the idea that these characteristics are not noise to be eliminated from diagnosis and treatment of disorders. Instead, such complexity and heterogeneity arises from intrinsic features of brain function and, therefore, represent opportunities for computational models to provide a more accurate biological foundation for diagnosis and treatment of psychiatric disorders. The challenges to be addressed by a computational framework include the following. First, it must improve the search for risk factors and biomarkers, which can be used toward primary prevention of disease. Second, it must help to represent the biological ground truth of psychiatric disorders, which will improve the accuracy of diagnostic categories, assist in discovering new treatments, and aid in precision medicine. Third, to be useful for secondary prevention, it must represent how risk factors, biomarkers, and the underlying biology change through the course of development, disease progression, and treatment process.

Introduction

Understanding and treating the enigmatic symptoms of psychiatric disorders has placed the field of psychiatry at an impasse (see Gordon and Redish, this volume). Key issues that must be addressed to advance psychiatry include: highly comorbid diagnoses and heterogeneity of patients grouped under a single diagnostic label; lack of understanding of the causal biological mechanisms, which form the basis of disease etiology and progression, and their impact on treatment decisions; and the absence of biomarkers and clear risk factors that can be used to predict and prevent disease. These challenges likely arise because of the complexity and heterogeneity present in the brain, in the environment, and in our collective attempt to assign patients to diagnostic groups. Moreover, sources of complexity and heterogeneity are dynamic and change over time with natural development and in response to the disease process itself. For instance, the severity of symptoms and the presence of specific symptoms change across various illness stages, which makes diagnosis and treatment decisions difficult. Treating psychiatric disorders will require improving their biological grounding through an understanding of the brain, an organ that is itself a dynamic and complex system. Computational models are well placed to build a bridge between the initial patient self-report and behavioral observations and the complex, dynamic neurobiological and neurocomputational state of the patient.

In this chapter, we propose that computational models can be used for multiple purposes. They can be used as tools of prediction in psychiatry and as as a method for increasing the biological grounding and quantitative, mathematically formalized framework with which disorders can be understood. We begin with an overview of the issues of complexity and heterogeneity that hinder progress in psychiatry. Thereafter we identify opportunities for computational models to address these issues and the data available for use in models. We close by outlining some examples in which computational models may be used as tools or as methods for improving our understanding of the etiology and progression of psychiatric disorders. We will demonstrate how these tools and methods might be used to improve diagnosis, identify biomarkers and risk factors, and prevent disease through treatment or removal of risk factors. By identifying the sources of complexity and heterogeneity present in psychiatry, we hope to provide a starting point for dialogue between psychiatrists, neuroscientists, and computational scientists. Throughout our overview, we use clinical vignettes to illustrate, from a practical viewpoint, the complexities that challenge the field of psychiatry.

Sources of Complexity in Psychiatry

Clinical Vignette 1: Peter, an 18-year-old high school senior, is brought to the local mental health crisis clinic by his parents. They report that for the past two weeks he has isolated himself in his bedroom, taped up the windows with foil, and refused meals with his family, eating only packaged food that he prepares in the microwave. He has had several angry outbursts when his family has attempted to coerce him to go to school or to get help. The police were called to the house by a neighbor, due to the shouting that occurred, and he reluctantly agreed to come to the clinic voluntarily because the officer told him that he would take him into custody if he did not. Up until about a year ago Peter was performing with average to above average grades in school, playing in his school jazz band, and socializing regularly. At that time, he had an episode of anxiety and depersonalization after smoking cannabis for the first time. The episode lasted for several days and led to a visit to his family doctor. During the visit, he described intermittent feelings of being detached from things, that things were not real, that there were strange shadows outside his windows at night, and that he felt at times as if someone was looking in at him although, when he checked, no one was there. The family physician prescribed an antidepressant for "panic attacks." However, three days after starting this drug he became increasingly irritable, restless, religiously preoccupied, argumentative, and he began to stay up all night playing video games with online partners; this led the physician to discontinue the drug. At that time, Peter's grades started to deteriorate and he became increasingly withdrawn from friends and family. Over the next few months, he spent more time alone in his room and on the Internet. By his report, he was concerned initially that he was being singled out at school by a group of peers with whom he had an argument over rival sports teams. He noticed that his computer became very slow to boot up and was infected by a virus; he became concerned that he had been hacked and that other students in the school, and eventually the teachers, were involved. He began hearing whispers in the background audio of music videos and then, several weeks later, began hearing several voices commenting on his behavior and making derogatory remarks. He stopped showering and his room was piled with dirty clothing. He began to notice that his food smelled different and at one point he had an epiphany: his parents and teachers were illuminati and were trying to poison him. When he was seen in the clinic, he was visibly terrified, scanning the room, trembling and unable to sit still.

Clinicians are confronted by an extremely complex set of variables when they must diagnose a patient and make a treatment decision. One level of complexity is intrinsic to the patient's narrative and the symptoms that they report, and this can lead to comorbid diagnoses. For instance, Peter is irritabile and restless, staying up all night playing video games; he could be diagnosed with bipolar disorder. But Peter also meets criteria for schizophrenia, with prominent symptoms of social isolation and psychosis. Indeed, one-third of individuals diagnosed with bipolar disorder are also diagnosed with schizophrenia (Gonzalez-Pinto et al. 1998). Yet the psychiatrist must select a diagnosis and this decision will then guide treatment. This complexity that confronts the clinician, therefore, challenges their ability to deliver an accurate diagnosis, predict the time course of disease, select treatment, and monitor treatment response.

In addition to the individualized complexities that arise as a result of each patient's unique narrative, other highly individualized complexities occur at the level of biology and also influence psychiatrists' ability to accurately diagnose and treat the patient. At the level of the individual such complexities include genetic endowment, environmental influences, longitudinal changes in the brain (due to both natural neurodevelopment and neuroadaptation to disease), and interaction with stable traits. At the population level, this overlap between patients leads to comorbidity and to fluctuating diagnoses over time. For example, patients' symptoms will change and cause a diagnosis to fluctuate between, say, a diagnosis of anorexia nervosa and bulimia nervosa (Tozzi et al. 2005). Furthermore, illnesses have temporal components; they can be chronic, relapsing/remitting, or a combination thereof. The onset of an illness can also occur over multiple timescales. It can begin in a punctate manner (e.g., triggered by a trauma) or express itself in a punctant manner (e.g., a suicide attempt or a decision to use a drug), or it can progress longitudinally over a lifetime. Adaptive or nonadaptive processes can alter the course of disease (see chapters by Krystal et al. and Huys, this volume). Differences in treatment history cause diverse disease trajectories, even across patients with the same diagnostic label or underlying disease process. The evolution of Peter's symptoms, from social isolation and weird behavior to frank paranoia, seem to reflect some underlying temporal progression. An additional source of complexity that must be considered is that all of these temporal aspects of illness interact with stable traits and propensities, such as personality, temperament, and genetic endowment.

The heterogeneities that we have identified have real-world implications for both basic research and practical matters of effective diagnosis and treatment by clinicians. For instance, comorbid and temporally fluctuating diagnoses result in heterogeneous patient populations, which present a challenge to discovering common biological correlates and risk factors. Jonathan Flint has compared the search for genetic risk factors in psychiatry with searching for a genetic risk factor for a diagnosis of fever, which would sample a disease-predisposing genetic makeup from a highly heterogeneous group of patients with autoimmune disease, various infections, cancer, and many other conditions (Ledford 2014). What can be done to improve the accuracy of diagnosis? Clearly, we need a greater understanding of the biological ground truth underlying psychiatric disease. The lack of correspondence between diagnostic labels and biology has hindered the search for biomarkers, etiological mechanisms, and risk factors that can be mitigated through preventative medicine. In fact, a recent attempt at discovering genetic correlates for depression was successful, whereas others have not been, because the study was designed to reduce heterogeneity across the population of subjects which shared the common diagnostic label of depression (CONVERGE Consortium 2015).

Treatment decisions are also hindered by the mismatch between diagnostic labels and the underlying biology. For example, clinical trials designed to test existing and novel therapies typically fail to take into account differences across subjects that are due to the biological stage of the illness or the biological effects of treatment history. For Peter (the patient in the clinical vignette who early on seems to have met criteria for both mood and psychotic disorders), the clinician must decide whether to either administer an antidepressant for a mood disorder, a mood stabilizer for bipolar disorder, or a preventative treatment for prodromal schizophrenia. His initial treatment by his family doctor was an antidepressant, which unfortunately precipitated his symptoms of mania and hypomania. Could this have been avoided had the physician taken into consideration the complex temporal progression of his disease? There is an emerging recognition that schizophrenia can be conceived of as having "predromal," "prodromal," and "syndromal" phases (Lieberman et al. 2001; Fusar-Poli et al. 2014; Krystal et al., this volume). Given that there are biological differences at each disease stage, a greater understanding of disease stage could lead to better treatments and, perhaps, even prevention. There is also evidence that treatment history, itself, can alter the biological factors underlying disease and must, therefore, be taken into account during diagnosis, treatment decisions, and clinical trial design. For example, patients enrolled in clinical trials for schizophrenia treatment may have already received dopaminergic antipsychotic drug treatment, which could alter the efficacy of the novel treatment being tested. Importantly, basic experiments performed in animals have provided direct support for this hypothesis by demonstrating that long-term administration, and then subsequent withdrawal of antipsychotics, affects the behavioral and neurophysiological response to a novel GABAergic treatment (Gill et al. 2014).

Next we will outline the current state of knowledge about the biological factors underlying psychiatric disease. Central to our proposal for the usefulness of a computational framework in meeting the challenges that face psychiatry is the idea that, although the complexity of the underlying biology leads to inaccurate diagnoses and treatment decisions, it reflects the intrinsic nature of the biology and this "noise" must be harnessed rather than avoided. We believe that a computational framework provides the ability to make sense of the noise. We begin by illustrating the biological complexities present at the genetic level, and their intraction with environmental and stable trait factors, using another clinical vignette.

Genetic Complexity

Clinical Vignette 2: Wendy's first stint in rehab was at the age of 26. She began drinking at age 12 when her mother, a heavy drinker, was diagnosed with breast cancer and died. Having initially surreptitiously sampled unfinished drinks at her parents' parties, she started sneaking drinks from the liquor cabinet before going to sleep at night. This became a habit during her college years, when she would come home after studying or social activities and have several drinks before going to sleep. During exams she would use Adderall to overcome the effects of the night before. She graduated from college with excellent grades and began working as a clerk in a law firm while preparing to take the qualifying exam for law school. She was punctual, courteous and managed to perform her duties in a manner that was satisfactory to the attorneys for whom she worked. She was in a long-term relationship with her boyfriend who occasionally binge drank with her at parties, but he was unaware of her nightly drinking. At times she felt that she had problems with her memory, and she was frequently hung over in the morning. She often thought about quitting drinking. Each day, as her workday wore on, she would begin to think about what she would drink that night and plan her social activities such that she would be able to get back to her apartment and drink. She would often decide that she was not going to drink after all, just as she entered her apartment, but would later abruptly decide to go to the store to buy alcohol. Her alcohol use dramatically increased after her father received a cancer diagnosis. She called in sick for several days and was then arrested driving her car erratically in the early hours of the morning. She had no recollection of how she got there; she had a blood alcohol level four times the legal limit along with amphetamines in her system. On her way home from her court-mandated treatment, she felt ashamed and determined to confront her substance use problem before it cost her lifelong dream to attend law school and become an attorney. As she passed her local supermarket she decided to stop and purchase nonalcoholic beer. She left the store with three bottles of her favorite wine.

Wendy's story illustrates many of the complexities inherent in psychiatric disease. There are often clear predisposing factors, such as, in this case, genetic endowment and family history (her mother was a "heavy drinker") that help explain why she might be more susceptible to addiction than others, such as her boyfriend, who "occasionally binge drank with her at parties." Wendy's alcohol consumption increased during times of stress. Stress is often thought to play a role in precipitating psychiatric symptoms. Might also more stable traits, such as her gender or personality, be additional risk factors? A synthesis of research demonstrates that these factors all contribute and interact. Drug consumption in males tends to be driven by the hedonic, reward-related aspects of the drug, whereas in females it is driven by negative hedonic states and stress (Li et al. 2005; Fox et al. 2006; Hyman et al. 2008; Potenza et al. 2012). These data may explain why her boyfriend binge drank at parties, whereas she consumed larger amounts during times of stress. Such observations about human behavior are bolstered by gender differences in reward-related neurophysiological activity. In comparison to women, alcohol administration increases striatal dopamine release to a greater extent in men, which may drive increased appetitive reward-seeking behavior in men (Berridge 2006; Urban et al. 2010). Finally, both gender-specific hormones and genetic endowment make independent contributions to drug/reward consumption and habitual drug-seeking behaviors (Quinn et al. 2007; Barker et al. 2010; Seu et al. 2014). These data indicate that genetic endowment, stable traits, hormones, and environmental factors may all interact to contribute to the expression of psychiatric disorders.

Even if we solely look at genetic factors, the picture remains quite complex. In general, psychiatric disorders are known to be highly heritable. Schizophrenia, for instance, is thought to be 70% heritable (Lichtenstein et al. 2009). Familial inheritance studies and twin studies have encouraged decades of research into genetic risk factors and their potential as causal pathophysiological mechanisms (LaBuda et al. 1993; Sullivan et al. 2003; Mitchell et al. 2011). Puzzling out the nature of genetic risk, however, has proved formidable, in part because there are multiple kinds of genetic risk factors and multiple genes of each kind. Copy number variants (CNVs), which are deletions or duplications spanning multiple genes, appear prevalent in schizophrenia, autism, and attention-deficit/hyperactivity disorder (Sebat et al. 2007; Abrahams and Geschwind 2008; Walsh et al. 2008; Xu et al. 2008; Gilman et al. 2011; Levy et al. 2011). Dozens of CNVs have been identified, each accounting for a relatively small proportion of patients with a given disorder, but each having a relatively large effect on risk (on the order of 5- to 30-fold increases). On the other end of the spectrum, genome-wide association studies have provided a list of hundreds of common single nucleotide polymorphisms (SNPs) for schizophrenia. Each of these common risk alleles are seen in large numbers of patients (and healthy controls), and all have small effect sizes (typically much less than 1.5-fold increases in risk) (Raychaudhuri et al. 2009; Stefansson et al. 2009). Similar risk structures may underlie autism (Wang et al. 2009; Gaugler et al. 2014).

Adding to this complexity, even a single one of these hundreds of risk genes can be associated with multiple psychiatric disorders (Cross-Disorder Group of the Psychiatric Genomics Consortium et al. 2013; Zhu et al. 2014) (Figure 3.1). For example, the translocation of DISC1 is highly hereditary (although it has irregular expression and affects carriers' phenotypes to different degrees) and strongly predicts a number of psychiatric disorders including schizophrenia, bipolar, and depression (St. Clair et al. 1990; Millar et al. 2000; Blackwood et al. 2001; Chubb et al. 2008; Jaaro-Peled et al. 2009). Similarly, several CNVs predispose to both schizophrenia and autism.

Therefore, multiple types of genetic variation—from hundreds of SNPs, each with a minor effect, to CNVs or other mutations of large effect size, and likely gray areas in between—can interact to give rise to psychiatric phenotypes, and any given gene can raise the risk for multiple different phenotypes. From this perspective, it is clear that genes must be considered part of a network with multiple, interacting genetic pathways to a disorder. It may be

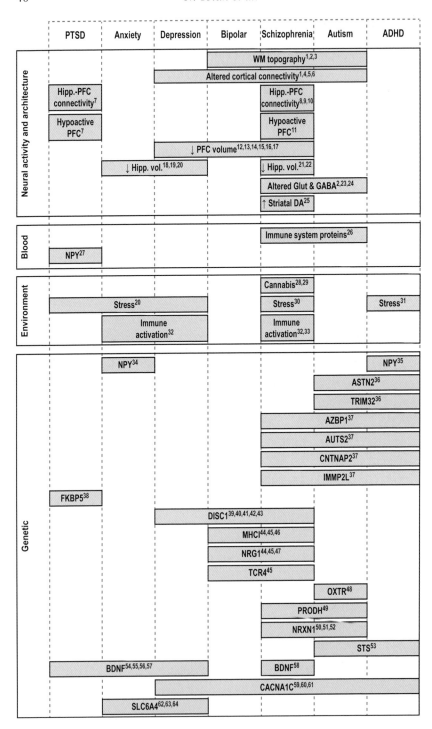

Figure 3.1 The genetic, environmental, anatomical, and physiological factors that are associated with a variety of psychiatric disorders. This non-exhaustive list conveys the breadth of factors and their overlap across disorders. References are numbered.

1. Bernard et al. (2015)	33. Brown and Derkits (2010)
2. Nelson and Valakh (2015)	34. Kaabi et al. (2006)
3. Sussmann et al. (2009)	35. Lesch et al. (2011)
4. Downar et al. (2014)	36. Lionel et al. (2011)
5. Just et al. (2004)	37. Elia et al. (2010)
6. Koshino et al. (2008)	38. Binder et al. (2008)
7. Fani et al. (2012)	39. St Clair et al. (1990)
8. Meyer-Lindenberg et al. (2005)	40. Millar et al. (2000)
9. Lawrie et al. (2002)	41. Blackwood et al. (2001)
10. Ford et al. (2002)	42. Jaaro-Peled et al. (2009)
11. Weinberger and Berman (1996)	43. Chubb et al. (2008)
12. Pantelis et al. (2003)	44. Intl. Schizophrenia Consortium et al. (2009)
13. Smieskova et al. (2010)	45. Stefansson et al. (2009)
14. Wright et al. (2000)	46. McGuffin (1979)
15. Ward et al. (1996)	47. Harrison and Weinberger (2005)
16. Rajkowska et al. (1999)	48. Gregory et al. (2009)
17. Drevets (2000)	49. Guilmatre et al. (2009)
18. Phillips et al. (2002)	50. Autism Genome Project Consortium (2007)
19. Velakoulis et al. (2006)	51. Walsh et al. (2008)
20. Sapolsky (2000)	52. Rujescu et al. (2009)
21. Bois et al. (2015)	53. Kent et al. (2008)
22. Nelson et al. (1998)	54. Felmingham et al. (2013)
23. Lewis (2014)	55. Soliman et al. (2010)
24. Volk and Lewis (2014)	56. Verhagen et al. (2010); *males only
25. Howes et al. (2009)	57. Tocchetto et al. (2011)
26. Chan et al. (2015)	58. Neves-Pereira et al. (2005)
27. Morgan et al. (2000)	59. Cross-Disorder Group of the Psychiatric Genomics Consortium et al. (2013)
28. Moore et al. (2007)	60. Green et al. (2010)
29. Semple et al. (2005)	61. Ferreira et al. (2008)
30. Anda et al. (2006)	62. Lesch et al. (1996)
31. Van den Bergh and Marcoen (2004)	63. Hariri et al. (2002)
32. Markham and Koenig (2011)	64. Caspi et al. (2003)

necessary, then, to analyze genetic factors as a causal network, rather than as direct causes (Raychaudhuri et al. 2009; Barabási et al. 2011; Gilman et al. 2011). Computational methods for analyzing effects in complex networks may be helpful in this area of research (Boccaletti et al. 2006; Bullmore and Sporns 2009; Barabási et al. 2011). Other work has used computational methods, such as machine learning, to demonstrate that schizophrenia is predicted by interacting combinations of genes that affect multineuron population activity associated with working memory, an impaired cognitive function in individuals diagnosed with schizophrenia (Nicodemus et al. 2010). This finding highlights the importance of considering the effects of *interacting* genetic mutations at the level of neural circuit function.

Neurobiological Complexity

Clinical Vignette 3: Jennifer woke up in the intensive care unit of a regional medical center. She had been in the hospital for three days following an acetaminophen and benzodiazepine overdose that had caused some liver damage but was not going to be life threatening. At 36 years, this was an unfamiliar, shocking, and embarrassing experience. Since the economic downturn six years ago, she and her family had struggled financially. Her marriage of 13 years had been stressed and her school-age children were having academic and other problems at school. Her oldest son (aged 12) had been suspended for fighting with classmates. Always a worrier, over the past months she had increasing difficulty getting to sleep and staying asleep, leaving her tired and irritable during the day. She ruminated constantly about her family problems and blamed herself for them. She felt increasingly irritable, sad, and empty. She also felt tired and disinterested in pleasurable experiences, including food and sexual activity. Initially, she recognized that she was depressed and sought self-help on the Internet; however, over time she increasingly felt that there would be no help for her, that she was a burden on her family, and that they would be better off without her. During the two weeks before her suicide attempt, she saw her family doctor to complain about fatigue and insomnia and was prescribed the benzodiazepine. She also researched suicide on the Internet, including the potential lethality of the medication on which she overdosed. She left a note apologizing to her family for letting them down, stating that they would be better off without her.

Jennifer clearly suffered from depression, but the causes of her depression are, well, complex. Were there genetic factors? Probably. But whatever genetic predisposition to depression lay in her genes, this predisposition was filtered through multiple downstream events: economic troubles and stressful family issues seemed to play a grave role in her illness, which progressed slowly over months but was then punctuated by a terrible event—her suicide attempt. How do we understand her illness at a biological level?

We must start with a fundamental understanding about brain disease: whatever the fundamental causes, genetic or environmental, these precipitants contribute to behavioral phenotypes only when filtered through neural circuit development, cellular physiology, and neural circuit activity, each of which adds further complexity to the picture. Thus, for causal events to be useful as biomarkers or etiological mechanisms in psychiatry, they must be integrated with information about gene expression, cellular physiology, and neuronal circuits.

Once again to simplify, consider the case of a disease gene. One recent study has demonstrated that the number of neurons expressing a disease-related mutation is surprisingly low (Cai et al. 2014). This critically important result demonstrates that a mutation can affect the cellular physiology of a relatively confined group of neurons. Well, then, might we just study the small number of cells that are affected directly by this mutation? This approach will probably not be fruitful. Buzsáki and Mizuseki (2014) have proposed that behavior, cognition, and perception depend not only on an active minority assembly of neurons, but on their coordination with a global majority, in order to provide the optimal trade-off between fast, yet accurate thought and behavior. Multiple neurophysiology studies have demonstrated that a small group of neurons can easily affect a large population of neurons (Cardin et al. 2009; Li et al. 2009; Thiagarajan et al. 2010; Kwan and Dan 2012; Logothetis et al. 2012; Olsen et al. 2012; Beltramo et al. 2013). For instance, a single mouse visual cortex neuron can drive spiking of other neurons within a 100 μm radius with different degrees of drive on different cell types (Kwan and Dan 2012), giving rise to complex, nonlinear dynamics in large networks of neurons. In other words, complex genetic networks (operating within neurons) impact similarly complex neural networks (operating between neurons). The challenge is to build a mechanistic bridge between these genetic and neural networks that offers insight into human thought, mood, perception, learning, and memory (Figure 3.2).

Given that multiple genetic pathways can converge at the level of particular cell types or neuronal circuits, it is imperative that we increase our understanding of how behavior and cognition arise from the activity of these neuronal networks. Neural activity and neurotransmitter systems have been tied to many cognitive functions and behaviors that are impaired in disorders, such as impulsivity (Robbins 2002), working memory (Goldman-Rakic 1996; Arnsten 2011), reward expectation and value (Schultz 2007; Roesch et al. 2010), and fear extinction (LeDoux 2000; Letzkus et al. 2011). In some cases, the neuronal activity underlying these behaviors has been further focused on various cell types interacting as microcircuits or broadened to examine macrocircuit interactions across brain regions. Based on knowledge gained from animal neurophysiology and neuropsychopharmacology studies, testable hypotheses about human neural circuit dysfunctions have been proposed and tested in animals. For example, Homayoun and Moghaddam (2008) have suggested that the orbitofrontal cortex (OFC) is a site of convergence for both dopaminergic antipsychotics (established schizophrenia treatments) and metabotropic glutamate agonists and positive allosteric modulators (novel treatments) to normalize OFC activity that has become aberrant via altered excitatory-inhibitory

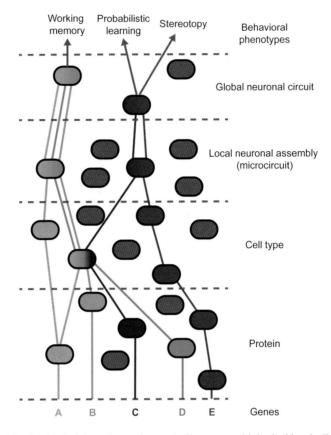

Figure 3.2 Multiple, interacting pathways to disease span biological levels. Each level provides unique pathways to disease. Genes A and C are part of the neuroligin family, which is associated with autism and schizophrenia (possibly through interaction with neurexin) (Carroll and Owen 2009; Sun et al. 2011; Kenny et al. 2014). Gene A is neuroligin-2, which drives changes in the synaptic targeting between excitatory, glutamatergic cortical pyramidal neurons and inhibitory, parvalbumin-containing interneurons (Gibson et al. 2009). Gene C is neuroligin-3, which affects excitatory, glutamatergic synaptic formation on D1 dopamine receptor expressing nucleus accumbens neurons that are involved in learning-motivated behaviors (Rothwell et al. 2014). Gene B effects cell division cycle 42 (cdc42) mRNA expression, which is reduced in schizophrenia and has been shown to reduce dendritic spines on layer 3 prefrontal cortex (PFC) pyramidal neurons, thus effecting prefrontal cortical excitatory-inhibitory balance in a layer-specific manner that is critical for cognitive faculties such as working memory (Goldman-Rakic 1995; Hill et al. 2006; Lewis et al. 2012). Gene D is neuregulin 1 (NRG1), which is associated with bipolar disorder and schizophrenia and is associated with PFC activation during working memory tasks (Nicodemus et al. 2010). Its effects are mediated by interaction of its protein product with an enzyme (γ-secretase), which further regulates gene expression via intracellular signaling. Gene D highlights an example of genes interacting with genes. One consequence of this pathway is altered excitatory synaptic transmission (Fazzari et al. 2014), Finally, Gene E is the dopamine transporter (DAT) gene, which affects dopamine neurotransmission in striatum by altering reuptake.

balance. In other circuit models, Grace (2006, 2010) has suggested that aberrant interaction between the prefrontal cortex (PFC), limbic subcortical areas, and neuromodulatory brainstem structures during stress could provide the basis of comorbid mood disorders, schizophrenia, and addiction. Dysfunctional circuits have, to some degree, also been discovered in humans. A hyperactive subgenual cingulate leads to dysregulation of a network of brain regions associated with numerous symptoms of depression and direct electrical stimulation of this cingulate node normalizes activity in the rest of the network and can cause remission or lessen symptoms (Mayberg et al. 2005). Additionally, individuals with posttraumatic stress disorder are thought to have inadequate PFC excitatory control over inhibitory interneurons in the intercalated nucleus of the amygdala, which provides local inhibition of the amygdala neurons that trigger automatic expression of behavioral responses to fearful stimuli. A lack of PFC excitation to the locally inhibitory cells of the intercalated nucleus removes the internal dampening of fear-related amygdala activity that occurs during learning to extinguish stimulus-evoked fear responses (Parsons and Ressler 2013).

The challenge of understanding how neurons and neural circuits give rise to behavior will require not only acquiring additional data, but interpreting data from a computational perspective. Rieke et al. (1997) have described the computations performed by neurons that encode perception. According to their overview, the neuron's only glimpse of the world that we perceive, a world which is full of random and dynamic stimuli, comes from discrete spikes emitted by sensory receptors which must be continuously decoded by neuron upon neuron (and so on) to provide a continuous readout of the world. The authors describe the computations that neurons could perform to represent the world using single, discrete spikes. *One challenge for computational neuroscience is to translate ideas such as these beyond perception and into how neurons represent mood, memories, values, and so on. Further, these ideas must be*

Figure 3.2 (continued)   Note the overlap between the pathway formed by gene B and gene D, which both effect PFC working memory related neuronal activity, possibly by both effecting a common microcircuit parameter (cortical pyramidal neuron dendrite formation). Moreover, note that gene A also effects synaptic targeting, albeit in a different class of cell types, thus modulating microcircuits in a different manner. Finally, there is overlap between gene C and gene E, which may both affect glutmatergic input to nucleus accumbens and its modulation by dopamine, which is a critical component of learning from positive reinforcement (impaired in schizophrenia; Strauss et al. 2011) and controlling goal-directed movements (impaired in autism; Rothwell et al. 2014). Note that the striatal microcircuit affected by genes C and E differ from the prefrontal microcircuit affected by genes A, B, and D, although they may all contribute to symptoms of schizophrenia by affecting different global neuronal circuits. Fault tree analysis (see MacDonald et al., this volume) might be helpful in organizing these data. Tools of studying complex, nonlinear dynamic systems may also be useful in interpreting these data (see text).

integrated with our understanding of the genetic networks that regulate the neurophysiological characteristics of a neuron and how neurons wire together.

Finally, we must of course recognize that the brain is not static: it evolves over time, through neural development as well as through experience. Genes can raise the risk for illness by conferring susceptibility to environmental risk factors, which are experienced at particular points in time, like the polymorphisms in the serotonin transporter promoter that raise the risk for depression only within the setting of early childhood stress (Caspi et al. 2010). Other environmental events only raise the risk for psychiatric phenotypes when they occur at particular times in development, presumably during critical periods of growth and change in the physiology and anatomical connectivity of the brain (Lodge and Grace 2008; Hradetzky et al. 2012). Similarly, the treatments that reverse the effects of time-sensitive insults can have greater effects if they are administered during specific time windows (Du and Grace 2013). These findings have been interpreted within the familiar concept of developmental critical periods, namely, by proposing that genetic factors interact with developmental changes to increase susceptibility to stress-induced psychiatric disorders (Lodge and Grace 2008). In summary, neurodevelopmental trajectories are yet another source of complexity in the etiology of psychiatric disorders.

An additional type of trajectory, which we call "neuroadaptive," is derived from the disease process itself and is thought to contribute further temporal complexity to the etiology and primary prevention of psychiatric disorders. For example, perinatal gene–environment interactions during development might shift the normal gene-guided pruning of excitatory glutamate synapses into a presyndromal disease state of overly reduced glutamate synapses (Feinberg 1982; Weinberger 1987; Lieberman et al. 2001; Lewis and Levitt 2002; Fusar-Poli et al. 2014). This disease state of hypoexcitation will, itself, evoke neuroadaptive changes, such as the reduction of inhibitory GABA that serves to rebalance the reduced excitatory glutamate synapses (see Krystal et al., this volume; Volk and Lewis 2014). These types of neuroadaptive changes—driven by the disease process itself—will also have a trajectory that affects genes and brain circuits throughout the course of the disease. Thus, there are complex gene–gene, gene–environment, and brain circuit interactions that dynamically change in relation to the timing of both neurodevelopmental periods and neuroadaptive periods. Moreover, neuroadaptive processes occur in the context of treatments that are provided to the patient. Recall, for example, that administration of an antipsychotic dopamine antagonist alters the subsequent response to novel antipsychotic medications that primarily affect GABA in animal models of schizophrenia (Gill et al. 2014). It is highly likely that treatment alters the course of neuroadaptive processes during disease. *Methods are needed for representing these complex data sets and their interacting temporal dynamics.* Computational models are appropriate for addressing this complexity by functioning as tools to identify fundamental factors (or combinations of factors) that contribute to disease.

Summary of the Complexity Challenging Psychiatry

Psychiatry must confront these sources of complexity and heterogeneity in its attempt to become ever more biologically grounded. Psychiatric disease is due to the dysregulated function of neural circuits, and this disruption arises from the temporally dynamic interplay of genetic, environmental, developmental, and neuroadaptive factors. Each of these factors, rather than being viewed as noise to be eliminated from psychiatry, are opportunities to provide a more biological foundation to psychiatry. The missing piece is a framework—perhaps probabilistic or mathematically explicit—that represents this complexity in a way that benefits psychiatry. In the next section, we turn to the potential for computational approaches to provide this much needed framework.

How a Computational Approach Can Be Useful

Psychiatry desperately needs greater neurobiological understanding, better diagnostic accuracy, and improved treatment and prevention strategies. Here we discuss how computational approaches might be used to address these challenges because they are suitable for representing and drawing inferences from the complexities discussed above. We outline two potential ways that computational psychiatry can be useful: (a) by providing a sophisticated set of tools and techniques (from fields such as machine learning and nonlinear dynamical systems) to analyze data in ways that are more powerful than customary statistical approaches, and (b) by providing a formal framework for theory and model development in psychiatry (Maia 2015).

Reducing Complexity of Presentation: Improving Diagnoses

Diagnosis in psychiatry is based on clustering enigmatic symptoms, rather than biological ground truths. This diagnostic framework unfortunately lends itself to symptoms that span multiple disorders, which produces a large amount of comorbid diagnoses. A second consequence of this diagnostic framework is that the label given to a group of patients may cluster patients with different disorders, which have different biological causes or risk factors. Computational approaches clearly have a role in helping to construct more useful diagnostic symptoms (see Flagel et al., this volume).

One approach to improving diagnoses is to ground observations in a more objective framework. Some objectivity may be obtained by simply reconceiving enigmatic, self-reported symptoms (e.g., delusions) as objective cognitive variables that can be measured in behavioral tasks. For instance, computational methods have been used to reconceptualize anhedonia, which is present in both schizophrenia and depression and contributes to comorbid diagnoses between these disorders. Theoretical models of reinforcement learning have been used to place anhedonia into a biological framework by relating it to underlying cognitive processes. Viewing anhedonia from the perspective of reinforcement learning has shifted the emphasis of study away from the severity of anhedonia and toward the cognitive operations that might be associated with the experience of pleasure. For example, these models have suggested that anhedonia in schizophrenia should be reconceived as dysfunction of the cognitive process by which subjects decide to explore their environment (Strauss et al. 2011). In this interpretation of schizophrenia symptoms, anhedonia is actually a reduced tendency to explore actions that could improve their status quo, rather than an altered experience of reward or reward learning. Similarly, an important meta-analysis of studies has also reconceived of anhedonia within a framework of cognitive operations during major depression (Huys et al. 2013). This work has revealed that major depression and anhedonia were more strongly associated with the appetitive aspects of reward (i.e., the experience of reward) than with reductions in the rate of reward-related learning. Reductions in other dimensions of reward were also noted and have been studied computationally using reinforcement theory models (Chen et al. 2015). As a whole, consideration of the work by Strauss et al. (2011) on schizophrenia and Huys et al. (2013) on depression suggests that patients with a comorbid diagnosis who self-report anhedonia might be diagnostically reclassified using the model parameters derived from fitting their behavior on computerized tasks that dissociate exploration, appetitive behaviors, and learning rate. Indeed, there is growing evidence that these models of reinforcement learning apply to anhedonia as a transdiagnostic dimension of psychiatric disorders (Whitton et al. 2015).

Even without a formal, conceptual model for behavior, more descriptive (yet still quantitative) formalisms (e.g., machine learning or dynamic causal modeling) may be useful. With sufficient data, tools might be developed for predicting the chance of suicide by using data to classify patients into diagnostic groups or dimensional clusters, mapping the landscape of causes and inferring endophenotypes, weighing gene–environment interactions for use in prevention, selecting appropriate treatments, and diagnosing and predicting response to treatment. For instance, a model could predict how the co-occurrence of anhedonia and lack of social support increase the chance of a depressive episode. In addition, by including task performance measures of anhedonia that focus on specific parameters, such as sensitivity to reward or tendency to explore for new rewards, these models may provide more accurate task-based predictors of disease course and treatment response, which are the formal purpose of diagnoses. As noted above, these descriptive models may be combined with theoretical models that quantify task performance and generate parameters, and can then be used as data in descriptive models to predict illness or treatment response.

As a specific example, consider again the clinical vignette illustrating the case of Peter, who expressed symptoms of bipolar disorder and schizophrenia. A clinician would greatly benefit from a single number that would express the likelihood that Peter had schizophrenia or bipolar disorder. Functional and anatomical connectivity differ between schizophrenia and bipolar disorder (reviewed by Krystal et al., this volume). Connectivity is an example of a complex system, which can be analyzed and quantitatively described using mathematical methods, such as graph theory and information theory (Boccaletti et al. 2006; Bullmore and Sporns 2009; Quian Quiroga and Panzeri 2009). Although mathematical theories of complexity in the brain are extremely nascent, some attempts have been made using these theories to provide a single variable that uses EEG or fMRI signals to quantify arousal or level of consciousness into a single number (Tononi 2004; Gosseries et al. 2011). The same approach could be applied in diagnosis of psychiatric disease. One such study used information theoretic methods to generate a quantification of complexity which was predictive of later development of autism (Bosl et al. 2011). Thus, one could imagine quantifying the complexity of connectivity (or some other predictive measure) in Peter, and determining whether that quantity was more similar to patients with bipolar or schizophrenia, to arrive at the probability of either diagnosis.

An alternative diagnostic approach might be fault tree analysis (FTA) (see MacDonald et al., this volume). One important goal of psychiatry is to integrate data across complexities and make predictions. For instance, changes of small effect size interact and accumulate in ways that are not simple to visualize and quantify. FTA provides a probabilistic framework that can compute the probability that genetic, environmental, and physiological evidence, as well as evidence of stable traits, impacts behavior and symptoms. In the case of suicide, many genetic, environmental, and behavioral/cognitive observations predict suicide. In the FTA framework, these observations can include protective or resiliency factors (e.g., personality traits, being in a romantic relationship). Critically, an emergent property of FTA is that multiple observations can interact in a probabilistic manner to produce dimensional symptoms that overlap across disorders. Three limitations of the FTA framework are: (a) it addresses discrete (categorical) variables, (b) it can only combine variables according to Boolean operators, and (c) it assumes that the relations are known *a priori*. Psychiatry, however, often addresses (a) interval variables that can (b) exhibit complex probabilistic relations for which (c) we do not know the relations *a priori*. Probabilistic graphical models provide an alternative to FTA (Koller and Friedman 2009; Pearl 2009b). These models have a richer framework that can address multiple types of variables (including categorical and interval); this allows different types of conditional probability relations to be specified (i.e., is not limited to Boolean operators) in which the relations between variables can be automatically learned from data. Both FTA and probabilistic graphical models hold the promise of diagnoses that represent breakdowns in specific elements (or combinations of elements) within the complex system that is the brain.

Improving Treatment

Improving how patients are assigned a diagnostic label will also improve treatment selection and monitoring of treatment response. For example, re-conceptualization of anhedonia within a reinforcement-learning model framework has led experimentalists to study the neural mechanisms associated with depression using behavioral tasks that have been informed by these models. These efforts have drawn attention to circuits previously implicated in reward (Pizzagalli et al. 2005; Kumar et al. 2008) and the mechanisms of antidepressant treatment (Herzallah et al. 2013). Subsequently, these circuits have been probed with regard to compromised functional connectivity in depression and response to circuit-based treatments (Downar et al. 2014). Thus, by rendering symptoms into constructs that can be studied with computational and experimental approaches, models of reinforcement learning have provided new windows into studying neurobiological mechanisms and potentially designing new treatments.

Similarly, models might also serve as a way to improve existing treatments. Learning models, for instance, permit one to reverse engineer the causal factors in cognitive behavioral therapy, enabling efficacy predictions of efficacy against depression (see Huys, this volume). Advances in methodology and the increased availability of computer power has allowed increasingly complex models to be tested and validated (Huys et al. 2012, 2015a). The resultant models suggest that specific psychotherapeutic interventions might be efficacious in particular types of patients, classified by their relative impairments in specific processes implicated by these models. Huys (this volume) reviews how these computational approaches have been used for assessing the need for, and response to, cognitive behavioral therapy.

Understanding Neurobiological Complexity

Another component of complexity is the multitude of genetic, environmental, neurodevelopmental, and neuroadaptive factors that cause disease. Here, too, computational models could be tremendously helpful for understanding how these factors dynamically interact and mechanistically cause psychiatric disorders.

Gene–gene interactions are now being studied in the context of complex networks (Barabási et al. 2011). The goal of these computational models is to define simple mathematical rules that describe how components of the network interact and use those mathematical formalisms to predict how perturbations of the network will change it. This type of computational approach has been studied in many different contexts, from gene–gene interactions, to neuronal interactions, to social interactions (Boccaletti et al. 2006; Bullmore and Sporns 2009; Barabási et al. 2011). Future work can build on these gene- and neuron-based networks by integrating them together and including environmental information. The goal of such network models could be, for example, to predict how perturbations, like stress, occurring at a particular neurodevelopmental period, affect gene expression and neuronal activity.

The above approach will require the integration of large amounts of data that span multiple levels (genes to environment to neuronal circuits). The same approach has been used in the context of computational models of reinforcement learning to improve the accuracy with which they predict behavioral performance (Luksys et al. 2009). In their work, Luksys et al. used a simple reinforcement-learning model that learns the value of various states and actions; however, they insightfully incorporated additional variables that modulated the model's learning. These modulatory variables spanned multiple levels, from genes to neuronal circuits: they reflected changes in arousal, attention and learning rate, and the tendency to explore new choices. Importantly, these modulatory variables were configured using the subject's (mouse) genetic endowment; arousal- and attention-related neurotransmitter (norepinephrine) levels; stable traits for anxiety, responsivity to stress, responsivity to novelty, and arousing situations; and task performance history and experience. Critically, in comparison to traditional reinforcement-learning models, this broader model, which incorporated data from many biological levels, was able to better predict the behavioral task performance. In summary, the biological cause(s) of symptoms and disorders emerge frothe complex and rich interactions between genes, environment, and neural circuits. By incorporating all of these data together into a computational model of reinforcement learning, we may obtain models that more accurately fit human task performance during learning. Moreover, in line with the discussion of diagnoses above, the parameters generated by the model fit may help to reconceptualize some enigmatic symptoms in a cognitive or decision-making framework, which could be useful for splitting patients into different diagnostic groups.

Taking Advantage of Temporal Complexity

One of the needs of psychiatry is the ability to predict the onset and stage of a disease using biomarkers. A wealth of genetic, anatomical, neurophysiological, and behavioral data is available to apply computational approaches to build predictive models. One recent study used machine-learning methods to classify and separate controls from individuals diagnosed with schizophrenia (Pettersson-Yeo et al. 2013). Furthermore, the method was able to distinguish correctly the stage of the disease, in that it separated prodromal individuals from syndromal (first episode psychosis) individuals. The data set highlights the usefulness of combining multiple types of data, including cognitive task performance. In this work, Pettersson-Yeo et al. (2013) used anatomical data obtained using structural and functional MRI, white matter topography data obtained by diffusion tensor imaging, genetic SNP data, and cognitive task performance. The inclusion of cognitive task data in analyses of this type is critical because other studies have demonstrated improved patient classification when the data set includes parameters derived from fitting theory-driven (e.g., reinforcement-learning) models to the behavioral data (Wiecki et al. 2015). These studies highlight the usefulness of computational methods for determining the onset and stage of disease, specifically, by incorporating a wide range of multilevel genetic, neural, and behavioral data.

Similar approaches on broad data sets that span multiple biological levels may assist efforts toward primary and secondary prevention of disease. For example, machine learning has been used as a tool to produce a potential biomarker that predicts adolescent drug use (Whelan et al. 2014). It is conceivable that this tool could be used to select vulnerable individuals for targeting efforts aimed at primary prevention of disease. Whelan et al. (2014) conducted a longitudinal study that incorporated personality traits, task performance, environment, and genetic endowment. Their work revealed that life experience (e.g., romantic relationships) in combination with neurobiological and personality characteristics can predict the emergence of future adolescent drug use. Therefore, machine-learning methods could be used to screen for vulnerability to future drug use, and primary prevention efforts could focus on removing this risk factor to lower the incidence of mental health disorders in college students.

Another recent study that used a novel computational approach to predict the transition from the prodromal stage to psychosis should be highlighted here: Bedi et al. (2015) note that one of the prominent symptoms of schizophrenia, which begins in the prodromal stage, is aberrant speech. Clinicians often pick up on aberrant speech during patient interviews. Bedi et al. (2015) used computational methods to decode informative patterns in speech from prodromal (ultra high risk) individuals and found that this method predicted transition to psychosis better than a psychiatrist, and better than biological signals obtained from fMRI. Although many computational approaches are focused on the biology (genes, physiology, and neural circuit connectivity and activity), this example highlights some uniquely human symptoms, such as language, which may be particularly amenable to computational approaches that aid in diagnosis or tracking the stage of a disease.

Moving Toward Prevention

A key unmet goal in psychiatry is the transition from treating disease to preventing it. There are two chief forms of prevention: primary and secondary. Primary prevention aims to stop individuals from contracting an illness. Secondary prevention aims to stop an illness from progressing. Neither is currently possible for most, if not all, psychiatric diseases; either would be tremendously beneficial to patients. Computational methods could, in principle, help with identifying methods for either primary or secondary prevention.

To achieve primary prevention in psychiatry is to mitigate the effects of the above-mentioned genetic, environmental, and stable trait-risk factors.

Approaches aimed at understanding the interactions between these factors, therefore, might reveal mechanisms to reverse or compensate for risk. It is important, however, to note that these factors do not always confer "risk." The exact same factor can provide risk in some contexts, whereas it is actually protective against disease in other contexts. For example, animal models have demonstrated that stress can have both maladaptive and protective effects on neural circuits and the propensity to develop psychiatric disorders (Ladd et al. 2005; McEwen 2006). Thus, efforts to prevent psychiatric disorders must take into account the ability of a single factor to have opposing contributions to disease, depending on the other factors that are present.

Efforts toward primary prevention are also stymied by an unclear picture of what constitutes the formal onset of the disease. The onset of anxiety disorders (Beesdo et al. 2009), schizophrenia (Eaton et al. 1995; Levine et al. 2011a), and autism (Zwaigenbaum et al. 2009; Brian et al. 2014) is variable across individuals. In schizophrenia, for example, the predromal (before symptoms) and prodromal (some signs of dysfunction but no frank psychosis) stages could reflect the presence of risk factors which may or may not guarantee development of schizophrenia symptoms. Without an accurate time point for disease onset, it is thus not clear what risk factors should be the focus of primary prevention efforts, nor is it clear when preventative efforts should be made. Here again, computational approaches might be helpful, if formal models, which characterize diease progression from a biological standpoint, can be developed to predict disease stage and progression.

Efforts toward effective secondary prevention will also require understanding how biological systems change with disease onset and progression. The prodromal stage of schizophrenia contains subsyndromal symptoms, such as abnormal thoughts and perceptions, anxiety and irritability, cognitive problems, and social withdrawal (McGlashan 1988; Seidman et al. 2010; Giuliano et al. 2012). This collection of symptoms has been termed "attenuated psychosis syndrome" or APS (Fusar-Poli et al. 2014) and clear biological correlates of APS have been reported (Howes et al. 2009; Egerton et al. 2012; Fusar-Poli et al. 2013a; Bernard et al. 2015; Chan et al. 2015), which suggests that there are biological targets for prevention of continued progression of the disease to the chronic stage. Furthermore, there are other biological changes that occur later in the syndromal and chronic stages of the disease (Krystal et al., this volume).

Although it is a source of complexity, disease stage-specific biology provides an opportunity for secondary prevention efforts that target stage-specific biological processes. Indeed, there is some evidence that stage-specific treatments succeed during the prodromal stage of schizophrenia. For example, cognitive behavioral therapy and omega-3 polyunsaturated fatty acid supplementation appear to be effective interventions precisely during the prodromal stage (Addington et al. 2011; Morrison et al. 2011; van der Gaag et al. 2012). Secondary prevention, using treatments that target stage-specific biological processes, will rely on searching for biological correlates and testing mechanistic hypotheses about the etiology and progression of disease; for example, animal models can test hypotheses about how perinatal dietary intake of omega-3 polyunsaturated fatty acids affect glutamate synapses (Bondi et al. 2014). In summary, the ability to enact secondary prevention to halt the progression of disease will rely on detecting the onset of disease and determining stage-appropriate treatments.

There are already areas where computational models have suggested therapies aimed at secondary prevention. These computational models fall into two categories: biophysical and theoretical (e.g., reinforcement learning). For example, biophysical models have suggested that mGluR2/3 agonists could prevent the progression of schizophrenia if administered at the right phase of illness (see Krystal et al., this volume). Another example comes from Parkinson disease research, where reinforcement-learning models combined with biophysical models have suggested that adenosine drugs are useful, during the early stage of Parkinson disease, for preventing the aberrant motor learning that underlies some of the chronic motoric impairment (Beeler et al. 2012). In these studies, dopamine D2 antagonists were used to model Parkinson disease by inducing direct motor performance deficits in rodents, but they also induced aberrant learning, which interfered with subsequent performance, even after drug washout. Computational models of the basal ganglia simulated this pattern via biophysical modeling of the effects of D2 antagonists on the excitability and plasticity of neurons that represent action costs, which were modeled by a reinforcement-learning (theory-based) model. The computational modeling study suggested that the blockade of adenosine receptors could reverse the plasticity underlying aberrant learning without affecting direct motor performance. This computational interpretation implies that adenosine antagonists might be fruitfully applied during the early disease stage to prevent further aberrant learning and progression of symptoms. In addition to suggesting biological pathways for secondary prevention, this computational interpretation provides novel testable hypotheses for animal models of action and learning in health and disease.

Finally, we highlight another type of computational approach, a hidden Markov model, which has not been extensively used to model disease progression, but may be useful for predicting the transition from one disease stage to another by taking into account the underlying biological changes. In psychiatry, it would be helpful to predict the transition from abstinence to relapse in addiction or the transition from the predromal, prodromal, syndromal, and chronic stages of schizophrenia. Current data suggests that intermittent and attenuated psychotic symptoms predict psychosis in only 30% of patients, whereas the remaning patients do not transition to psychosis or are subsequently diagnosed with bipolar disorder instead of schizophrenia (Fusar-Poli et al. 2013a). The benefit of hidden Markov models is that they can be used to model states that you cannot directly measure in humans (e.g., glutamate synaptic dysfunction, GABA deficit, synaptic downscaling, and atrophy). These states are assigned a probability based on what can be measured (e.g., using EEG, MRS). These models may allow inference of the neurobiological states and the corresponding disease stage of the patient.

Predicting Punctate Events

Finally, we wish to draw attention to the potential utility of models that predict punctate events. Such events include taking a drug, self-harm, suicide, sexual offense, and dangerousness to others. Between 1999 and 2009, suicide accounted for an average 34,523 deaths per year in the United States (CDC 2014). Furthermore, while 41,149 deaths occurred due to suicide in 2013, there were 1,028,725 attempts (USA Suicide 2013). Preventing suicides would have a huge impact on society. How accurately might the clinician predict the likelihood of a suicide attempt, based only on the information available in a typical case, as illustrated in the clinical vignette of the patient Jennifer? A decision to release a patient from a secure psychiatric facility is based on predicting whether they will attempt suicide in the next few hours or days, not over the next five years. Yet studies that have estimated the accuracy of suicide prediction suggest that our abilities to predict acute risk are woefully inadequate. Suicide is associated with factors such as psychiatric disorders, history of suicide attempts, insomnia, and self-reports of suicidal ideation (e.g., answering yes to "have you felt that life is not worth living?"). Jennifer meets criteria for diagnosis with major depressive disorder, she has insomnia, and she has suicidal ideations. Furthermore, she appears to have stable traits ("she was always a worrier") that could contribute to her predisposition for mood disorders and suicide. Although a suicide attempt appears likely in the long run for Jennifer, it is extremely difficult to predict the timing of that attempt over the short term. Thus, the decision to release her or not is a seemingly impossible scenario. One study that used the above factors as predictors of suicide attempt had only a 55% sensitivity (Pokorny 1983). Using self-report of suicidal thoughts alone to predict a suicide attempt is also likely inadequate. Approximately 9.3 million adults in the United States reported thoughts of suicide in 2013, yet there were considerably less (approximately 1 million) attempts (USA Suicide 2013; Substance Abuse and Mental Health Services Administration 2014). An alternative approach for prediction is to use biological factors, rather than self-reports and symptoms. Genetic polymorphisms and stress-related hormones (both available in saliva or blood samples) have some demonstrated usefulness as potential biomarkers that predict suicide, but not enough to warrant being used as a decisive criterion for holding a patient (Caspi et al. 2003; McEwen 2015). Here, computational approaches could help weigh various factors and make a prediction. For example, Bayesian methods based on biological factors (immune and inflammatory proteins) have improved sensitivity in comparison to predictions made using self-report and symptoms alone (Amsel and Mann 2001; Mann et al. 2006).

Theoretical models based on decision theory and reinforcement-learning models have also been proposed as a means of predicting suicide. For instance, game theoretic models have examined suicide attempts from the perspective of signaling others (e.g., family, physicians, bystanders) and influencing their behavior (Rosenthal 1993). Decision theoretic models can also examine suicide from the reinforcement-learning perspective of death being a "goal" to be obtained by a self-harming action. The goal, albeit an abnormal one, is associated with probabilistic risks, costs, and values, just as any other goal would be; in addition, the actions, in this case self-harming, can be associated with probabilities of obtaining the goal and amount of control over impulsively taking a self-harming action. These types of variables and their ability to predict a decision are commonly used in models of decision making (for a review, see Frank as well as Huys, this volume). Furthermore, these variables are quantifiable using behavioral tasks and have been found to be altered in individuals who attempt suicide. Suicide has been associated with impulsivity (Nock et al. 2009), altered emotional processing (Pisani et al. 2012), and altered valuing of future events (Courtet et al. 2011). We propose that models of reinforcement learning and agent-based choice models, which take these factors into account, could learn to "avoid" the suicidal goal or maladaptively learn that suicide has the highest utility out of all available choices.

How could these reinforcement-learning models be used to predict suicide? Models which "pathologically" learn to choose suicide could be fit to the behavior of patients measured in computerized decision-making tasks. The extent to which a patient's behavioral task performance is fit by a reinforcement-learning model, with the goal of attempting suicide, could provide a quantitative measure for their actual propensity to make an attempt. The model would incorporate the patient's valuation of suicide and other outcomes (such as seeing a loved one) and their ability to control impulsive actions that are self-harming. There is evidence that latent variables, such as impulse control and emotional state, can be measured in behavioral tasks and used to predict suicidality. For example, Nock and Banaji used computerized cognitive tasks to demonstrate that suicidal individuals bias their responses to emotional, suicide-related stimuli, as reflected in reaction times (Nock and Banaji 2007; Cha et al. 2010; Nock et al. 2010). Quantitative measures, such as reaction times, also allow potentially inaccurate self-reports of intent to harm to be avoided. Nock and Banaji have proposed that self-reports are not useful as a clinical criterion of suicidality, because individuals who are suicidal may lack the insight and reflection to report their intentions accurately, or may attempt to conceal their plans. Agent-based decision-making models could be fit to data in tasks using emotional, suicide-related stimuli so as to provide some model parameters that could be predictive of suicide. Bayesian hidden Markov models are another model type that could be used to infer individuals' beliefs about the utility, risk, and probability of a suicide attempt being successful. Models such as these have been used to infer subjects' beliefs about task structure in other types of decision-making tasks and have accurately predicted reaction times in those tasks (Paulus et al., this volume; Shenoy and Yu 2011; Ma and Yu 2015). Perhaps similar computational methods can be used to infer individuals' beliefs about the utility and cost of a suicide attempt based on reaction time, eye gaze pattern, and autonomic arousal data from tasks using emotional, suicide-related stimuli.

Limitations and Requirements of Models

Computational models provide a substantial arsenal for addressing issues of heterogeneity and complexity that occur in the computations performed by the brain and in the presentation of symptoms and the time course of disorders. For this endeavor to succeed, it is imperative that psychiatrists, researchers who study biology and neuroscience, and computational researchers communicate in a nonexpert manner. To this end, we have set out some requirements of models from the perspective of interactions between computational modeling research and psychiatry.

Successful use of computational approaches will require that behavioral tasks and data collection are designed with a particular computational model in mind. Models intended to disentangle various multilevel factors that could be implicated in mental illness will be far more useful if the data collected are informed by the model in the first place. For example, one might want to know whether a patient's seemingly maladaptive choices reflect changes in decision making or aberrant reward learning (Collins and Frank 2014). For simplicity, consider the most basic reinforcement-learning model that might be used to fit behavioral data collected to answer this question. This model has two parameters:

1. A learning rate that scales the impact of unexpected outcomes on future reward estimates
2. An exploration parameter that scales the degree to which the model either chooses what it perceives to be the best action or engages in some amount of random exploration

Depending on the behavioral task given to the subject, these model parameters can either be separable or colinear. For instance, if the task is deterministic (each choice always leads to the same outcome), a shallow learning curve could be explained by a low learning rate, a low choice exploration parameter, or both. However, if the task includes choices with multiple levels of reward probabilities, and if the task has sufficient duration to allow learning curves to reach an asymptote, then model-fitting task performance can reveal separable influences of learning and choice parameters. The model parameters will then be useful for answering questions about how a patient's choices are reflected in their decision-making and reward-learning processes. Thus, it is critical that the task manipulates experimental factors that are most identifiable to the model (or preferably, class of model) being employed. Simulations can be run before any data are collected to optimize the task design. Although it is understandable that tasks which accommodate patients can be difficult to design, it is worth increasing our efforts to design tasks in collaboration with modeling researchers. Importantly, this effort should also be made in modeling task performance in animals, since they are often a source of mechanistic insight into disease etiology and treatment. Finally, if models are to be useful for understanding the etiology of disease and developing novel treatments, they must be robust and validated and capable of generating testable hypotheses for experiments.

In the development of computational models, attempts should be made to build on existing models and integrate models of various types (biophysical, connectionist, etc.), which has been done to a great extent already. They should also continue to integrate vertically across levels (e.g., incorporate genes for the dopamine reuptake transporter and dopamine synthesis in reinforcement-learning models). Finally, models should also be user-friendly and the methods communicated clearly. For example, overfitting impacts reliability and should therefore be communicated when the methods used to design the model are explained.

Recommendations

We close with a set of specific recommendations to guide how computational models can address pressing issues in psychiatry. It is crucial that models are used to provide a biologically grounded and formal mathematical framework to our understanding of psychiatric disorders. The goals of this framework must be to identify critical biological factors and risk factors that predict disease risk, to offer differentially diagnostic criteria, and to define treatment and monitor its efficacy. The discovery of critical factors should be used to refine experimental hypotheses about the etiology and progression of disease. In turn, models should accept iterative updating to generate testable hypotheses in the laboratory setting.

As the field of computational psychiatry matures, the tools and novel schemata that it contributes to diagnosis, treatment selection, and evaluation of treatment response must be realized in practice. This practical impact will depend on outreach to health services providers and incorporating models into how mental health services are provided. We recommend focusing outreach, not only on psychiatrists, but specifically on nonmental health service providers (e.g., primary care physicians, school counselors and teachers, social workers) as these people often make the initial identification of mental illness and treatment decisions.

By focusing on practical applications of computational psychiatry, a complementary, but critical need in psychiatry—management and reduction of costs—may be addressed. For instance, a recent study in England found that the total cost to society of only schizophrenia alone amounts to 11.8 billion GBP annually; this equates to 60,000 GBP per person annually (Andrews et al. 2012). One report from England reports that the total annual costs (in GBP for year 2007) for other psychiatric disorders have been estimated as 7.5 billion (depression), 8.9 billion (anxiety), 4.0 billion (schizophrenia), 5.2 billion (bipolar disorder), and 14.9 billion (dementia) (McCrone 2008). The practical impact of computational approaches on mental health services will hopefully address the urgent need to manage ballooning healthcare costs by determining the most clinically and cost-effective interventions. Even a slight improvement in treatments or ability to predict illness emergence and progression would reduce health costs and improve lives.

Acknowledgments

We thank all of our colleagues at the Ernst Strüngmann Forum for insightful and energizing conversations. Additionally, we especially appreciate the contributions made by Cameron Carter (for clinical vignettes) and Michael Frank (for details on some exemplar computational models). N. Totah thanks Ms. Frederike Klein for proofreading and discussions.

4

What Does Computational Psychiatry Need to Explain to Capture Mechanisms of Psychopathology?

Facts, Almost Facts, and Hints

Deanna M. Barch

Abstract

This chapter provides specific research examples on the neurobiology of mental illness—using psychosis as a case in point—that may begin to rise to the level of "facts," or at least "almost facts" or strong "hints," about important etiological mechanisms that need to be explained to capture key components of at least some facets of mental illness. These examples are then used to illustrate where computational psychiatry approaches may help. In particular, there is an opportunity to provide links across different levels of analysis (e.g., behavior, systems level, specific circuits and even genetic influences) in ways that can lead to a more unified framework for understanding the apparent multitude of impairments present in psychosis, which may in turn lead to the identification of new treatment or even prevention targets. This chapter also discusses some of the known conundrums about the etiology of mental illness that need to be accounted for in computational frameworks, including the presence of heterogeneity within current diagnostic categories, the vast degree of comorbidity across current diagnostic categories, and the need to reconceptualize the dimensionality versus categorical nature of mental illness.

Introduction

One way to define the field of *computational psychiatry* is to view it as the attempt to *use computational theories of cognition and neuroscience to build computational models of mental disease and injury.* To do this, these theories and models need to capture known features of mental illness at multiple levels of analysis, including information about neurobiological mechanisms, behavior, clinical presentation, course, outcome, and treatment response. This is a tall order, as most psychiatrists and psychologists have a very short list to present when asked to provide the known "facts" about any specific mental illness. Often such facts are at the level of epidemiology or links to specific environmental events (Tandon et al. 2008; Van Os et al. 2008; Brown and Derkits 2010; Vassos et al. 2012); when it comes to neurobiological mechanisms, much less is known. This may, in part, reflect the limits of our currently available technologies, which do not provide the level of *in vivo* examination of neurobiological mechanisms in humans that is achieved in animal models. It may also reflect the real complexity and heterogeneity of the causes of mental illness, and the diversity of pathways that may lead to what appears to be a similar set of outcomes. Further, this state of affairs may arise partly because of our reliance on a set of categories and conceptualizations about the various "types" of mental illness that do not map cleanly onto clear distinctions at the neurobiological level. This latter issue, recognized as a particular crisis point, has led to the emergence of the Research Domain Criteria (RDoC) initiative (Insel et al. 2010; Morris and Cuthbert 2012; Cuthbert and Kozak 2013; Cuthbert 2014a), which is attempting to develop a psychiatric nosology based on variation in known neural systems that link to core aspects of behavior.

In this chapter I provide a few examples of research on the neurobiology of mental illness—using psychosis as an example—that may begin to rise to the level of "facts," though a more conservative stance might frame these as "almost facts" or strong "hints" about important etiological mechanisms that need to be explained to capture key components of at least some facets of mental illness. Following this, I discuss some of the known conundrums about the etiology of mental illness that also need to be accounted for in computational frameworks.

There Is Something Going on with Dopamine in Schizophrenia

The field of psychiatry has long hypothesized a critical role for dopamine in the pathophysiology of schizophrenia, though much of the early evidence for this was based on serendipitous treatment findings and the effects of drugs that stimulate the dopamine system on the emergence of psychotic symptoms. Over the years, numerous researchers have variously argued that dopamine dysregulation is or is not a key feature of the etiology of psychosis. However, in the past several years, the accumulating literature has clearly solidified an important role for dopamine dysregulation in the pathway to psychosis (Howes and Kapur 2009; Bonoldi and Howes 2013; Kambeitz et al. 2014; Howes et al. 2015). It is clear that this is not the only mechanism. It is also clear that there is variability in the degree to which dopamine abnormalities are present across individuals with psychosis and within individuals with psychosis over time. The nature of dopamine dysregulation associated with psychosis is multifaceted. Perhaps the most consistent finding among individuals with psychosis is the presence of increased presynaptic striatal dopamine synthesis capacity (Howes et al. 2012; Fusar-Poli and Meyer-Lindenberg 2013). This increased presynaptic striatal dopamine synthesis capacity is present both in individuals at clinical high risk for psychosis (Howes et al. 2009, 2011b; Allen et al. 2012; Egerton et al. 2013) and in individuals with diagnosed schizophrenia (Lyon et al. 2011), with the most reliable results in the dorsal as compared to the ventral striatum. There is mixed evidence as to whether similar results are present in individuals at genetic high risk for psychosis: at least one study found increases (Huttunen et al. 2008) whereas another did not (Shotbolt et al. 2011).

Research has also shown that this increased presynaptic striatal dopamine is associated with the severity of symptoms during the prodromal phase (Howes et al. 2009), that individuals who go on to develop manifest psychosis had greater presynaptic striatal dopamine availability at baseline (Howes et al. 2011b), and that the severity of increased presynaptic striatal dopamine availability deteriorates as individuals worsen in their psychotic illness (Howes et al. 2011a). Interestingly, there is also evidence that individuals who respond to antipsychotics are more likely to have increased presynaptic striatal dopamine availability than those who are treatment resistant, again pointing to potential heterogeneity in causal pathways (Demjaha et al. 2012). Importantly, this increased presynaptic striatal dopamine availability has been indirectly linked to altered salience attribution and increased attribution of motivational salience to irrelevant features of stimuli (Roiser et al. 2013).

In addition to the evidence for increased striatal dopamine availability, there is also evidence for increased striatal dopamine release following amphetamine administration (Kambeitz et al. 2014). This is present in both medication naive (Abi-Dargham et al. 2009) and unmedicated individuals with schizophrenia (Laruelle et al. 1996, 1999), as well as in individuals with other clinical manifestations related to schizophrenia, such as schizotypal personality disorder (Abi-Dargham et al. 2004). However, these dopamine alterations may no longer be present in individuals in remission (Laruelle et al. 1999). There is also some evidence of increased occupancy of D2 receptors by synaptic dopamine, again with the most robust evidence for dorsal versus ventral striatum (Kegeles et al. 2010).

Dopamine Dysfunction, Behavior, and Brain Function

There is now a robust behavioral, neuroimaging, and computational literature in healthy individuals that provides strong links between various aspects of dopamine function and a number of different components of cognition, learning, motivation, and effort- and value-based decision making (Hazy et al. 2006; Niv et al. 2007; Schultz 2007; Dayan 2009; Samson et al. 2010; Aarts et al. 2011; Cools 2011; Dayan and Walton 2012). As such, one would expect to find evidence that dysregulated dopamine function in psychosis is directly related to impairments in one or more of these aspects of behavior and function. However, it is not clear exactly what one would predict in terms of the direction of these impairments, given the nature of the dopamine dysfunction present in this illness, in that it is less clear how one would expect the interaction of enhanced dopamine availability, increased dopamine release, and increased D2 receptor occupancy to combine to change behavior. As such, this is a domain in which formal modeling could help to make principled predictions and to link across levels of analysis in a way that might clearly point to novel treatment targets. Surprisingly, there is essentially no research that has provided a *direct* link between any measure of learning, motivation, or decision making in humans with psychosis and indices of either increased presynaptic dopamine availability, increased dopamine response to amphetamine, or increased D2 receptor occupancy in schizophrenia. There is, however, indirect evidence that can inform modeling efforts.

Dopamine, Reward Learning, and Motivation in Schizophrenia

There is, of course, evidence that individuals with schizophrenia show impairments in reward processing, learning, and motivation domains that have been strongly associated with dopamine function, though, as discussed in more detail below, it is not clear whether there is a core mechanism that provides a unified account of these impairments. Further, some of the evidence is mixed in terms of impaired versus intact behavior and, as noted above, it has not been directly linked to dopamine function. For example, one might predict that individuals with schizophrenia should show impairments in learning mechanisms supported by dopamine in the striatum, such as the ability to learn what cues predict reward and to update stimulus response associations via striatal-learning mechanisms. However, the evidence suggests surprisingly intact performance on a range of tasks in which learning is either relatively easy or relatively implicit (Elliott et al. 1995; Hutton et al. 1998; Joyce et al. 2002; Turner et al. 2004; Tyson et al. 2004; Jazbec et al. 2007; Waltz and Gold 2007; Ceaser et al. 2008; Heerey et al. 2008; Weiler et al. 2009; Somlai et al. 2011), though with some exceptions (Oades 1997; Pantelis et al. 1999). Further, for the most part, individuals with schizophrenia show intact learning rates on the weather prediction task, a probabilistic category-learning task frequently used to measure reinforcement learning, though with overall impaired performance (Kéri et al. 2000, 2005a, b; Weickert et al. 2002; Beninger et al. 2003; Weickert et al. 2009). There is some evidence that reinforcement learning may be more intact for patients on atypical than typical antipsychotics, though it has been found in those on typicals as well (Beninger et al. 2003; Kéri et al. 2005b). Of course, people can accomplish tasks in many different ways, and intact performance at the level of broad behavioral metrics might arise from varying strategies. Again, this is a domain in which formal modeling might be able to help clarify whether performance is indeed intact using metrics that presumably index a particular approach to performing the task.

In contrast, when the reinforcement-learning paradigms are more difficult and require the explicit use of representations about stimulus-reward contingencies, individuals with schizophrenia show more consistent evidence of impaired reinforcement learning (Waltz et al. 2007; Morris et al. 2008b; Koch et al. 2009; Gold et al. 2012; Yilmaz et al. 2012; Cicero et al. 2014). Interestingly, these impairments may be greater when individuals with schizophrenia must learn from reward versus from punishment (Waltz et al. 2007; Cheng et al. 2012; Gold et al. 2012; Reinen et al. 2014), though some studies also find impaired learning from punishment (Fervaha et al. 2013a; Cicero et al. 2014). Further, recent work suggests that working memory impairments may make a significant contribution to reinforcement-learning deficits in schizophrenia (Collins et al. 2014). Such findings are consistent with the larger literature, which suggests altered cognitive control function in schizophrenia, and are also consistent with the growing basic science literature that suggests important interactions between what have been referred to as "model-free" learning systems (e.g., dopamine in the striatum) and "model-based" learning systems that engage prefrontal and parietal systems, which support representations of action-outcome models (Gläscher et al. 2010; Daw et al. 2011; Doll et al. 2012; Lee et al. 2014; Otto et al. 2015). Given the large body of evidence for altered dopamine function in the striatum in schizophrenia, it is quite puzzling as to why, at the behavioral and functional neuroimaging level, the deficits appear to be more in the realm of the model-based components of learning, which are thought to be supported by cortical systems. This is precisely an area where formal modeling may help us to understand this apparent conundrum.

One might also predict alterations in schizophrenia in the neural signals thought to reflect dopamine-mediated functions, such as reward anticipation or reward prediction error responses. Again, it is not entirely clear what direction of alteration one might predict in schizophrenia, given the nature of the dopamine abnormalities found in this illness. A number of studies have reported reduced ventral striatum activity to reward cues in schizophrenia, both in unmedicated (Juckel et al. 2006b; Schlagenhauf et al. 2009; Esslinger et al. 2012; Nielsen et al. 2012b) and in medicated individuals (Juckel et al. 2006a; Schlagenhauf et al. 2008; Simon et al. 2009; Walter et al. 2009; Grimm et al. 2012). There are some hints that these deficits might not be present in individuals on atypical medications, but these data are from small samples (Kirsch et al. 2007). Some work has found reduced ventral striatal responses to anticipation cues in antipsychotic-naive schizophrenia patients, which improved following atypical treatment (Nielsen et al. 2012a, b). Further, there is evidence that the magnitude of these impairments may vary as a function of the severity of specific types of symptoms in schizophrenia, such as negative symptoms (Juckel et al. 2006a; Simon et al. 2009; Waltz et al. 2010).

Other studies in schizophrenia have also examined prediction error responses using functional neuroimaging: an increase in striatal (potentially dopaminergic) responses to unexpected rewards and a decrease in striatal responses when predicted rewards do not occur. Several studies have found altered prediction error responses in schizophrenia, manifesting as either reductions in responses to unpredicted rewards and larger than expected responses to predicted rewards (Murray et al. 2008; Morris et al. 2012; Schlagenhauf et al. 2014). Gradin et al. (2011) found reduced prediction error responses in the caudate, but increases in the ventral striatum. Waltz et al. (2009) found evidence for reduced positive prediction error responses in a range of regions that included the striatum (dorsal and ventral) as well as insula, but relatively intact negative prediction errors in these same regions. In contrast, Walter et al. (2009) found intact prediction error responses in the striatum for both positive and negative prediction errors. Thus, this literature is quite mixed. There is again suggestion that medication may have a key influence. Insel et al. (2014) found that individuals with chronic schizophrenia taking higher doses of medication showed smaller prediction error responses. However, the fact that reduced prediction error responses have also been seen in unmedicated individuals (Schlagenhauf et al. 2014) argues against such abnormalities resulting only from medication effects in schizophrenia. Again, there is evidence that the magnitude of these impairments may vary as a function of the severity of specific types of symptoms in schizophrenia, such as negative symptoms (Waltz et al. 2009).

There is also a growing literature on altered effort-based decision making in schizophrenia, another function often associated with the dopamine system. The animal literature provides strong evidence that dopamine plays a key role in regulating physical effort allocation and vigor (Niv et al. 2007). For example, dopamine blockade, especially in the accumbens, reduces physical effort allocation (Salamone et al. 2009, 2012; Farrar et al. 2010; Salamone and Correa 2012), and increased D2 receptor expression in the nucleus accumbens of adult mice increases physical effort expenditure (Trifilieff et al. 2013). In humans, Treadway et al. (2012) found that increased dopamine release in response to d-amphetamine in the left striatum and the left ventromedial prefrontal cortex (PFC) was associated with increased willingness to expend physical effort. Based on these data, one might predict that individuals with schizophrenia might not show impaired effort-based decision making or that they might even be willing to expend greater levels of physical effort (at least those not on medication) if schizophrenia is characterized by hyperdopaminergic function. Surprisingly, the opposite has been found.

The majority of the literature on effort in schizophrenia has used physical effort tasks that involve finger tapping (Treadway et al. 2009), a balloon-popping task (Gold et al. 2013), or grip strength as metrics of physical effort allocation. Studies of finger tapping have consistently found a specific pattern of reduced effort allocation in schizophrenia: they do not differ from controls at low levels of reward or low levels of probability of receiving the outcome; they also do not show the same increase in effort allocation as either reward or probability increase (Fervaha et al. 2013b; Gold et al. 2013; Barch et al. 2014; Treadway et al. 2015). The two studies using grip strength showed differing results: one found reduced effort allocation in those with schizophrenia rated clinically as having higher apathy (Hartmann et al. 2015), whereas the other study found no significant differences in schizophrenia (Docx et al. 2015). Two studies have also examined cognitive effort. One study used a progressive ratio task and found evidence for reduced effort allocation in schizophrenia, although the design of the task was such that cognitive effort was confounded with physical effort (Wolf et al. 2014). In contrast, Gold et al. (2015) found little evidence of reduced cognitive effort in schizophrenia across three studies, though these studies did suggest that individuals with schizophrenia had difficulty detecting variations in cognitive effect among conditions.

In summary, the behavioral and neuroimaging literatures do suggest some consistent evidence for impairments in reward processing, reinforcement learning, and motivational functions putatively associated with dopamine in schizophrenia. However, it is not clear how easily the observed patterns map to what one might expect, based on the normative literature about dopamine's role in these functions and the impact that one would expect increased versus decreased dopamine metrics to have on behavior or brain function. Further, none of these studies have directly linked such impairments to variations in dopamine function in schizophrenia. As noted above, this is clearly a domain in which formal modeling, which can take into account the complex interactions among different components of the dopamine systems, could help to make rational predictions about behavior and neural activity, and help to bridge the levels of analysis. Some examples of the types of models that could be relevant are nicely outlined by the contribution of Frank (this volume).

Dopamine, Cognitive Control, and Working Memory in Schizophrenia

Although much of the literature on the role of dopamine in behavior has focused on reward or motivationally related functions, there is also a robust literature on the role of dopamine in other domains, such as cognitive control and working memory. Such theories have argued that dopamine is critical for modulating active maintenance of representations in PFC, potentially by providing a gating signal that indexes the need to update maintained representations (Braver 1997; O'Reilly et al. 1999; Braver and Barch 2002; O'Reilly and

Frank 2006), or to modulate local network activity in a way that regulates the maintenance of representations in PFC (Camperi and Wang 1998; Seamans and Yang 2004; Wang et al. 2004; Vijayraghavan et al. 2007).

There is a huge literature documenting that individuals with psychosis experience deficits in working memory and cognitive control (Barch 2005; Lee and Park 2005; Forbes et al. 2009), with evidence that deficits are present at first episode in unmedicated individuals and in individuals at familial risk for psychosis (Agnew-Blais and Seidman 2013; Bora and Murray 2014; Fatouros-Bergman et al. 2014). Critically, there is also a body of research indicating that cognitive impairment in schizophrenia is a critical determinant of quality of life and function, potentially more so than the severity of other aspects or symptoms of schizophrenia, such as hallucinations and delusions (Nuechterlein et al. 2011; Lepage et al. 2014).

One of the major challenges to understanding the nature of cognitive function in schizophrenia (or psychosis more broadly) is that at least on the surface, individuals with this illness appear to have deficits in a wide array of domains, not just in working memory and cognitive control. These domains include language function, episodic memory, processing speed, attention, inhibition, and sensory processing (Forbes et al. 2009; Mesholam-Gately et al. 2009; Sheffield et al. 2014), with such deficits clearly present even in unmedicated individuals (Fatouros-Bergman et al. 2014). It is unlikely that a single mechanism or model will be able to account for all of these impairments. At the same time, it seems equally problematic, and definitely not parsimonious, to develop different theories or models about the causes of impairments in each of these domains independently. Instead, there are likely several core mechanisms that each contribute to impairments in a number of cognitive deficits in psychosis.

In the spirit of identifying core mechanisms of cognitive dysfunction in schizophrenia, I and others have argued that one such mechanism is a deficit in the ability to actively represent goal information in working memory needed to guide behavior, and that this deficit reflects impairments in the function of the dorsolateral prefrontal cortex (DLPFC), its interactions with other brain regions (e.g., the parietal cortex, the thalamus, and the striatum), and the influence of neurotransmitter systems such as dopamine, GABA and glutamate (Barch et al. 2009; Edwards et al. 2010; Lesh et al. 2011). This framework is based in part on computational modeling work by Cohen and colleagues, which put forth the hypothesis that intact function of dopamine in DLPFC was responsible for the processing of context, and that a disturbance in this mechanism could account for a range of cognitive deficits in schizophrenia (e.g., Braver et al. 1999; Cohen et al. 1999; Barch et al. 2001). We have suggested that impairments in working memory, attention, inhibition, and language processing in schizophrenia can all be understood in terms of a deficit in goal representations, as each of these domains requires the active representation of such context information for effective function (for full discussion, see Braver et al. 1999; Cohen et al. 1999).

In more recent years, the role of context processing in cognition and in schizophrenia has been reconceptualized somewhat more broadly as the function of proactive cognitive control (Braver et al. 2007, 2009; Haddon and Killcross 2007; Edwards et al. 2010). This reframing builds upon concepts of context processing to argue for flexible mechanisms of cognitive control that allow humans to handle the range of challenges faced in everyday life. One such theory, termed dual mechanisms of control (Braver et al. 2007, 2009; Edwards et al. 2010), makes a distinction between proactive and reactive modes of cognitive control. Proactive control can be thought of as a form of "early selection," in which goal-relevant information is actively maintained in a sustained or anticipatory manner, before the occurrence of cognitively demanding events. This allows for biasing of attention, perception, and action systems in a goal-driven manner. By goal information, we mean information about what one needs to accomplish in this particular task situation or the intended outcome of a series of actions or mental operations. In real-life settings, such goals may include the main points you wish to communicate in a conversation or the need to organize a shopping trip so that you can make sure to get everything you need. In contrast, in the reactive mode, attentional control is recruited as a "late correction" mechanism that is mobilized only when needed, such as after a high-interference event is detected (e.g., you encounter unexpected distracting stimuli and need to retrieve the topic of your conversation). Thus, proactive control relies on the anticipation and prevention of interference before it occurs, whereas reactive control relies on the detection and resolution of interference after its onset.

A number of prior studies have provided support for these hypotheses concerning context processing, goal representation, and proactive control deficits in schizophrenia using a range of paradigms (for reviews, see Barch and Braver 2007; Barch and Ceaser 2012; Barch and Sheffield 2014, 2016). Further, there is robust evidence that individuals with schizophrenia have difficulty with components of working memory which one might closely link to the mechanisms of proactive control (Barch 2005). Specifically, individuals with schizophrenia show impairments on working memory tasks with all different material types (e.g., verbal, spatial), and there is relatively little evidence for selective deficits with one material type over another (Lee and Park 2005; Forbes et al. 2009). In addition, they consistently show deficits on tasks designed to measure a range of functions ascribed to the "central executive" component of working memory, including manipulation (Kim et al. 2004; Horan et al. 2008), interference control, and/or dual-task coordination (e.g., Smith et al. 2011) as well as information updating and temporal indexing (e.g., Galletly et al. 2007).

Despite this wealth of evidence for impairments in proactive control, working memory, and other "executive" type tasks, there are also some important conflicting data points. First, the literature on task switching is not seamlessly consistent with ideas about proactive control and prefrontally maintained task representations. In task-switching paradigms, people often show worse performance when they have to switch between tasks and update rules. This effect, however, is reduced when they have a longer time between task cues and stimulus, allowing them to use anticipatory mechanisms to reconfigure task sets prior to applying them to a stimulus. In such situations, one might expect that individuals with schizophrenia would show increased "switch" costs and less of a benefit from a longer time between task cue and stimulus. However, only a handful of task-switching studies have shown evidence for increased task- or response-switching costs in schizophrenia (Elvevåg et al. 2000; Meiran et al. 2000; Franke et al. 2007). Many others have not shown any evidence of increased task-switching costs (Barton et al. 2002; Manoach et al. 2002; Karayanidis et al. 2006; Kieffaber et al. 2006; Greenzang et al. 2007; Jamadar et al. 2010; Manoach et al. 2013), at least not when overall longer reaction times are taken into account. Further, individuals with schizophrenia seem to show as much benefit as controls when there is a longer period between the task-switch cue and the trial (Meiran et al. 2000). The fact that individuals with schizophrenia seem able to use this cueing time to prepare in advance is not consistent with a deficit in proactive control. Ravizza and colleagues have shown task-switching impairments in schizophrenia when the task involved more complex rule switching, but not in a more perceptually based task (Ravizza et al. 2010; Wylie et al. 2010). They argue that task-switch deficits will be more apparent when the rule that needs to be updated is the more complex.

A second conflicting data point is that one would predict that individuals with schizophrenia should have challenges using predictive cues to help them select task-relevant versus task-irrelevant information for encoding into working memory. However, studies by both Jim Gold's and Ed Smith's groups found evidence that individuals with schizophrenia had an intact ability to use predictive cues to guide management of the contents of working memory (Gold et al. 2006; Smith et al. 2011). Gold's group, however, has found strong evidence for impaired working memory capacity (Gold et al. 2003, 2010), and robust evidence that individuals with schizophrenia have difficulties inhibiting the impact of salient distractors in working memory (Hahn et al. 2010). As with the task-switching literature, it is not quite clear why there is not more evidence for impaired predictive control over working memory in schizophrenia, though one could again speculate as to whether introducing a prepotency manipulation might reveal susceptibilities in these mechanisms.

As with the literature on reward processing and reinforcement learning described above, there is almost no evidence directly linking specific types of cognitive impairments to dopamine function in schizophrenia, though there is clearly indirect evidence. For example, a number of studies have linked manipulations of the dopamine system to performance on tasks tapping proactive control in both humans and nonhuman animals (Barch 2004; Chudasama and Robbins 2004; Barch and Carter 2005; Barch and Braver 2007; Cools and

D'Esposito 2011). This is consistent with the literature discussed above on a potential role for dopamine in gating representations in working memory that are relevant for maintaining cognitive control (Braver 1997; O'Reilly et al. 1999; Braver and Barch 2002; O'Reilly and Frank 2006). Nevertheless, while such evidence is consistent with the idea that there may be a role for dopaminergic abnormalities in cognitive impairment in psychosis, more direct evidence is needed, as well as formal modeling efforts that can link across levels of analysis.

How Does Dopamine Dysfunction Relate to Other Neurobiological Abnormalities Present in Psychosis?

As described above, there is now consistent evidence for specific types of dopamine abnormalities in individuals with psychosis. However, there is also consistent evidence for other types of neurobiological impairments in psychosis, and it will be important to develop models that allow us to understand the full range of neurobiological factors that may contribute to the emergence of psychosis. As an example, meta-analyses indicate that individuals with schizophrenia show robust evidence for reduced activation in the DLPFC during a range of cognitive tasks, especially those tapping into cognitive or executive control (Minzenberg et al. 2009; Ragland et al. 2009). This consistent meta-analytic evidence for altered activation extends to dorsal parietal and anterior cingulate regions as well, though it is actually increased activity in anterior cingulate (Minzenberg et al. 2009). There is also meta-analytic evidence for a range of alterations in brain structure in schizophrenia, including reduced hippocampal volumes (Vita and de Peri 2007; Adriano et al. 2012), reductions in insula, anterior cingulate, thalamic, and caudate volumes, and increased ventricular volumes (Ellison-Wright et al. 2008; Glahn et al. 2008; Adriano et al. 2010; Bora et al. 2011). There is also some evidence of progression to these brain volume changes, with decreases in whole brain volume and increase in ventricular size over the illness course (Olabi et al. 2011; Vita et al. 2012).

There is also at least some evidence for alterations in other neurotransmitter systems besides dopamine in schizophrenia, with literatures regarding both GABA and glutamate. The evidence for GABA impairment has recently been reviewed by Taylor and Tso (2015) and is nicely articulated by Krystal et al. (this volume). One example provided by the authors was that good data from postmortem studies indicates that certain types of GABAergic interneurons are reduced. Specifically, Taylor and Tso argue that there is consistent evidence from postmortem studies for reductions in 67-kDa isoform of GAD67, localized to parvalbumin (PV)-positive interneurons. This has been found across a number of different brain regions. However, they also note that the *in vivo* studies of GABAergic function in schizophrenia have not yet provided strongly consistent evidence in this regard. Taylor and Tso note that there are at least two types of PV-positive interneurons that could be particularly functionally relevant to understanding cognition and behavior in schizophrenia. One set are the fast-spiking PV-positive basket cells that have been associated with cortical gamma oscillations (Bartos et al. 2007; Sohal et al. 2009), which has in turn been associated with working memory and proactive control (Cho et al. 2006; Lewis et al. 2008a; Minzenberg et al. 2010). The other are chandelier cells, which may play a role in depolarizing and exciting pyramidal cells when they are less active, but potentially inhibiting active ones (Woodruff et al. 2011).

The evidence for glutamate impairment in schizophrenia has recently been succinctly summarized by Howes et al. (2015), with much of the evidence coming from either ketamine studies in healthy adults (which elicit psychotic-like symptoms) or from magnetic resonance spectroscopy studies. Hypotheses around glutamate in schizophrenia tend to focus on hypofunction of the NMDA receptor, though as Howes et al. (2015) note, the evidence for this in postmortem studies is not consistent. A recent meta-analysis examined studies using ^{1}H magnetic resonance spectroscopy (^{1}H-MRS) to examine glutamate and glutamine (precursor to glutamate) (Marsman et al. 2013). This meta-analysis found reduced glutamate in the frontal cortex, but increased glutamine. Marsman et al. suggest that this could reflect a deficit in glutaminase, which normally converts glutamine into glutamate. Howes et al. (2015) also note the growing literature that suggests important links or interactions between dopamine and glutamate dysfunction in the etiology of schizophrenia. The hypothesis is that dopaminergic hyperactivity might be secondary to glutamate dysfunction in regions such as the hippocampus; this, in turn, leads to disinhibition of dopamine neurons in the striatum (Lodge and Grace 2006, 2007, 2011a). As an aside, there is also a literature on modeling of glutamate/GABA interactions with dopamine contributions to working memory that might also inform formal modeling and help us to understand the nature of working memory impairments in schizophrenia (Wang 1999; Seamans et al. 2001; Yasumoto et al. 2002; Lapish et al. 2007; Rolls and Deco 2015).

As yet, relatively little work has been done to link the diverse neurobiological impairments found in schizophrenia. However, there are some intriguing hints. For example, presynaptic striatal dopamine availability has been linked to altered PFC activity during cognitive performance in individuals at clinical high risk for psychosis, with evidence for links with both increased inferior prefrontal activity (Fusar-Poli et al. 2011) and decreased middle frontal gyrus activity (Fusar-Poli et al. 2010). In addition, other work has shown a negative correlation between hippocampal glutamate levels and striatal dopamine in individuals at clinical high risk for psychosis (Stone et al. 2010).

How Does Dopamine Dysfunction Relate to
Other "Facts" about Psychosis?

A number of consistent epidemiological factors have been associated with the development of psychosis. These factors include, among others, the evidence that pregnancy and birth complications with hypoxia are associated with a higher risk of psychosis in the developing fetus (Cannon et al. 2002; Miller et al. 2011). In addition, other prenatal and perinatal adversities (including stress, infection, malnutrition, maternal diabetes, or other medical conditions) have been linked to psychosis (Brown 2011). Work has also linked season of birth to the rates of schizophrenia (Brown 2011), potentially because season of birth marks the potential risk for the exposure to maternal influenza in the fetus. Intriguingly, offspring with older paternal age are at a greater risk for schizophrenia (Malaspina et al. 2001; Stilo and Murray 2010). How does one integrate such factors with the evidence described above for impairments in dopamine function, glutamate, brain activity, or brain structures? An elegant framework for incorporating these factors into a unified understanding has been made available by Howes and Murray (2014), who have used a neuro-developmental framework to integrate these seemingly diverse factors. More specifically, they have argued that a number of the early childhood factors associated with increased risk for schizophrenia are ones that are also known to lead to alterations in the dopamine system and at least some aspects of brain structure, such as hippocampal volume. This includes evidence that *in utero* inflammation and exposure, as well as a variety of stress and social risk factors, are associated with altered dopamine function, including stress-related sensitization of the dopamine system. Although these links are primarily conceptual at this stage, they do provide an intriguing way to begin to try to unify our understanding of the diversity of impairments present in psychosis, in a way that may be relatively amenable to formal modeling.

Additional Challenges and Considerations
for Computational Psychiatry

Above I very selectively reviewed some "facts," "almost facts," and "hints" about impairments that are present in psychosis, and which may be part of the pathophysiology of this illness. While not at the same level of analysis as the types of mechanisms and impairments described above, there are other "facts" about psychiatric disorders and the presence of varying types of neural impairments that need to be taken into consideration when trying to develop and apply computational frameworks to help us understand etiological mechanisms.

The first is something alluded to at the start of this chapter: we clearly do not have our diagnostic categories quite right, or maybe not even close to right.

This is reflected in the fact that there is huge heterogeneity among individuals with the same diagnoses in terms of the types of symptoms with which they present, with accompanying heterogeneity in the severity of behavioral and neural impairments. Further, there is massive comorbidity across psychiatric disorders, with the same individuals often meeting criteria for many different disorders, and with both symptoms and behavioral/neural deficits shared across putatively different diagnostic boundaries.

The second is that the manifestations of psychiatric disorders vary across the lifetime of the individual, even after the development of manifest illness. Some of this may reflect treatment-related changes, but it may also reflect the interaction of illness factors with normal developmental and/or aging mechanisms. If so, then information about developmental changes in neural mechanisms need to be incorporated into our models so that predictions can be made about factors that may influence emergence, presentation, and treatment at different stages of life.

Computation

5

Computational Approaches for Studying Mechanisms of Psychiatric Disorders

Zeb Kurth-Nelson, John P. O'Doherty, Deanna M. Barch,
Sophie Denève, Daniel Durstewitz, Michael J. Frank,
Joshua A. Gordon, Sanjay J. Mathew, Yael Niv,
Kerry Ressler, and Heike Tost

Abstract

Vast spectra of biological and psychological processes are potentially involved in the mechanisms of psychiatric illness. Computational neuroscience brings a diverse toolkit to bear on understanding these processes. This chapter begins by organizing the many ways in which computational neuroscience may provide insight to the mechanisms of psychiatric illness. It then contextualizes the quest for deep mechanistic understanding through the perspective that even partial or nonmechanistic understanding can be applied productively. Finally, it questions the standards by which these approaches should be evaluated. If computational psychiatry hopes to go beyond traditional psychiatry, it cannot be judged solely on the basis of how closely it reproduces the diagnoses and prognoses of traditional psychiatry, but must also be judged against more fundamental measures such as patient outcomes.

Introduction

The human mind is the most complex known phenomenon. Mental illness is, almost certainly, correspondingly complex. To treat mental illnesses effectively, we need a deep understanding of its mechanisms, not just a description of its surface properties. Yet these mechanisms likely can only be described as sophisticated interactions between many moving parts that both function and fail in complex, nonlinear ways. A promise of computational psychiatry is to provide the tools that are naturally suited for describing this complexity, so as to capture the essence of both function and failure.

Although computational methods address complexity, they should be viewed as a way of making things simpler, not more complex, by providing insight and understanding that transcends what can be gleaned from experimental observation alone. In the end, this understanding should be communicable without equations.

A recurring theme in this chapter is that computational tools can be applied at many different points within the science and clinical practice of psychiatry. A crucial mechanism may be at the level of protein folding and ion channel kinetics, but may equally be at the level of structure hidden within patterns of treatment response across patients. Computational psychiatry does not require reductionism but seeks to apply computational tools wherever they might yield insight, often with the result of linking domains of knowledge in a parsimonious way.

The Space of Computational Psychiatry

At play in psychiatric illness is the whole organism: from genes to molecules, cells, circuits, brain systems, behavior, and social influences. To begin to attack this host of processes with computational methods, first we need to organize both the processes and the computational methods. On the computational side, the way that we think about the problem can be divided into levels of analysis, called "computational," "algorithmic," and "implementation," after Marr (1982). In this section we outline the levels of the organism and levels of analysis, and then, within this organizational framework, show where computational methods can be applied to psychiatry.

Levels of Biology and Psychology

Biological processes relevant to psychiatry have been studied extensively across many levels of scale (Figure 5.1). The smallest level of scale relevant for biology is arguably the molecular. (Genes are often situated below molecules because their code is unpacked into molecules.) Proteins and protein complexes are the building blocks of biology, carrying out the various processes that permit neurons, the cells of the nervous system, to develop, survive, signal, learn, grow, and senesce at the appropriate times and places. The cell, most centrally the neuron, is the next level of organization. The neuron (and perhaps the glial cell) forms the elemental information-processing unit, integrating synaptic and other signals in its dendrites and soma and passing this information, transformed according to its own special calculations, downstream

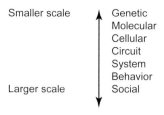

Figure 5.1   Levels of biology and psychology.

via an axon to its various partners. Multiple neurons and their connections form circuits, which act cooperatively and dynamically to increase the complexity and robustness of these neuron-based computations. Circuits combine to form distributed neural systems; activity in these systems guides behavior.

Psychiatric disorders can perhaps best be understood as spanning multiple levels of scale. For example, consider the case of the 22q11.2 microdeletion syndrome. Individuals with this microdeletion have cognitive and neuropsychiatric deficits and often meet criterion for schizophrenia (Karayiorgou et al. 2010). At the genetic level is the root cause of the disorder, a 1.5 to 3 Mb deletion knocking out 2–3 dozen genes. In mice, at least, one of the many consequences of the microdeletion at the molecular level is the mislocalization of an enzyme regulator in the axons (Mukai et al. 2015). At the cellular level, axon-branching deficits can be seen in cortical projection neurons; at the circuit level, neural transmission is reduced between the hippocampus and the prefrontal cortex (PFC). At the systems level, neural synchrony between the hippocampus and PFC is reduced (Sigurdsson et al. 2010), while at the behavioral level, mice carrying the microdeletion have deficits in spatial working memory (Stark et al. 2008).

Each of these individual deficits exists within a given level of scale. Yet a thorough understanding of the syndrome requires integration across scales. Are these observations connected? Does the molecular level deficit in enzyme localization cause the cellular level deficit in axon branching? A computational approach has the potential to enable integration by demonstrating, in a mathematically rigorous way, how phenomena on one level impact phenomena on another. Thus, one can construct a formal, mathematically quantifiable model of the axon-branching process, incorporating the location of the enzyme in question as a variable, and then test the effect of mislocalization on branching. The model predictions can be used to predict the results of experimental manipulations to provide further evidence in support of causal connections across scales.

It should be noted that the linear organization of levels presented here is simplified. There are fuzzy boundaries: between molecules and cells, there are organelles like mitochondria, macromolecular complexes such as the postsynaptic density, and signal processing pathways. There are also backward causal connections: neuronal activity shapes gene expression, circuit connectivity induces cellular specializations (such as synapses that recruit further changes in gene expression), and behavior itself strongly influences activity and plasticity across multiple levels. There are clear instances of level jumping: molecular changes can affect circuit formation and behavior, and systems activity can alter gene expression patterns. Finally, environmental influences appear at every level. Environment affects gene expression, neuronal growth and integrity, circuit plasticity, functional connectivity, and, of course, behavior.

These complexities notwithstanding, the differentiation of biological and psychological processes into levels of analysis can be useful as a means to identify how approaching a phenomenon at one spatial and temporal scale might inform understanding at other spatiotemporal scales. This differentiation can also serve as a starting point for the computational modeler in guiding the selection of a modeling approach that is appropriate to the type of phenomenon being studied.

Levels of Analysis in Computational Modeling

Next, we seek to organize the problem space of psychiatry from the point of view of how it can be described with computational models. David Marr famously described three levels of analysis for computational neuroscience (Figure 5.2; Marr 1982). The "computational level" specifies the goal of a system. For example, a computer program might be charged with sorting a list of numbers in a descending fashion. There are, however, many strategies for tackling this problem, such as sequentially searching for the nth-largest number or repeatedly swapping pairs of numbers that are out of order. These strategies live at the "algorithmic level." Finally, a given algorithm must ultimately be realized in software and hardware, with details such as where in memory to store the array. This is the "implementation level." Computational models of brain and behavior can potentially link to psychiatry at all three levels.

Take, for example, reward learning in the brain: A reinforcement learner's computational goal is to maximize their sum of future rewards. One (of several) algorithm that attempts to achieve this goal is the actor-critic algorithm. Here, the critic learns to predict the expected value of particular environmental states, and errors in these predictions (i.e., reward prediction errors) are used in two ways: (a) to improve future estimates and (b) to adjust weights in the actor, which selects among available actions (for further discussion, see chapters by Frank, Huys, and Montague, this volume). Those actions which yield the largest reward prediction errors in the critic are more likely to be selected, increasing experienced rewards. Implementation of this algorithm has been linked to the basal ganglia and dopaminergic system, with the ventral striatum learning the predicted values of states, the dopaminergic neurons signaling reward prediction errors (together with the ventral striatum, this is the critic), and the dorsal striatum playing the role of the actor (O'Doherty et al. 2004). One distinguishing feature of the actor-critic model is that it predicts no preference

More abstract

More detailed

Computational
Algorithmic
Implementation

Figure 5.2 Levels of analysis.

for actions that have yielded unexpected rewards over actions that avoid potential losses but have yielded no reward, since both have elicited positive prediction errors relative to a baseline of expected outcomes. Healthy controls and schizophrenia patients with high negative symptoms are well characterized by this pattern, whereas schizophrenia patients with low negative symptoms are better described by an alternative reinforcement-learning model (Gold et al. 2012; Palminteri et al. 2015).

A few words of caution are warranted. As with levels of biology, Marr's levels are inevitably fuzzy. One person's computational goal could be another person's algorithm in service of a broader computational goal. Also, the use of the word "computational" in Marr's levels is sometimes confusing. (A better name might be the "problem" level; that is, which problem the system is designed to solve.) Throughout the rest of this chapter, we will use "computational" much more generally to mean the application of sophisticated mathematical and theoretical tools to complex biological systems. Finally, we note that Flagel et al. (this volume) distinguishes between "normative" models and "process" models: the former is roughly equivalent to the computational level and the latter is roughly equivalent to algorithmic and implementation levels.

Any level of analysis can be applied to any level of biology. For example, we can ask what "goals" a gene network is set up to accomplish (e.g., maintaining balance in expression levels of two proteins), what algorithm it uses (e.g., feedback inhibition), and how it is biologically implemented (e.g., binding of protein products to promotor regions).

One of the strengths of computational modeling is that it naturally draws out connections between levels of analysis by forcing us to think in detail about what algorithm achieves a computation, and how that algorithm may be implemented. These connections can lead to insights in psychiatry.

Different levels of analysis can and do inform each other. Sometimes biological implementation informs understanding of the algorithmic level. Biophysically detailed models can simulate the differential contributions of D1 and D2 striatal neurons and predict both neurophysiological recordings and effects of dopamine manipulations on behavior. The properties of these models can be summarized at the algorithmic level by a modified actor-critic called OpAL ("opponent actor learning") in which the actor is divided into two components that, via nonlinear learning rules, come to specialize on representing the benefits of alternative actions (in D1 neurons) and the costs of these actions (in D2 neurons) (Collins and Frank 2014). This framework allows the model to simultaneously capture the effects of a variety of manipulations across levels (including pharmacology, genetics, optogenetics, and behavior). This provides a clue as to the computational function of these opponent processes; indeed a normative analysis provided some evidence that they improve performance compared to classical reinforcement-learning algorithms. Thus, this example illustrates how algorithmic and computational considerations have informed interpretation of the biology and allowed for testable predictions, but also how mechanistic constraints can reciprocally inform the algorithmic level.

Principles for Applying Computation to Psychiatry

With this multidimensional spectrum of biology and computational approaches, what principles can guide our attack on psychiatric illness? The first principle, sometimes overlooked (Markram 2012), is that computational models must be targeted carefully to the questions we want to answer. A model can, *by design*, answer some questions but not others. For example, an architectural model of a building that is made of cardboard can be used to ask questions about general aesthetics, aspect ratios, and visual impact of the building, but not structural questions such as whether the roof will sustain a pool. The questions one would like to answer with the model should prescribe the level of description at which the model is designed, and what levels of description it can safely abstract over. For instance, if one would like to ask whether different methods for detoxification (inpatient, outpatient, etc.) might be more or less effective in preventing relapse to the addictive substance, a model at the level of systems and behavior might be more useful than one specified at a biophysical level of detail. In contrast, if the goal is to develop targeted gene therapy, detail at the biophysical level may well be needed.

The second principle is that models can provide insight into complex systems, but the insight itself need not be complex. For instance, consider the hypothesis that dopamine neurons implement a reward prediction error which is used for reward-related learning (Montague et al. 1996; Sutton and Barto 1998). This insight arose from theoretical models of reinforcement learning, yet in the end, the principle that it revealed is quite simple. This is a strength of computational models, not a weakness. The gleaning of simple principles from complex neural and behavioral data is the goal. The model is a tool for sharpening our thinking and for formalizing hypotheses, not an end unto itself.

The third principle is that disease mechanisms and mechanisms of healthy function are both productive targets for computational methods in psychiatry. Directly seeking to explain behavioral and genetic abnormalities in disease are obvious targets. However, applying computational methods to enrich our understanding of healthy function can also lead to translational results. For example, electrophysiological work that described dopamine effects on persistent neural activity in PFC (Seamans and Yang 2004) and the role of dopamine on working memory-related persistent activity (Goldman-Rakic et al. 2004) informed computational models (Durstewitz et al. 2000; Durstewitz and Seamans 2002). These models, in turn, helped shape expectations about what D1 agonist pro-cognitive drugs might achieve (Rosell et al. 2015) and why PFC D1 deficits in schizophrenia might be associated with working memory dysfunction (Abi-Dargham et al. 2002).

The fourth principle is that both postdiction (i.e., explaining existing data) and prediction are useful. In some cases, models show how a simple set of principles can explain a broad range of existing data, which is valuable in clarifying our understanding and setting future directions. In other cases, models make novel, testable predictions that were difficult or impossible to make without the model. It is not uncommon to start by explaining a body of existing data and then make new predictions.

What Is the Toolkit That Computation Brings to Psychiatry?

In this section we outline the range of computational tools available to attack this problem space. It is useful to organize these tools both from the perspective of the biological or psychological systems to which they directly pertain as well as in terms of the mathematical framework from which they originate.

Organizing Computational Tools by Level of Biological Scale

Computational tools may be distinguished by the level of abstraction and biological scale they address (outlined in Table 5.1), moving from models that address detailed biophysical or biochemical processes to those that describe behavioral principles at an abstract level.

Table 5.1   Organizing models by level of biological scale. The left column lists some broad classes of computational models that target different features of biological systems. The right column maps the levels of biology and psychology to which these models are most often linked.

Type of model	Levels of biology
Biophysical models	Molecular
	Cellular
	Circuit
Connectionist models	Circuit
	System
Reinforcement learning	System
	Behavior
Bayesian inference	Circuit
	System
	Behavior

Biophysical Models

Biophysical models attempt to faithfully capture real biological details, such as the temporal evolution of membrane potentials or the temporal and voltage-dependent behavior of ionic conductances. The advantage of these models for psychiatry is that they provide a close link to pharmacology and genetics by explicitly describing drug targets and gene products. Biophysical models exist at many levels of abstraction. At the most detailed level, a biophysical model may capture the whole spatial extent of neurons with all their axons and dendrites (compartmental models), with the gating behavior of a large array of ionic conductances, and perhaps even intracellular molecular cascades; in short, any biochemical or biophysical process that can be expressed in terms of differential equations (Koch and Laurent 1999). Intermediate-level models may reduce this structure to just a few spatial compartments (e.g., one for soma and one for dendrite) and retain just a few ionic currents essential for the questions at hand (e.g., Durstewitz and Gabriel 2007). At the most abstract level, a biophysical model may consist simply of one or a few differential equations for the membrane voltage and for variables which capture the lump effect of many ionic currents (e.g., Hertäg et al. 2012). In general, the most abstract is therefore the class of models most suitable for addressing questions about how, for instance, specific pharmacological agents or, more generally, genetic, molecular, or physiological factors impact on network dynamics, as variables measured experimentally at this level can be translated into the models with none or only few additional assumptions or simplifications. For instance, in Durstewitz et al. (2000), changes in several currents due to D1- or D2-class receptor activation as measured *in vitro* were implemented in compartmental models, which then were used in a level-bridging approach to investigate the implications of these current changes for network dynamics, and ultimately working memory and cognitive symptoms in schizophrenia (Durstewitz and Seamans 2008; Frank 2015).

Connectionist Models

One key approach to understanding mechanism has been the use of neural network models, also referred to as connectionist or deep learning models (McClelland et al. 2010). These models bridge between cells and behavior by explaining how groups of cells encode information in ways meaningful to behavior, and are often applied to circuit- or systems-level neural phenomena. A first generation of neural network modeling, beginning as early as the 1960s but flourishing in the 1990s and 2000s, provided powerful insights into the computational mechanisms underlying complex patterns of clinical dysfunction in language disorders and certain forms of dementia (Plaut and Shallice 1993; McClelland and Rogers 2003). Neural networks were also applied, beginning in this period, to account for impairments of cognitive control and working memory function in schizophrenia and other disorders, implementing some of the first computationally explicit proposals concerning the role of dopamine in psychiatric pathophysiology (Cohen and Servan-Schreiber 1992). Innovations in neural network methods combined with the advent of faster computers, capable of running large-scale simulations, have recently triggered a new wave of neural network research, allowing more direct validation of these models as accurate representations of neural information processing (Afraz et al. 2014), and application to richer bodies of data. Another feature of this new wave of research is the development of neural network models that also include constraints from biological data, while retaining close contact with behavioral phenomena (Hoffman and Cavus 2002).

A key feature of connectionist models is learning representations of the inputs that are useful for generating outputs. Even very complex representations learned by these models can be strikingly similar to real neurons (Yamins et al. 2014), suggesting relevance for psychiatry. This may include examples such as disordered perceptual representations as in schizophrenia as well as disordered representations of abstract decision-related variables that could underlie many phenomena including posttraumatic stress disorder (PTSD). These models can also potentially explore the time evolution in psychiatric disorders and model the effects of treatments at a network level.

Reinforcement Learning

Reinforcement-learning models quantify the dynamics of learning and decision making over time. These models, often cast at the level of systems and behavior, can be used to specify precisely hypotheses about how information obtained at one point in time affects beliefs and behavior at another. Reinforcement-learning models are concerned with learning to predict future rewards and punishments (as in Pavlovian or classical conditioning) and learning to select actions that would maximize future reward (as in operant or instrumental conditioning). The name "reinforcement learning" suggests an emphasis on learning dynamics; however, the models can also be used in steady state, after learning has achieved equilibrium, to make predictions and test hypotheses about decision making and action selection in different situations. Importantly, by fitting free parameters of these models to time series of behavioral data, one can precisely quantify different aspects of learning and decision making in individual patients. Relevant parameters might include the learning rate for appetitive and aversive outcomes, the degree of exploration versus exploitation, and the extent to which patients generalize across instances and stimuli. These parameter values can, in principle, be used as a diagnostic tool to characterize the different ways in which decision making can break down and to quantify individual differences (Moran et al., this volume). Indeed, reinforcement-learning parameters systematically vary as a function of disease, genetics, and pharmacology in ways that match predictions from decades of systems neuroscience (Frank and Fossella 2011).

In terms of Marr's levels, one of the strengths of reinforcement models is that they link from the computational level of optimizing future reward, through the algorithmic level of temporal difference learning, to the implementational level of dopamine-dependent plasticity in corticostriatal synapses (Barto 1995; Montague et al. 1996; Schultz et al. 1997). These neural and behavioral systems are also largely preserved in phylogeny, and thus reinforcement-learning models can be applied to humans and animals alike, even insects (Montague et al. 1995). The simplicity and transparency of these models allows one to give semantic interpretation to every construct of the model. However, one might argue that this is at the expense of allowing properties to "emerge" from the model in a way that sometimes occurs with models that embody more complex dynamics. This aspect of reinforcement-learning models can be seen as a feature rather than a flaw. The fact that these models do not have many moving parts allows one to easily form an intuitive understanding of the behavior of the model even from simply observing the model equations. Thus these models are most useful for specifying and sharpening hypotheses regarding learning dynamics in both healthy and clinical populations.

It is important to note that although reinforcement learning is highly prominent and promising for psychiatry, there are other classes of models, which we do not discuss here, that similarly provide process models for behavior that are linked to neural substrates. These include sequential sampling models, of which drift diffusion models are the most familiar (see Frank, this volume; Gold and Shadlen 2007).

Organizing Computational Tools by Mathematical Framework

Computational tools could also be organized according to mathematical frameworks and methodological toolkits. Specific computational models may, for instance, rely on mathematical tools from areas such as probability and statistical theory, nonlinear dynamics, or information theory. These areas of mathematics provide general frameworks for addressing computational questions at any level of abstractness or biological organization. Other mathematical frameworks that are applied across many levels of biology and psychology include statistics and machine learning, dynamical systems theory, and Bayesian methods as well as, to a lesser degree, information theory and optimal control theory.

Machine Learning

Machine-learning tools come out of statistics and computer science and were originally used mainly in the context of pattern recognition applications. In general, they use various types of mathematical principles and specific algorithms to analyze data so as to make predictions about existing and future data.

Machine-learning tools can be either supervised (e.g., you have a specific categorical or dimensional variable guiding analysis) or unsupervised methods for characterizing data. Examples of supervised tools include support vector machines (Cortes and Vapnik 1995), which try to learn optimal decision boundaries for predicting class labels. Some examples of unsupervised tools include factor analysis, independent components analysis, and clustering approaches. Common to these unsupervised methods is that they attempt to detect structure within or suitable reductions of the data space without explicit advance knowledge of what that structure may be. All these approaches can be augmented by Bayesian methods to incorporate prior knowledge. These tools can either be used with a single type of data (e.g., behavior, neuroimaging, or genetic) or can be expanded to include several types of data or levels of data, such as in multimodal fusion approaches (Sui et al. 2012). Machine-learning tools can be used to identify novel structures in psychopathology, whether they might be dimensional, categorical, or a hybrid. For example, these tools could help identify categories or dimensions in high-dimensional data. They have already been used in this way in the psychopathology field, as a means to develop new models of the meta-structure of psychopathology based on phenomenological data (Krueger and Markon 2006; Wright et al. 2012, 2013). They can also help to integrate from one level of analysis to another (e.g., imaging to behavior, gene to imaging). In such cases, one could train on level A and predict on level B, without starting from strong hypotheses about how these transformations happen. For example, if we can identify structure directly in complex patterns of brain activity, this structure could inform our theories of the computations underlying behavior. As such, these general-purpose tools may help us identify additional types of data needed to understand the nature or mechanisms of such transformations. They can also help us to predict risk (Paulus et al., this volume) for the development of various forms of psychopathology based on different types of biomarkers. Further, they might even be able to be used to identify biomarkers that predict the success of different treatments based on similar cases, where similarity metrics are determined by the specific machine-learning method that is used. These tools can also be used more generally as data analytic tools for a variety of types of data (and are actively being used in this way), such as analysis of fMRI or connectivity data

For example, many risk factors for mental illness involve complex interactions between genes, the brain, and environmental influences which develop dynamically in naturalistic contexts (Kaddurah-Daouk and Weinshilboum 2015; Michino et al. 2015). To address this, we need data that is not only "big" but also multimodal. Ongoing studies are currently collecting functional neuroimaging, real-time/real-life data collected via smartphone, and geographical mapping. Mapping can link real-time smartphone data to specific locations where we have data about neuropsychiatric risk factors, such as urbanicity, pollution, sociodemographics, etc. Machine-learning techniques may aid the identification of patterns that predict risk-related neuroimaging markers.

Further extensions of this concept in longitudinal study designs (accelerated longitudinal data acquisitions covering critical age ranges of neurodevelopmental disorders) may aid the identification of the dynamics of the neural correlates over time (discussed further below).

The current fields of genetics, genomics, and epigenetics provide abundant data for machine learning. Publicly available databases provide hundreds of thousands of control and patient subjects throughout the world,[1] and these genetic data are paired with categorical and continuous phenotypic data. For a smaller subset of individuals, there are also physiology and neuroimaging data. Computational methods across available big datasets will almost certainly allow deeper understanding of connections across levels of analyses from genetics, to epigenetics, circuits and behavior.

Dynamical Systems

Dynamical systems theory is a field in mathematics that addresses systems described by sets of equations which dictate the evolution of variables over time and/or space. It specifically addresses nonlinear systems for which some of the more conventional mathematical techniques (analytical approaches to equation solving) break down. A central concept in dynamical systems theory is that of a state-space (i.e., the space spanned by all dynamical variables of the system). A point in this space captures the current state of the system, and the evolution of this state across time is given by a trajectory meandering through this space. The course of this trajectory is determined by various geometrical properties of this space, like for instance the existence of attractor states, such as stable orbits (limit cycles), which give rise to nonlinear oscillations. Nonlinear dynamics provides a set of tools to characterize the flow of these trajectories (and thus the system's evolution in time and space) and analyze their behavior. Since essentially all neural and behavioral phenomena can be cast in terms of variables that evolve dynamically in time, nonlinear dynamics provides a very general framework for describing and analyzing computational models.

Dynamical systems may also provide ways of capturing phenomena that may be central to understanding the mechanisms of breakdown in psychiatric conditions. For example, NMDA receptor dysfunction has been implicated in schizophrenia (Barch, this volume). We can begin to understand the mechanics of this dysfunction with dynamical systems theory. As NMDA conductance steadily increases, both in real cells and in biophysically plausible models, the system suddenly jumps from quiescence into a bursting mode, then jumps again from regular bursting into chaotic irregular activity, and finally from chaos into regular steady single spiking (Durstewitz and Gabriel 2007). Although the changes in the underlying system parameter (NMDA conductance) are

---

[1]   National Center for Biotechnology Information, http://www.ncbi.nlm.nih.gov/genome/ (accessed June 15, 2016).

gradual, the neuron's spiking modes change abruptly. These abrupt jumps between different operating regimes are called "phase transitions." There are also numerous examples of phase transitions at the level of neural populations (Durstewitz et al. 2010).

The idea of abrupt or critical transitions between operating regimes can also arise at very different levels of scale. For example, there may be enough resilience in the brain and in behavioral or social coping mechanisms to allow underlying biological changes to occur without obvious psychiatric symptoms, until some critical point is reached and there is an abrupt shift to a different regime such as depression or psychosis.

Bayesian Methods

Bayesian inference describes how one can use probability theory to infer the state of variables we are interested in, given prior knowledge and noisy observations. Bayesian approaches start with a hypothesis H, and some observed data O. For example, H could be the hypothesis that a patient has lung cancer, and O could be a positive blood test. It is relatively easy to measure $P(O|H)$ (i.e., the probability of getting a positive blood test given that one has lung cancer), by measuring the proportion of the population with cancer that have positive blood tests. This can be used to compute, using Bayes's rule, the more important quantity of $P(H|O)$; that is, the probability that a patient with a certain blood test result has lung cancer. According to Bayes's rule, $P(H|O)$ is proportional to $P(O|H) \cdot P(H)$, where $P(H)$ is the prior probability of having lung cancer (i.e., the prevalence of lung cancer in the general population). This general framework can also be used to include multiple observations or to predict new observations. It can also be used to build hierarchical representations: "H" can play the role of the observation, "O," for another, higher-order model. Finally, Bayes's rule can capture temporal prediction or temporal evolution of a state, with $P(H)$ corresponding to knowledge from the past and $P(H|O)$ representing the new knowledge updated by observations. The utility of this framework for psychiatry is highlighted by Flagel et al. (this volume).

Bayesian models can be used as "normative" descriptions of brain function. Bayesian belief updating represents the optimal solution to many problems under a very broadly applicable set of constraints. Thus, it is reasonable to posit that in many cases the brain may be attempting to calculate $P(H|O)$ or some good approximation thereof. This may be true of the computations of individual neurons (e.g., if their firing rates encode beliefs about perceptual features, and their synaptic inputs encode new evidence about these features), circuits, systems, or the entire organism (e.g., how an individual reaches their beliefs about others' intentions). This view is very powerful because it then allows us to explore the mechanism by which the brain calculates beliefs based on observations, or how the calculation might go wrong (Huys et al. 2015b).

The brain constructs highly hierarchical representations of its environment. For example, visual areas construct increasingly "meaningful" representations (from local contrast to contour, basic shapes, and objects) of the visual world. To do so, both feedforward connections (sending information from sensory area to "higher level" area) and feedback connections (sending information prior expectations to the sensory area) are essential. For instance, detecting a tree requires integrating feedforward sensory information ("green," "tall,"etc.) with prior knowledge ("I am in a forest"). Since both feedforward and feedback connections are excitatory in the human brain, such highly recurrent excitatory circuits could not work properly on their own. Sensory information would be sent up the hierarchy, generating expectation, then reverberated back, combined with themselves, then sent back up as if they were new sensory evidence, in an endless cycle. Such a system would suffer from an extreme amount of "circular inference," making us "see what we expect" or "expect what we see." It would also learn "fake" causal relationships between completely uncorrelated events, simply because their neural representations are correlated through the network dynamics. To function properly and generate an accurate belief system, the brain needs to combine excitatory (E) feedforward and feedback connections with strong, balancing inhibition (I), whose goal is to cancel all predictable (reverberated) excitatory inputs in the network. Such tight E/I balance is a widely observed phenomena in cortex. It could be that imbalances in E/I (involved in a wide range of mental illness such as bipolar disorder, schizophrenia, or autism) causes circular inference, leading to the formation of aberrant beliefs (overconfidence, hallucinations, delusions, alien control). New experiments confirm that the behavior of schizophrenic patients in probabilistic inference tasks was well described by such a "circular inference" hypothesis (Jardri and Denève 2013).

This method can also be used as a tool to organize and generalize from complex, high-dimensional and noisy data in any domain. As such, Bayesian inference can provide a useful tool for diagnosis and treatment of mental illness. A well-known example is "Bayesian causal models," widely used to interpret imaging data, but applicable, in general, to any type of data (for further detail, see Moran et al., this volume).

Bridging Levels

We have considered various levels of modeling and how they can be used to ask different sorts of questions, from biophysical to normative. We have also emphasized that no single level of analysis is sufficient to make the connections between mechanism and behavioral symptoms relevant for psychiatric illness; the complementary values of each level implies that an-all-of-the-above strategy is useful. Informally, one can also interpret modeling endeavors at one level in terms of the other. For example, a biophysical model of dopamine modulation of attractor dynamics and flexibility in PFC can be summarized by analogous functions in connectionist networks and their application to cognitive tasks (Cohen et al. 2002). More formally, one can also quantitatively map the properties of one model onto another. This affords a richer testable prediction that leverages the utilities of both levels (Frank 2015). Typically, two approaches are taken. The first is to derive exact mappings. For example, Ma et al. (2006) showed how spiking models, including probabilistic population codes, can precisely implement Bayesian inference in a sensory cue combination task, building on work from Zemel et al. (1998), who developed the idea of spikes as encoding probabilistic information. Bogacz and Gurney (2007) showed how an optimal model of evidence accumulation during decision making can be mapped onto the anatomy of the basal ganglia.

The second approach is not to assume that the mapping between levels is exact, but rather that it is approximate, and instead to fit the behavioral output patterns of complex network quantitatively using a higher-level algorithmic description. This leverages the advantage of the algorithmic models: because the behavioral data can be fit quantitatively with few free parameters and the same strategy (that these models use) can be applied when fitting to empirical data, a determination can be made as to which of several alternative algorithmic models best describes the behavior of the system. Thereafter, an estimate can be made as to the impact of biological manipulations in the network on higher-level algorithmic parameters, which in turn can guide empirical experimentation. For example, Ratcliff and Frank (2012) showed that the outputs of a network model of the basal ganglia are well approximated by drift diffusion models (DDMs). In these models, evidence for each of two or more options is accumulated noisily over time until one option reaches a decision threshold and is chosen. Ratcliff and Frank also found that parametric modulations of the subthalamic nucleus (STN) affect the decision threshold (as opposed to other decision parameters), particularly in the face of choice conflict. This prediction was tested empirically by recording and manipulating STN function and estimating its impact on drift diffusion parameters, based on choices and response time distributions (and EEG data). Indeed, subsequent fMRI studies provided evidence that STN activity is related to decision threshold adjustment during choice conflict (Frank et al. 2015). STN manipulation in Parkinson disease reduced the decision threshold for these choices (Cavanagh et al. 2011; Green et al. 2013), providing a novel interpretation for how impulse control disorders can arise in these patients. This is just one example of how computations at one level can afford analysis at another, allowing falsifiable predictions. One can also further bridge these levels with machine-learning tools to classify or cluster patients based on fitted model parameters as well as to identify which parameters/mechanisms contribute most strongly to classification (Wiecki et al. 2015).

Another example for the scale-bridging approach is provided by the "dual-state model" of PFC dopamine function (Durstewitz and Seamans 2002). In this biophysically anchored theory, slice-electrophysiological observations on the dopamine D1- and D2-class receptor modulation of a range of different voltage-gated and synaptic currents were linked through dynamical systems tools to alterations in prefrontal attractor dynamics, which in turn could be related to changes in working memory function and cognitive flexibility (see Frank, this volume).

Applying Computational Methods to the Evolution of Systems over Time

A central feature of mental illness is that it is not static; it evolves both developmentally and in adulthood with prodromal stages and subclinical antecedents, through episodic or slowly changing patterns of symptoms, to remission and often relapse. Adding the dimension of time creates significantly more complexity, which provides an entry point for computational methods.

As an example, PTSD is unique among psychiatric disorders in that one key component of its etiology (i.e., the traumatic event) is known. What remains uncertain is how the initial clinical manifestations of acute stress (e.g., hyperarousal, re-experiencing, avoidance) may progress in some individuals to the constellation of symptoms and associated social dysfunction that characterizes the disorder. It is noteworthy that only a relatively small proportion of individuals who are evaluated in the emergency room following a traumatic stressor (e.g., a motor vehicle accident or assault) are diagnosed with PTSD at 6 months to one year following the trauma. A prime application of computational methods would be to enhance prediction of symptom progression from acute stress reactions (as seen in an emergency room setting).

Another example pertains to the long-term course of depressive episodes of major depressive disorder, which can be a highly recurrent illness marked by discrete illness episodes and periods of relative stability (Thase 2013). Depressive episodes can be sorted into categories based on specifiers such as melancholic features (e.g., minimal mood reactivity, early morning awakening, diurnal variation), atypical features (e.g., mood reactivity, hyperphagia, hypersomnia), and psychotic features (e.g., delusions, hallucinations), which have state-dependent neurobiological correlates. However, individuals often switch unpredictably from one category to another between episodes (Oquendo et al. 2004), which poses challenges in implementing treatment strategies for relapse prevention. Moreover, proper clinical decision making requires predictions. Issues such as how long to continue a medication after a patient achieves remission, or whether to continue with a partially effective treatment or to switch to a new one are crucial for treatment providers. Computational models that take into account dynamics over time would be immensely helpful in making such decisions.

Change over time may be driven by spontaneous internal changes, but also by changing environmental factors, including treatments. Furthermore, internal and external variables interact in complex ways. Individuals with depression, for example, are especially vulnerable to variations in mood according to seasons as well as to hormonal perturbations related to pregnancy, the postpartum period, menstrual cycle variation, and use of hormonal contraceptives.

The emergence of the complex neurobiology of chronic or recurrent mood disorders may be viewed as having progressed through a number of stages. Before the first episode of depression, vulnerability exists at genetic, physiological, and environmental levels (Lupien et al. 2011). Chronic and acute stress (allostatic load) perturb homeostatic mechanisms at multiple organismal and neural levels, such as mood, sleep, appetite, and motivation (McEwen 2003). Factors contributing to the transition from "having a bad month" to developing "depression" include elevations in circulating glucocorticoid and inflammatory cytokines. These factors have numerous consequences for the brain. One mechanism that has received attention is the compromise of the glial capacity to transport glutamate, resulting in elevations in extrasynaptic glutamate levels. These elevations suppress point-to-point synaptic functional connectivity in circuits regulating mood by inhibiting glutamate release via stimulation of presynaptic mGluR2 receptors and by causing the retraction of dendritic spines. In the long term, this reduces dendritic complexity due to excessive stimulation of extrasynaptic GluN2B-containing NMDA receptors and reductions in the level of trophic factors (reviewed in Krystal et al. 2013). The disruptions in structural and functional connectivity, combined with many other neuroplastic mechanisms (including alterations in reward and social learning), may make it impossible to "bounce back."

Modeling Time

Data which describe trajectories over time come in many forms and include both behavioral assessments and physiological markers. Can this multivariate time series of data be used to learn more about the mechanisms of the disease? Can practical predictions be made about the future disease course or the effects of treatments and interventions?

Formally, there are several useful general-purpose approaches. First, autoregressive–moving-average (ARMA) models express current observations as a weighted linear combination of previous observations plus noise. Here, forward prediction is straightforward using the estimated weight parameters. Nonlinear variants of these models also exist, such as threshold or piecewise linear autoregressive models.

Second, state-space models include latent (i.e., unobserved) as well as observed variables. Latent variables capture underlying causes, such as neural activity, which cannot be directly measured but nonetheless have effects on the observed data. The time evolution of the latent variables can be described mathematically as an ARMA process, or as a discrete set of states with transition probabilities, or a combination thereof as in the class of switching state-space models (Ghahramani and Hinton 2000).

Third, there is a large toolbox from nonlinear dynamical (NLD) systems theory. NLD methods usually start from a state-space representation of the observed system. This is the space spanned by all the dynamical variables of the system (e.g., the firing rates of a set of recorded units). A point in this space specifies the current state of the system, and the movement of this point through the space, as time passes, yields a trajectory. In physical and biological systems, these trajectories are described by geometrical objects within these spaces like "attractor states." Based on such representations, NLD theory offers various methods for prediction and assessing the effect of interventions in these spaces (Lapish et al. 2015).

Understanding the dynamics of mental illness is not only crucial for their diagnosis, management, and treatment but also for bringing some light to the underlying mechanisms. For instance, when episodes occur in an approximately periodic fashion (as, e.g., in bipolar disorder), they might be described in dynamical systems terms through an underlying periodic or chaotic oscillator. This, in turn, may offer a way to study when in the cycle it would be best to intervene therapeutically. An acute episode recruits compensatory processes, which persist after the episode is finished. However, these compensatory processes themselves may be regulated through other feedback loops with the environment, as is common in biology, which in turn can cause another episode. In contrast, when episodes occur erratically without prior warning and sudden onset, compensatory processes might be better described through metastable states or bifurcation mechanisms giving rise to an instability. In this case, the healthy state is fragile (e.g., due to weakened homeostasis). The brain state can be temporarily thrown out of this state, jumping to a pathological state.

Is There a Use for Computational Approaches without Understanding "Fundamental" Mechanisms?

Computational neuroscience brings a powerful set of conceptual tools for understanding complex systems. However, we must be cognizant that we may never fully understand every level of mechanistic detail in the path from molecule to behavior. Still, there are many ways in which computational approaches can be used to enhance our understanding of behavior as well as approaches to treatments and interventions.

It is tempting to look to the most detailed or microscopic level for the most "fundamental" understanding, but this is often a mistake. For example, a liquid only exists as the interaction between atoms. Some argue that the fundamental level for a given phenomenon is the most detailed level at which the phenomenon exists. Others argue that it is the level at which the phenomenon is most parsimoniously captured. There is broad agreement that for phenomena which live mostly at higher levels, it is most useful to study them at these levels. Crucially, we *do not know* at which level most psychiatric illnesses are most usefully described and studied. Even if a gene conferring risk for an illness produces a malformed ion channel, it is possible that behavior at a cellular or circuit level (e.g., synaptic plasticity) might look essentially normal and that pathology may only appear when investigating properties of the brain at a higher level of organization.

Furthermore, the best level of description is related to the use that one makes of the description. If one is interested in etiology, for instance, then genetics may be particularly important, but if one is interested in developing pharmacologic treatments, then an understanding of cells and circuits are important. Even with a single type of use, say medication development, and desired endpoint, to alleviate a disorder, different treatment mechanisms will be developed to target different levels of description (e.g., to correct an abnormal protein, to correct synaptic or circuit dysfunction, or to correct a behavior).

In this section we outline three ways that computation offers a benefit without necessarily reaching the most detailed level of explanation. First, some approaches allow us to characterize some aspects of mechanism without requiring an understanding of fundamentals. Second, whether or not we can achieve mechanistic understanding of the disease, it is useful to obtain mechanistic understanding of other related phenomena (e.g., recovery and resilience). Third, we can eschew mechanism entirely and use computational methods to optimize treatment directly.

Characterizing Mechanism at a High Level

It is possible to extract knowledge and impact treatments using computational approaches, even without understanding the fundamental mechanisms behind the illness. We might have a very useful understanding of how the system behaves and misbehaves that is in some sense mechanistic, but without reference to deeper mechanisms. Neuroimaging, for example, can be used to identify which areas of the brain are activated during hallucinations. These areas could then be targeted with, say, a 1 Hz transcranial magnetic stimulation (TMS), which decreases the activation of the targeted area, to reduce hallucinations over a time period of several weeks. In service of treatment, this leverages a partial understanding of brain regions and disturbances in excitability (balance of excitation and inhibition) that might be corrected, through TMS, without requiring a complete picture of the underlying neural signaling disturbances or the impact of the TMS on these disturbances (Hoffman and Cavus 2002; Hoffman et al. 2007). Likewise, there are very effective treatments at a purely behavioral level that rely on some understanding of mechanism (e.g., from psychology) at this level, without understanding anything about the brain.

Mechanisms of Resilience and Recovery

Without fully understanding the original causes of a disease, we may begin to understand mechanisms of resilience to the disease or recovery from it. Resilience may simply constitute lower vulnerability to disease. On the other hand, resilience may be a more active phenomenon. Within a certain range, neural systems can return to their original homeostatic states. But if stretched too far, a system may "break," resulting in a discontinuity, such as a pathological response (e.g., PTSD, anxiety disorder). However, in some cases, the organism achieves a new stable state that not only adapts to the current stressor but can better withstand subsequent stress, thus resulting in enhanced resilience (Friedman et al. 2014). The mechanisms of this reactive resilience are still poorly understood, but represent a prime target for dynamical systems models that capture such multistability.

Likewise, mechanisms of recovery are sometimes quite distinct from mechanisms of pathology. Most psychiatric treatments do not fix the underlying pathology (e.g., depression is not caused by a lack of electroconvulsive seizures). However, models can potentially be used to understand the mechanisms of treatment and recovery. An understanding of learning theory suggests ways to make extinction permanent (e.g., fear in PTSD or phobia, compulsive behavior in obsessive-compulsive disorder, craving in addictive behaviors). Computational learning theory implies ways to optimize behavioral therapy or computer apps without understanding the underlying neural and molecular mechanisms of the original disorder.

Computational Methods to Optimize Treatment Directly

In some cases, we may temporarily abandon the quest for mechanistic understanding of a disease process and use computational methods to analyze data directly and make predictions and recommendations about treatment. Advances here can be in the realm of descriptive nosology (see chapters by First, MacDonald et al., and Flagel et al., this volume), such as clustering of patients using computational algorithms based on current symptom/intermediate phenotype datasets, independent of underlying mechanistic knowledge (Borsboom et al. 2011; Borsboom and Cramer 2013). Computational approaches can also be very helpful in optimizing treatments and understanding of outcomes: from optimal timing and dosing of medication, when to start/stop treatment, and even optimal organization of psychiatric treatment flow in clinics. All can be improved with computational modeling to optimize current processes, agnostic to the underlying mechanisms of functioning of these approaches. This approach is being used, for example, to enhance treatment parameters with electroconvulsive therapy (Deng et al. 2013; McClintock et al. 2014).

A striking and non-obvious observation (and perhaps a deep principle in computational psychiatry) is that applying analysis methods to data, without explicitly trying to model the mechanism, may actually help *reveal the mechanism of the disease*. For example, suppose one were to attempt to model the trajectory of disease episodes in a patient with schizophrenia using a hidden Markov model. The goal of this model might be to predict the next episode so as to guide treatment. Yet as part of the process of establishing an optimal model fit to the clinical data, we may infer a number of states in the hidden Markov model. If this parameter is consistent between subjects, or consistently relates to some other important variables that have biological or psychological relevance, we may accidentally reap clues about the mechanism of the disorder itself.

Finally, notwithstanding the above, it is worth striving for a more fundamental mechanistic understanding. Because the mechanisms of hypertension are known, a physician usually will not prescribe another beta blocker if a patient is already on beta blockers; instead, the physician will try adding a drug with a different mechanism of action. Perhaps even more importantly, a deeper understanding of mechanism will help to achieve more complete remission and ultimately lasting recovery. This could be the difference between a treatment that works partially and temporarily versus a cure. Realistically, all of these approaches must be combined to create a versatile armamentarium.

How Can We Measure the Success of Computational Approaches?

In early computational psychiatry, theoretical approaches were sometimes judged by how well they could reproduce traditional approaches (e.g., whether clustering model parameters could reproduce diagnostic categories in the DSM). Since computational psychiatry may soon exceed the usefulness of traditional approaches, this correspondence should not be used as a primary metric. Instead we need to step back and think about how we can gauge, in a more fundamental sense, what is or is not working (Clementz et al. 2016).

Treatment Outcomes

In a clinical sense, the ultimate gold standard is to improve outcomes for patients. In the ideal case, computational psychiatry could come to be very explicitly and directly part of treatment, so that changes in the prevalence or incidence of the disease after the introduction of the techniques could be measured. Here we outline five primary vehicles toward this end:

1. Computational approaches might inform basic research that subsequently leads to improvements in patient outcomes (e.g., by identifying critical neural circuits or components of cognition).
2. Computation might help predict risk status, thus enabling more informed interventions. In bridging levels of biology, models may be

particularly well suited to develop an integrated understanding of risk factors across levels: from the known genetic risk architecture of mental illness (Gottesman and Gould 2003; Preston and Weinberger 2005; Cannon and Keller 2006; Meyer-Lindenberg and Weinberger 2006; Kendler and Neale 2010) to known neurobiological, behavioral, and environmental factors. Relatedly, predicting risk may also serve a role in forensic psychiatry, which includes risk to others—an area that could potentially have a tremendous impact on society (Buchanan 1999; Freedman 2001; Loza and Dhaliwal 2005; Odgers et al. 2005; Dahle 2006; Odeh et al. 2006; Hill et al. 2012; Chu et al. 2013).

3. Computation will reveal new treatments and treatment targets. Biophysical models could be used to identify novel molecular targets, greatly facilitating screening for new drugs. At the circuit level, we can identify neural circuits for brain stimulation interventions based on an understanding of these circuits' role in overall brain function, allowing us to fine-tune stimulation parameters (Gutman et al. 2009; Datta et al. 2012; Rotem et al. 2014; Li et al. 2015; Senço et al. 2015). Similarly, models that identify particular behavioral variables, and deficits in the same, might suggest novel types of psychotherapies aimed at addressing these behavioral pathways. For example, reinforcement learning and other learning paradigms have already had a great deal of impact at the level of informing certain forms of cognitive behavioral therapy; in particular, prolonged exposure therapy, use of virtual reality-based therapies, and the use of cognitive enhancers (such as d-cycloserine) to enhance the rate of extinction learning in combination with exposure (Conklin and Tiffany 2002; Rothbaum and Davis 2003; McNally 2007; Craske et al. 2008; Abramowitz 2013).

4. By understanding *how* individual treatments work, existing treatments can be repurposed to treat a different disease or to work more effectively. In the models of Parkinson disease discussed above, modeling suggested that the disease involves learned avoidance due to exaggerated learning in D2 neurons, and that this learning component can be rescued by adenosine antagonists which block plasticity in these neurons (Beeler et al. 2012). This may also explain the failure of such antagonists in clinical trials wherein they were administered to Parkinson patients in the advanced stage: it predicts that these treatments will be most effective during very *early* stages of the disease to prevent aberrant learning.

5. A particular strength of computational approaches is identifying exactly what data we need to collect to make effective predictions. We often have the potential to gather a huge array of data modalities about the patient from neuroimaging, cognitive tasks, questionnaires, genetics, hormone levels, etc., but we need better methods to determine which will actually provide the critical information to guide treatment.

Making Scientific Progress

Finally, in some views, basic scientific progress is an end unto itself. Psychiatric disorders may even be seen as fortuitous because they shed light on how the healthy brain works. Early progress in neuroscience was accelerated by observing the consequence of gunshot wounds in particular areas of the brain (suddenly plentiful after World War I). Similarly, elucidation of the nature of dysfunction in psychiatric disorders may facilitate progress in the understanding of fundamental brain mechanisms.

We wish, however, to end with a note of caution. The field of computational psychiatry is still in its relative infancy, and the problem to be tackled is immense. Thus, patience must be exercised, as we expect progress in fits and starts. The original promise of computational psychiatry may take decades to be fully realized.

6

Computational Cognitive Neuroscience Approaches to Deconstructing Mental Function and Dysfunction

Michael J. Frank

Abstract

Advances in our understanding of brain function and dysfunction require the integration of heterogeneous sources of data across multiple levels of analysis, from biophysics to cognition and back. This chapter reviews the utility of computational neuroscience approaches across these levels and how they have advanced our understanding of multiple constructs relevant for mental illness, including working memory, reward-based decision making, model-free and model-based reinforcement learning, exploration versus exploitation, Pavlovian contributions to motivated behavior, inhibitory control, and social interactions. The computational framework formalizes these processes, providing quantitative and falsifiable predictions. It also affords a characterization of mental illnesses not in terms of overall deficit but rather in terms of aberrations in managing fundamental trade-offs inherent within healthy cognitive processing.

Introduction

Understanding how any system, including the brain, can become dysfunctional first requires at least a general understanding of how it is functional. Arguably the main obstacle to progress in psychiatry is its historical inclination to "put the cart before the horse" in its efforts to link illnesses and higher-level symptoms to individual genes or molecular mechanisms, before understanding the relationship between the many intermediate levels of analysis. Cognitive neuroscience takes smaller (but still lofty) leaps of linking isolated cognitive processes to larger-scale mechanisms (often with coarse descriptors) to offer better explanations of neurocognitive processes, but with less immediate application to mental illness. One of the central goals of the computational psychiatry mission is to develop principled mechanistic models that formalize functional objectives within the domains of perception, action, and cognition, and to explore how aberrations in such mechanisms lead to corresponding changes in mental function. As such, in this chapter I discuss only selected computational cognitive neuroscience approaches and domains relevant for mental illness, focusing on high-level concepts rather than details of the formulations. For other treatments of these thematic ideas and related assumptions and practices of computational psychiatry, see Maia and Frank (2011), Montague et al. (2012), Huys et al. (2012), Stephan and Mathys (2014), Huys et al. (2015a), Wiecki et al. (2015), and Wang and Krystal (2014).

Systems and cognitive neuroscience inherently link levels of analysis across a continuum: from biological mechanism to cognitive and behavioral phenomena. One can study, for example, ion channel conductances and receptors, intracellular signaling cascades, synaptic plasticity, excitation-inhibition balance, or probabilistic population codes. Zooming out a level, entire neural systems can be studied, such as the sensory cortices, frontal cortex, hippocampus and basal ganglia, ascending neuromodulatory signals, and interactions among several of these systems. At the cognitive level, a different set of topics and principles are relevant, for example, concerning attention, working memory, decision making, cognitive control, reinforcement learning, and episodic memory. A key role for computational models is their ability to provide a unifying coherent framework that links these levels, specifying computational objectives of a cognitive problem and providing novel interpretation of underlying mechanisms, while also forcing one to be explicit about the assumptions being made.

Indeed, computational approaches encompass a huge range: from those that specify detailed biophysics to those that consider high-level goals of a functional system without regard for implementation. Biophysical models explore, for example, how combinations of ionic currents and their dynamics give rise to higher-level "behaviors," but where behavior here is often defined in terms of the changes in membrane potentials of individual neurons or distinct compartments within neurons, or in terms of synchrony of neural firing across populations of cells. Sometimes these models explore further the impact of particular ionic currents on attractor dynamics thought to be relevant for cognitive function (but usually without simulating realistic cognitive tasks). The field of computational cognitive science, on the other hand, considers how behavioral phenomena might be interpreted—independently of the neural implementation—as optimizing some computational goal, such as minimizing effort costs, maximizing expected future reward, or optimally trading off uncertainty about multiple sources of perceptual and cognitive information to make inferences about causal structure. In between, computational cognitive neuroscience considers how mechanisms within neural systems can approximate (or even directly implement) optimal solutions to computational problems, and how alterations in these mechanisms can lead to predictable changes in behavior. Even here there exist several levels of models: neural network models typically capture some aspects of electrophysiology and dynamics whereas higher-level algorithmic models can summarize the key processes with few free parameters and are more suitable to quantitatively fit behavioral data. By linking the mechanisms at the neural-level model to observable changes in higher-level model parameters (for a review, see Frank 2015), one can derive predictions for how changes in neural activity due to disease, medication, or brain stimulation results in changes in cognitive computations. This linking process also imposes mutual constraint relations between higher- and lower-level descriptions, and allows both levels to be refined and/or reinterpreted by the other.

Moreover, one of the central aims of computational cognitive neuroscience is to identify *computational trade-offs* inherent in cognitive problems and to examine how the brain mitigates these trade-offs at the systems level. As one classic example shows, in memory there is a trade-off between being able to separately store distinct events (e.g., the location of where I parked my car today compared to yesterday) versus being able to accumulate information across events into a coherent representation (e.g., determining the best parking strategy on average). The former process requires distinct neural patterns to encode distinct events, whereas the latter requires a shared population of neurons representing all the times a particular strategy was used to associate with outcomes. Computational cognitive neuroscience approaches have suggested that the brain solves this trade-off by incorporating multiple memory systems in the hippocampus and neocortex (McClelland et al. 1995). Moreover, within the hippocampus, subregions such as the dentate gyrus can enhance pattern separation to minimize interference among similar events, whereas area CA3 supports pattern completion to allow retrieval of memories given partial cues (McClelland et al. 1995). Such models have inspired decades of research and empirical data that have confirmed their key predictions, which may not have been tested in absence of guiding theory, and served to refine further model development.

In what follows I give a brief survey of computational cognitive neuroscience approaches to select problems, highlighting various trade-offs that may be informative for mental illness.

Working Memory and Prefrontal Cortex

The prefrontal cortex (PFC) is long known to be involved in working memory, that is, the ability to hold task-relevant information in mind over short periods of time and use that information to contextualize and guide subsequent actions. Biophysical models have shown that the ability to sustain stable, persistent PFC activation states related to working memory depends on recurrent excitation, intracellular ionic currents and NMDA receptors, modulated by dopaminergic input (Figure 6.1) (for reviews, see Durstewitz and Seamans 2008; Wang and Krystal 2014). These models have made precise predictions regarding the effects of NMDA and dopaminergic manipulations and how they affect attractor states needed for working memory, and have also been instrumental in guiding study on the neural basis of working memory impairments in schizophrenia, associated with NMDA and dopamine (DA) hypofunction in the PFC.

If PFC neurons can maintain information in working memory, a related question is: What is the representational content of this information? Traditional analysis focuses on the coding of individual stimulus dimensions (e.g., color) or basic rule representations. Recent computational models and analysis have instead highlighted that a large proportion of PFC neurons have mixed selectivity across sensory dimensions; that is, their representations are rather high dimensional. Modeling suggests that this "multiplexing" of multiple variables affords a computational advantage by increasing the repertoire of possible input–output mappings that can be read out of attractor states (Rigotti et al. 2013). Support for this interpretation came from an analysis of trials in which monkeys committed cognitive errors (responding according to the wrong rule, thus permitting the analysis of failure modes); in these cases, the dimensionality of PFC neurons collapsed while the simpler low-dimensional representations of individual cues remained intact.

However, computational approaches have also identified a key trade-off within working memory: while it is desirable to robustly maintain task-relevant information in the face of distracting interference (noise in neural firing, task-irrelevant environmental input), it is also important to be able to flexibly update working memory states when incoming information is relevant or when behavioral strategies and plans need to be adjusted. The ability to both robustly maintain existing attractors and rapidly update them are at odds with each other but can be solved by a gating mechanism that dynamically increases the influence of incoming information. Multilevel models have identified DA as a likely candidate for implementing this gating function: when incoming information is relevant, phasic increases in DA can shift the balance in PFC from a "D1 state" optimized for robust maintenance, which is difficult to destabilize, to a "D2 state" optimized for flexibility in terms of allowing for shifting representations (Cohen et al. 2002; Durstewitz and Seamans 2008). Biophysical models show how this pattern emerges due to differential D1 versus D2 receptor sensitivity to different levels of DA, and their differential effects on NMDA and GABA currents, and their resulting network effects on attractor dynamics within populations of cortical pyramidal cells and interneurons.

This example serves to highlight that more DA does not imply better overall function, but rather that a dynamic range of DA signaling is needed to dynamically modulate a functional objective that balances a trade-off: facilitating a switch from maintenance to updating in the service of relevant task demands. Such insights have informed a variety of empirical evidence that DA and PFC states trade off in tasks that demand cognitive stability versus flexibility (Cools

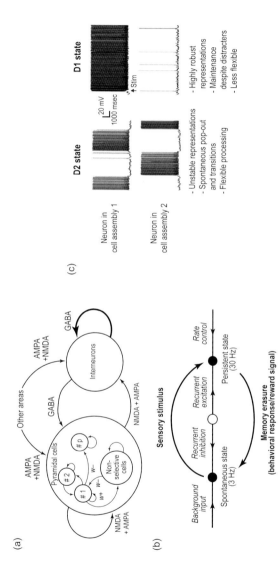

Figure 6.1 Prefrontal cortex and working memory. (a) Biophysical models of prefrontal cortex include multiple populations of pyramidal cells, their interactions with inhibitory interneurons, the differential effects of AMPA and NMDA currents, and recurrent connectivity between these cell populations leading to *attractor dynamics*. Reprinted with permission from Brunel and Wang (2001). (b) Schematic of attractor dynamics for working memory. Background input supports a spontaneous spike rate of prefrontal cells regulated by recurrent inhibition. A potent sensory stimulus provides sufficient excitation to elicit a higher spike rate, which is supported by recurrent excitation leading to a persistent state (*active maintenance*). Other factors can determine when such representations are no longer needed (e.g., task end state reached, behavioral response, reward), driving the population back to spontaneous state. Reprinted with permission from Wang (2001). (c) Biophysical models of dopaminergic influences show how dopamine modulates such prefrontal attractor dynamics. In the D2 state, dopamine levels are high and spiking in model neurons support multiple representations, facilitating updating of working memory. In contrast, the D1 state promotes the stable persistent activations in a dominant population which is robust against interference. Reprinted with permission from Durstewitz and Seamans (2008).

and D'Esposito 2011) and serve to inform the interaction between motivation and cognition more generally.

Finally, other models—from more detailed implementations to more algorithmic approaches—have suggested that in addition to direct DA input to PFC, the basal ganglia (BG) can act as a gating mechanism by disinhibiting thalamocortical input to selective PFC subregions, allowing more refined and selective control of working memory updating (Frank et al. 2001; Todd et al. 2009). Again, these models have framed and guided a variety of subsequent findings showing complementary roles of PFC and BG in working memory updating versus maintenance, and have facilitated a computational theory of the role of the BG that extends beyond its classical role in motor control. This framework and that of dopaminergic signaling within PFC reviewed above provide clear translational implications for patients with mental illness (e.g., in attention-deficit/hyperactivity disorder, schizophrenia and, more generally, other frontostriatal disorders), by providing a coherent set of mechanisms relevant for understanding distractibility, attentional focus, and the interactions between reward and cognition.

In sum, this example provides a target set of phenotypes to study in mental illness: the trade-off of flexibility versus stability inherent in PFC-BG networks, and the dynamic modulation of this trade-off by DA inputs to both PFC and BG according to task objectives and reward maximization. Incorporating other neuromodulators into these theories is an important line of work with some promise; for example, reduced serotonin function in orbitofrontal cortex has been related to getting "stuck" in attractor states and associated with obsessions (Maia and Cano-Colino 2015).

Reinforcement Learning and Motivated Choice in Corticostriatal Circuits

One of the most seminal contributions of computational work in understanding systems and cognitive neuroscience was the proposal that phasic activity in midbrain DA neurons signal reward prediction errors (RPEs) (Figure 6.2a), with increases in activity for positive RPEs (outcomes that are better than expected) and dips below baseline for negative RPEs (worse than expected) (Montague et al. 1996). This model found striking correspondence between dopaminergic patterns of activity during simple reward conditioning tasks and RPEs as reflected in the temporal difference model of reinforcement learning, which allows an agent to learn precise expected reward values of various states in the environment, and—when augmented to learn the value of its own actions as well—to take actions that can maximize its cumulative reward. The quantitative link between DA and RPEs (both positive and negative) has since received enormous degree of support across species and methods (Schultz 2013), largely overturning older theories about the roles of DA in motor function and/

or reward signaling per se. Moreover, subsequent rodent genetic engineering studies have confirmed the causal importance of dopaminergic RPEs for inducing both Pavlovian and instrumental learning in ways that conform to learning theory. Human studies show neural markers of RPEs in striatal BOLD signals which are amplified by DA manipulations and correlate with reward learning (Pessiglione et al. 2006; Jocham et al. 2011).

Early theories also proposed that the downstream mechanism by which these DA signals promoted learning involved modification of corticostriatal synaptic strengths (e.g., Doya 2000; see also Figure 6.2b). Subsequent models expanded this notion by examining how the biology of this system supported the existence of two opponent systems that differentially learn from positive and negative RPEs (i.e., when outcomes are better and worse than expected), as a function of differential DA modulation of D1- and D2-containing medium spiny neurons, which act to promote action selection and avoidance (Frank 2005; Figure 6.2c). This model was motivated by decades of systems neuroscience including electrophysiological, pharmacological, and behavioral data. It suggests differential roles of these pathways, but was developed as an attempt to explain data from human Parkinson patients, whereby dopaminergic drugs can sometimes impair and sometimes enhance cognitive function. Many studies across species, over the last decade, have provided support for the basic model mechanisms, showing that modulation of D1 and D2 corticostriatal pathways is both necessary and sufficient for inducing reward/approach and aversive/avoidance learning, respectively (Hikida et al. 2010; Kravitz et al. 2012). Incorporating this opponent process into a refined algorithmic reinforcement learning facilitated a formal analysis of its properties, allowing for quantitative fits to multiple datasets, and provided a normative account to explain why this system might have evolved in this manner (Collins and Frank 2014; Figure 6.2d). It also provided an explanation for the finding that antipsychotics (and indeed striatal DA denervation, more generally) can induce an aberrant learning process resulting in progressive motor deterioration, beyond the direct effects of DA depletion on motor performance. Thus it hinted at a different mechanism for potential therapeutics. This same model has been applied to explain differential sensitivity to positive versus negative decision outcomes (a different sort of trade-off) across a range of conditions induced by dopaminergic dysregulation, including Tourette syndrome, schizophrenia, attention-deficit/hyperactivity disorder, pathological gambling, and substance abuse (Maia and Frank 2011).

Various extensions of simple reinforcement-learning models have also been developed and relate to underlying biology. First, basic models often assume a fixed *learning rate*; that is, the degree to which RPEs are used to update action value estimates and hence behavioral adjustment. More sophisticated models show how this learning rate can itself be dynamically adjusted to take into account different forms of uncertainty, so as to integrate optimally the informativeness of the incoming RPE relative to current knowledge and

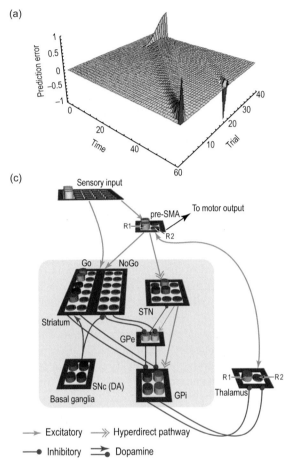

Figure 6.2   Dopamine, striatum, and reinforcement learning. (a) Temporal difference reinforcement-learning model showing phasic reward prediction error (RPE) signals initially when reward is delivered (end of the trial, final time step on *x*-axis) as well as with learning across trials (*z*-axis), where there is a lack of response to the reward itself; this prediction error signal is propagated back to the earliest predictor or reward (conditioned stimulus). On trial 20, the expected reward was withheld and a negative prediction error is observed. This pattern closely matches the phasic responses of midbrain dopamine (DA) neurons. Reprinted with permission from Schultz et al. (1997). (b) Models formalize how these DA RPEs are used to adjust the predicted values of sensory states V(s) and state-action pairs Q(s, a) in the striatum, with action selection dictated by comparison of action values in the downstream pallidum and subsequent disinhibition of the thalamocortical neurons coding for the most rewarding action. Reprinted with permission from Doya (2000). (c) Refined neural network of corticobasal ganglia circuit, with differential "Go" and "NoGo" striatal populations, representing positive and negative action values for given actions in pre-supplementary motor area (pre-SMA) and the current sensory state. Action selection is again governed by disinhibition (gating) of the corresponding column of thalamus, but where "NoGo" units provide evidence against a given action to prevent this disinhibition via the indirect pathway.

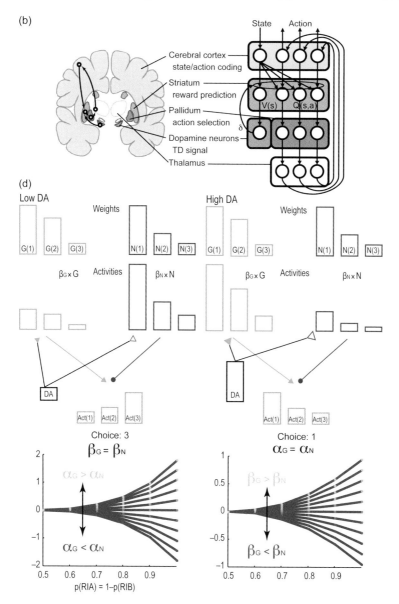

Figure 6.2 (continued) In the substantia nigra pars compacta (SNc), DA modulates excitability of "Go/NoGo" units via simulated D1 and D2 receptors, and phasic changes during RPEs drive opponent plasticity signals. The hyperdirect pathway from pre-SMA to subthalamic nucleus (STN) to globus pallidus internal (GPi) modulates the overall gating threshold by globally exciting GPi and making it more difficult for striatal "Go" signals to disinhibit the thalamus, thereby regulating impulsive choice. Individual neurons are cylinders, with instantaneous spike rate reflected by height and color. (*continued on next page*)

Figure 6.2 (continued) (d) Top: opponent actor learning (OpAL) model summarizing the core learning/choice computations of the neural network in algorithmic form, and capturing DA effects on both learning and incentive choice (Collins and Frank 2014). Separate G and N weights reflect learned propensities for each action to yield a positive or negative RPE. In the example, there are three actions and the corresponding learned G and N weights are shown in the top row. The middle row shows activity levels where DA levels during choice can be used to differentially amplify G or N weights via differential effects on excitability. In the "Low DA" condition, G weights are de-emphasized while differences among N weights are enhanced. Choice is governed by differences in activity levels; here the third choice which has the lowest cost is executed. In the "High DA" condition, for the same learned weights, the benefits are differentially amplified and Choice 1 is executed. Bottom: The OpAL model allows for asymmetry in the effects of DA on G versus N learning via the α parameters, and for asymmetry on their expression during choice, via the β parameters. Asymmetry in either set of parameters can produce differential sensitivity to probability of a choice leading to positive versus negative outcomes, and more so as the positive/negative probabilities become more deterministic.

to incorporate one's estimation of environmental volatility. Some evidence implicates the anterior cingulate cortex in tracking such volatility and adjusting learning rates (Behrens et al. 2007), while cholinergic interneurons may serve this function in the striatum (Franklin and Frank 2015). Recent studies suggest that this process is altered in individuals with high trait anxiety (Browning et al. 2015).

Second, rather than learn only values of simple stimulus-action pairs, *hierarchical reinforcement-learning* frameworks allow an agent to learn the values of, and select among, more abstract actions which themselves might involve multiple temporally extended sets of primitive actions (Botvinick et al. 2009). Once an abstract action is selected it can then carry out the sequence of state-action pairs that define it, allowing for more efficient reuse of previously learned actions (subgoals) that can be applied in the service of new goals. This hierarchical nesting of action selection and learning has been related to anatomical hierarchically nested rostral to caudal corticostriatal circuits, where anterior and lateral frontal circuits select actions which can then constrain the selection of lower-order actions in posterior loops.

Motivational Vigor and Incentive Choice

Although the learning theory of DA has been successful, some researchers prefer to emphasize the motivational aspects of DA in directly driving changes in vigor (speed with which actions are selected) and incentive choice (risky decision making). Computational models have simulated such effects as well by differential modulation of tonic (as opposed to phasic) DA, proposed to reflect the opportunity cost that would result from inaction (Niv et al. 2007). The opponent D1/D2 model ties together these DA roles on learning and choice: indeed, the mechanism by which DA modulates learning in this model is by altering the excitability of these populations, and hence also affects their relative expression (activity levels) at the time of choice, providing a mechanism to dynamically modulate the emphasis on costs versus benefits of alternative choice strategies (Collins and Frank 2014; Figure 6.2d). Optogenetic studies show that effective action values can be enhanced or suppressed for particular choices when stimulating D1- or D2-expressing striatal neurons, respectively. However, this modeling work has also further highlighted an issue already recognized by many in the animal-learning community; namely, that many findings in reinforcement-learning experiments which appear to result from differential modulation of learning could instead reflect differential modulation of incentive choice, or vice versa. Thus careful designs are needed to tease apart their differential contributions.

Pavlovian factors can also affect instrumental performance. Pavlovian-to-instrumental transfer is the phenomenon by which stimuli taking on Pavlovian values can invigorate or inhibit instrumental action (Liljeholm and O'Doherty 2012). Computational models have quantified these effects and how they interact with instrumental learning (Huys et al. 2011a), and have further suggested that they involve both dopaminergic and serotonergic components as well as ventral striatal value-based modulation of dorsal striatal action (Boureau and Dayan 2011).

Model-Based Learning

Lacking in all of the above discussion on learning is a consideration of actions that are "goal-directed" (i.e., taken with the purpose of achieving a particular outcome), which often involves planning and forward thinking. Indeed, the DA–RPE hypothesis belongs to a special class of reinforcement-learning algorithms referred to as *model-free* in the sense that it involves learning incremental associative values (whether positive, negative, or combined), reflecting the statistical probability that an action will result in a good or bad outcome—a sort of net "gut-level" value—often intended to explain habits rather than goal-directed behaviors (Daw et al. 2005; Liljeholm and O'Doherty 2012). In contrast, a *model-based* learner will represent the expected outcomes of their actions using a cognitive map of the environment (Figure 6.3). The outcomes of actions could be other states that have no intrinsic value in and of themselves but open up yet other potential actions and consequent states. The model-based agent then conducts a mental search using their cognitive map to decide which action to take based on their goals. A model-based agent is much more flexible in that it can plan which course of action to take based on its current valuation of particular states and shift the course of action when those values change, without having to reexperience RPEs and incrementally adjust values for all relevant state-action pairs. However, it is also much more computationally demanding and time consuming, requiring the existence of a model of state transitions and the ability to search through the future trajectories while planning,

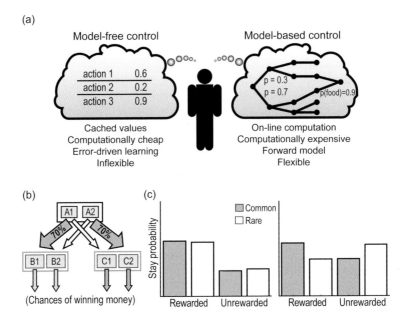

Figure 6.3 Model-based versus model-free reinforcement learning. (a) Model-free reinforcement learning as supported by dopaminergic RPEs allows a decision maker to learn a scalar value assigned to each action based on past reinforcement history; choices in this framework require only comparing such values without imagining their consequences for future choices. Such a system is efficient but inflexible when reinforcement contingencies of future states are altered and is typically invoked to explain habitual behaviors. In contrast, a model-based system reflects the predicted outcomes (subsequent states) that result from each action using a mental model of the environment, allowing a learner to make choices by searching this model and planning. Such a system is computationally demanding but flexible. Reprinted with permission from Smittenaar (2015). (b) Two-step task used to assess model-based versus model-free learning. Subjects choose between A1 and A2 which then yield subsequent B or C states. Choices at the bottom level yield different chances of obtaining rewards (money). Choice A1 results most commonly in the B state whereas A2 results in the C state, but rare transitions from A1 to C and A2 to B are also present. (c) Predicted probability of repeating the same top-level choice (A1 or A2) after rewards versus non-rewards, as a function of transition type from top to second level. A model-free agent will increase probability of staying (repeating the top-level choice) after rewards relative to non-rewards, regardless of the transition (it has no model of the transition structure). In contrast, a model-based agent shows an interaction: if rewarded on a rare transition, it knows that it will be more likely to reach the highly rewarding outcome on the next trial if it instead takes the top-level choice that most commonly transitions to that state. Reprinted with permission from Doll et al. (2012). Humans typically show a combination of model-based and model-free contributions, supported by interactive neural systems (see text).

thus involving working memory (and associated capacity limitations) and representations of the spatial environment. Indeed, rodent studies have shown striking evidence for forward projections of future states (in the hippocampus) and their values (in ventral striatum) as animals engage in decision making (van der Meer et al. 2010).

This example illustrates another key trade-off: the model-free system is efficient in allowing an agent to make choices rapidly based on accumulated values but is inflexible, whereas the flexible model-based system can allow for much more sophisticated choice strategies but is resource intensive. This trade-off suggests that the brain evolved to include both components, with various proposals debated about how they may compete or collaborate (Daw et al. 2005, 2011; Collins and Frank 2012). In rodents, model-based and model-free systems appear to depend on dissociable corticostriatal circuits, but in humans there is evidence for both competition and collaboration (Doll et al. 2012).

Thus multiple mechanisms are engaged in model-based reinforcement learning, including prefrontal-dependent working memory processes, hippocampal-dependent episodic and spatial memories, and an "arbitrator" for deciding which system should govern choice based on uncertainty or reliability (sometimes proposed to involve anterior cingulate cortex). Furthermore, the tendency to engage in model-based processing might itself be subject to motivational influences in addition to intrinsic capacities, and thus ventral striatum and anterior cingulate cortex modulation of cost-benefit trade-offs are equally relevant.

The interactions between model-based and model-free processes are only beginning to be uncovered. For example, the model-based system might train the model-free system so that it can engrain useful stimulus-response policies such that they no longer depend on cognitive resources. Reciprocally, model-free processes can be used to learn when to engage model-based systems. Model-based systems can also be used to learn representations of task structure—the variables in a task that matter, hidden causal states that govern contingencies—which can dramatically enhance the efficiency of model-free learning by collapsing across irrelevant features and facilitating generalization (Collins and Frank 2013; Wilson et al. 2014). Moreover, the complexity of model-based processing suggests that even when it is engaged, some shortcuts are often needed to prevent one from having to consider all possible courses of action; one such shortcut is a Pavlovian effect which drives subjects to avoid considering routes that elicit immediate negative states, even when this is suboptimal (Huys et al. 2012).

All of these mechanisms are ripe for further investigation into their disturbances in mental illness. Indeed, preliminary evidence suggests that a range of compulsive disorders are associated with reduced model-based processing (Voon et al. 2015). Further studies are, however, needed to examine the precise nature of these effects and their potentially dissociable underlying mechanisms. For example, model-based decision making can be impaired due to (a) impaired *learning* of the model (i.e., impairments in detecting sequential transitions that describe the environment), (b) reduced tendency to *use* the model when making choices (reluctance to engage in deliberative processing required for planning), and/or (c) reduced motivation to engage in either model-based learning or model-based choice depending on the motivational stakes, cognitive load, etc.

Exploration versus Exploitation

To optimize learning, it is not always best to take the action that has higher expected value based on previous reward histories. Indeed, one should sometimes explore actions that have potential to provide yet better outcomes than the status quo (Figure 6.4). There are two main strategies that have been studied for balancing this exploration–exploitation trade-off. The first is to simply add some noise into the choice function: rather than deterministically choosing the options with highest reward values, a typical reinforcement-learning agent will make choices stochastically, allowing it to explore the values of actions it does not know. The most common of such choice functions is called *softmax* and is a logistic function that effectively adds more stochasticity to choices when the perceived values are more similar to each other, with more deterministic exploitation for values that are further apart. Such an algorithm is relatively simple to implement, with various proposals suggesting that cortical norepinephrine can dynamically modulate the noise in the choice function, and is itself regulated by recent task performance—encouraging more exploration during periods of poorer performance (Cohen et al. 2007). However, exploration might also demand cognitive control and prefrontal resources for overriding the dominant striatal tendency to exploit (Daw et al. 2006).

The second, a more strategic model-based approach to exploration, is to direct exploration toward those actions that have the greatest potential to be informative about the value of the current policy. Behavioral studies have shown that humans use a combination of both random and directed exploration (Wilson et al. 2014; Figure 6.4b). Further, fMRI and EEG studies have shown that the degree to which humans engage in directed exploration toward uncertain options is accompanied by rostrolateral PFC activity, which dynamically tracks the potential gain in information (relative uncertainty) that would result from such exploration (Badre et al. 2012); genetic and pharmacological studies suggest, however, that this tendency is modulated by prefrontal catecholaminergic function (Kayser et al. 2015). Notably, deficits in such uncertainty-driven exploratory behavior are correlated with anhedonia in patients with schizophrenia (Strauss et al. 2011). This finding might imply that anhedonia is not actually related to hedonics (i.e., an inability to experience pleasure)—indeed much evidence in the schizophrenia literature rejects that notion—but could

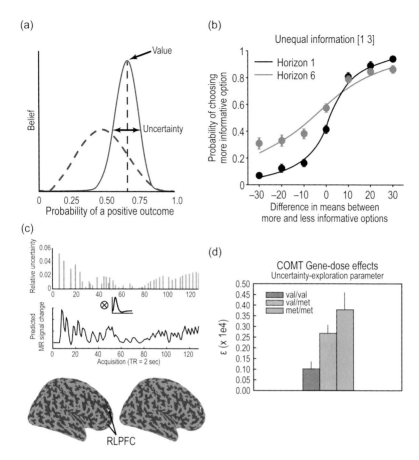

Figure 6.4 Exploration versus exploitation. (a) Adaptive reward-based choice requires not only comparing the values among alternative choices but considering the uncertainty in what those values are. Predicted rewards (or here, probabilities of obtaining a positive outcome) can be represented as entire belief distributions rather than single values. The solid curve represents an action that has high belief associated with value of 0.7 but other nearby values have similar belief levels. The dashed red line shows the belief distribution of an alternative action which has lower mean expected value but higher uncertainty (given limited experience), i.e., there is some possibility that its true value lies higher than that for the other action. Directed exploration thus takes informative actions that reduce this uncertainty. (b) Probability of choice increases as a function of mean difference in expected value based on limited samples, but there is a bias toward choice of the more informative action, particularly when the subject knows they will have the opportunity to make more choices (horizon 6) to capitalize on this information. Adapted with permission from Wilson et al. (2014). (c) fMRI study regressing trial-to-trial measures of information gain against BOLD activity reveals rostrolateral prefrontal cortex (RLPFC) region that tracks relative uncertainty about action values, and more so in explorers that use directed exploration behaviorally. Reprinted with permission from Badre et al. (2012). (d) Candidate gene affecting prefrontal catecholaminergic function modulates degree of directed exploration (Frank et al. 2009).

instead result from a reduced tendency to engage in those activities that could potentially improve their long-term situation.

Decision-Making Dynamics: From Simple
Choice to Inhibitory Control

Aberrant decision making can also arise from changes to the decision process itself. Above we have been assuming that choices are made using some sort of comparison process among the learned reward values. While the reinforcement-learning literature focuses on how these values are acquired, the decision-making literature focuses on how choices are made within a given trial in the face of competing sources of evidence for each alternative, which may fluctuate based on momentary changes in perception, attention, or memory. Again, multiple levels of modeling have been applied to understand the decision-making process. One of the more popular frameworks is the drift diffusion model (Figure 6.5), which has been widely used for several decades in mathematical psychology; it accounts not only for choice proportions (which choices are made, given differences in the evidence for each option), but also for the full distributions of response times of those choices (Ratcliff and McKoon 2008). Sources of evidence in this framework can be perceptual (e.g., make a choice to discriminate whether you see an animal or a man-made object on a screen with different levels of discrimination difficulty, contrast and/or distractors), memory (e.g., determine whether an object presented has been studied before, given different levels of encoding), or based on reward values having multiple attributes (such as taste vs. health, e.g., choose among an apple or a cookie). In all cases, the drift diffusion model can be used to extract decision parameters that govern the choice process. The most relevant here are the "drift rate," which quantifies the amount of evidence inherent in the stimulus itself (or in the neural representation thereof), and the "decision threshold," which reflects the degree of evidence in favor of one option over the other before a participant is willing to commit to a choice. (It also has a Bayesian interpretation: given the stimulus presented, it reflects the likelihood ratio in favor of one option relative to the other; Gold and Shadlen 2007.) While changes to the drift rate or the decision threshold can produce changes in response times or choice proportions, they can be disentangled by examining simultaneously the choice proportions and response time distributions: small drift rates imply slower and more variable choices, higher decision thresholds lead to slower but more consistent (and accurate) choices. Any prior bias to select one option over the other (perhaps due to differential expectations before stimuli are observed) can be captured by the starting point, or bias parameter. More refined patterns of choice data and response time distributions can also be captured by estimating the degree of cross-trial

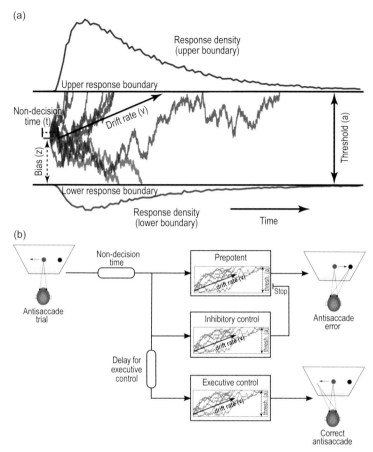

Figure 6.5 Decision-making dynamics. (a) The drift diffusion model is one of a larger class of sequential sampling models of simple decision making that can quantitatively capture choices and their response time distributions. It is commonly used to study neural mechanisms of the underlying process. Evidence for one choice over the other is noisily accumulated over time (*x*-axis) with average drift-rate *v* (reflecting the degree of evidence for one alternative over the other) until one of two boundaries (separated by threshold *a*) is crossed and a response is initiated. Trajectories of multiple drift processes (blue and red lines) are shown to illustrate the variability of this process across trials. The resulting response time distributions are shown for each choice. The non-decision time *t* accounts for time prior to (perception) and after (motor execution) the decision process; the bias (*z*) accounts for any predisposition toward one choice or the other. (b) Accumulator models show how different sources of evidence can accumulate independently to govern choices. This example shows how such models can be used to simulate dynamic processing within a trial supportive of inhibitory control, as in the antisaccade task, where an initial "intuitive" prepotent response drives accumulation toward one response boundary, an inhibitory signal accumulates to stop this process, and later information derived from cognitive rules can implement executive control to modify responding toward the correct action. Adapted with permission from Noorani and Carpenter (2012).

variability in these parameters (drift rate and starting point), which can reflect attentional lapses and/or changes in neural noise.

Beyond providing an abstract formalism that would allow psychologists to quantify distinct cognitive processes governing simple choice, many neuroscientific investigations have shown evidence for the otherwise unobservable internal processes of such models. Much of this evidence comes from perceptual decision-making tasks in which electrophysiological signals (e.g., spike rates in parietal cortex and striatum) exhibit the predicted ramping of decision variables quantifying the degree of evidence (or expected reward value) for one option over the other, with the slope of this ramping proportional to the signal-to-noise ratio in the stimulus; choices are executed once these neural signals reach a critical threshold (Gold and Shadlen 2007; Ratcliff and McKoon 2008). In reward-based decision making, the drift rate is proportional to relative difference in reward values among alternative options and is modulated by the subject's visual attention and reflected in striatum and ventromedial PFC (Lim et al. 2011). Moreover, choice values reflect multiple attributes: those involving impulsive urges (e.g., unhealthy foods) are encoded earlier in ventromedial PFC and are subsequently modified by long-term goals (Harris et al. 2013); competition among each attribute can occur at various hierarchical stages (Hunt et al. 2014). These models can also be modified to simulate decisions in inhibitory control such as the antisaccade task, where a prepotent response is elicited by an imperative stimulus, but where cognitive rule-based representations (e.g., in PFC) can modify the decision process later in time (Noorani and Carpenter 2012). More generally, these models can be used to estimate the separable contributions of mechanisms giving rise to fast automatic response tendencies, those associated with inhibiting such responses, and those associated with guiding behaviors toward controlled responses.

Mechanistic neural models also articulate how these different choice processes are implemented in cortical attractor networks (Wang 2012) and in the BG (Lo and Wang 2006; Bogacz and Gurney 2007; Ratcliff and Frank 2012). One critical variable that should be evident in terms of aberrant decision making is the decision threshold: if, as noted above, impulsive urges contribute to value signals early during the choice process, then a mechanism to regulate decision thresholds is critical to regulate decision making. Indeed, mechanistic neural models have specified how the decision threshold is malleable and subject to communication between frontal cortex and BG. In particular, when subjects experience "decision conflict" (i.e., when there are multiple decision alternatives or when the execution of a rule-guided action requires overriding an initial prepotent response), the frontal cortex signals this information to the subthalamic nucleus (STN) (see Figure 6.2c) which—by effectively raising the decision threshold—makes it more difficult for initial striatal valuation signals to impulsively govern choice. This role for the fronto-STN pathway in decision-threshold regulation and inhibitory control is corroborated by neuroimaging, intracranial electrophysiology, and deep brain stimulation studies (e.g., Hikosaka and Isoda 2010; Cavanagh et al. 2011). Indeed, deep brain stimulation of STN can induce impulsivity in patients' daily lives and prevent them from adjusting decision thresholds as a function of conflict. These studies provide an opportunity for developing phenotypes that identify when such mechanisms are aberrant and likely to be causing dysfunction. Moreover, this same STN mechanism could play a role not only in impulsivity (linked to a reduced decision threshold) but in the opposite scenario (when STN and/or its cortical inputs are overactive, linked to too high a decision threshold). It might also be implicated in perfectionism, obsessiveness, and deficits in proactive reward-based decisions.

This last example highlights that a particular phenomenon of impulse control disorders can arise from multiple mechanisms. Here, though, the emphasis is on the need to regulate the decision threshold to consider alternative goals given an impulsive urge, whereas earlier it was on the imbalance in the sensitivity to prospective positive versus negative decision outcomes due to changes in DA function. Indeed, the modeling framework and relevant task paradigms have facilitated the ability to dissect components of impulsivity related to insensitivity to adverse consequences (which are affected by DA medications given to patients with Parkinson disease and associated with pathological gambling), from those involved in disinhibition during conflict-based decision making that is affected by deep brain stimulation of STN (Frank et al. 2007b). Moreover, yet other forms of impulsivity may involve differential discounting of long-term versus immediate rewards (e.g., McClure et al. 2004), the neural mechanisms of which are also intensely studied.

Social Interactions

Although we have focused in this chapter on simple decision-making and learning tasks, computational cognitive neuroscience approaches have also examined social interactions. For example, game theory, traditionally developed in economics, has been recently applied to understand how humans interact with each other in various cooperative and competitive environments that require theory of mind and value representation of self versus other (Montague et al. 2012). This research program shows how computational approaches can be used to infer latent processes involved in social decision making, much as latent processes are involved in inferring task variables in complex reinforcement-learning environments alluded to above. For example, I may view the value of a particular choice in terms of its immediate outcome, but if that outcome also depends on another person's goals and intents, we can develop a model that incorporates their beliefs and goals and determine if that should influence our own decisions and associated values. This type of approach is beginning to see some useful application to explore how these processes are altered in disorders such as autism.

Conclusion

This selective and quite incomplete overview covers only a small portion of computational approaches to a restricted set of domains within the larger field of cognitive neuroscience. Nevertheless, it has highlighted how computational models at multiple levels of description have contributed to a richer understanding of the neural basis of cognitive function, including several examples for how these have been or could be capitalized to understand the failure modes of such functions in mental illness. Much more work is needed to study which individual mechanisms can be assessed using refined cognitive tasks, neural measures, and quantitative models, as well as how these mechanisms interact to form distinct functional profiles (Stephan and Mathys 2014; Wiecki et al. 2015).

7

How Could We Get Nosology from Computation?

Christoph Mathys

Abstract

Psychiatry has found it difficult to develop a nosology that allows for the targeted treatment of disorders of the mind. The historic inability of the field to agree on a nosology based on clinical experience has led it to retreat to diagnoses based on symptom checklists as laid down in the Diagnostic and Statistical Manual of Mental Disorders (DSM). While this has increased the reliability of diagnoses, hopes that biological findings would lead to the emergence of mechanistically founded diagnostic entities have not been realized despite considerable advances in neurobiology. This article sets out a possible way forward: harnessing systems theory to provide the conceptual constraints needed to link clinical phenomena with neurobiology. This approach builds on the insight that the mind is a system which, to regulate its environment, needs to have a model of that environment and needs to update predictions about it using the rules of inductive logic (i.e., Bayesian inference). The application of the rules of inductive logic is called Bayesian inference because Bayes's theorem is the most important consequence of these rules, prescribing how beliefs need to be updated in response to new information. Importantly, while Bayesian inference is by definition consistent with the rules of inductive logic, it can still be false (to the point of being pathological), in the sense of leading to false predictions, because the model underlying the inference is inadequate. Further, it can be shown that Bayesian inference can be reduced to updating beliefs based on precision-weighted prediction errors, where a prediction error is the difference between actual and predicted input, and precision is the confidence associated with the input prediction. Precision weighting of prediction errors entails that a given discrepancy between outcome and prediction means more, and leads to greater belief updates, the more confidently the prediction was made. This provides a conceptual framework linking clinical experience with the pathophysiology underlying disorders of the mind. Limitations of this approach are discussed and ways to work around them illustrated with examples. Finally, initial steps and possible future directions toward a nosology based on failures of precision weighting are discussed.

Introduction

The State of Psychiatric Nosology

Before DSM-III, the state of psychiatric nosology was widely seen as unsatisfactory. The main point of criticism was the lack of diagnostic reliability (e.g., Robins and Guze 1970). Given a dearth of biological or conceptual constraints on nosological speculation, clinical experience had to serve as the main guide in developing a nosology of the mind. Clinical experience came—and comes—in two forms: (a) each clinical practitioner has his or her own immediate experience with patients, but (b) he or she is also acculturated into the thinking of the field, whose collective clinical experience has been condensed into nosological categories that are passed on as a traditional body of knowledge. While neither of these sources of nosological insight is to be scoffed at, it is not surprising that, owing to the diversity of individual experience and nosological traditions, the inter-rater reliability of psychiatric diagnoses was low. DSM-III was a conscious effort to address this problem by shifting the focus of diagnosis to lists of easily observable or reportable symptoms. However, despite decades of efforts, and despite an increase in the reliability of diagnoses (Clarke et al. 2013; Freedman et al. 2013; Narrow et al. 2013; Regier et al. 2013), the state of psychiatric nosology is still widely held to be dire (Craddock and Owen 2010; Kapur et al. 2012). Critics focus mostly on the missing biological foundation of the existing diagnostic categories (Hyman 2012; Insel 2012; Owen 2014), and they express hope that directing research efforts toward the biological mechanisms underlying psychiatric disorders will enable a new, mechanistically grounded nosology and uncover new targets for pharmacological treatment. In mitigation, one might say that the current system produces reliable categories that have been useful for clinical care up to a point, but that to make further nosological advances which can help guide neurobiological research, improve predictions regarding what treatments will be most efficacious, and ultimately identify new treatment targets, computational approaches will be essential.

Constraints on Nosology from Systems Theory

There is relatively little appreciation for the fact that in addition to constraints on psychiatric nosology rooted in clinical experience and human biology, there are also constraints originating in systems theory. While systems theory is complicated, these constraints are simple, which makes them an important guide to the interpretation of the clinical and biological patterns observed in disorders of the mind. My aim here is to explain these constraints, their implications, and their simplicity.

The main systems theoretic constraint, from which all others follow, is the good regulator theorem (Conant and Ashby 1970): every good regulator of a system needs to be a model of that system. This forms the basis of the reasoning in this article. Setting out from this theorem, I argue that any mind striving successfully to preserve its existence fits the description of a good regulator and therefore needs to be, in a well-defined sense, a model of its environment. To go about regulating its environment, the model (i.e., the mind) performs inference, learns, and selects actions in line with the laws of probability, or in other words, according to Bayesian inference (Jaynes 2003). At the heart of Bayesian inference is Bayes's theorem (Bayes and Price 1763; Laplace 1774). It prescribes the only way that, given a model, beliefs (i.e., probability distributions) can be updated in response to new information without violating the laws of probability. It is important to note that the word "belief" is used here as shorthand for "probability distribution of a state or parameter of the mind's model of its environment." Beliefs need not be conscious or even consciously accessible. In most cases, including almost all interesting ones, the equation governing probabilistic belief updates given by Bayes's theorem has no closed-form solution, meaning that it is mathematically impossible to write down an equation giving the solution. For example, the very simple equation $2^x = x + 1$ cannot be solved for x in closed form (i.e., it cannot algebraically be transformed into an equivalent equation of the form $x = \ldots$). However, solutions exist ($x = 0$ and $x = 1$), but short of guessing them, the lack of a closed-form solution forces us to find them by using approximations, which always involve assumptions. Importantly, approximations do not diminish the complexity and richness of the models being used to perform inference. To the contrary, approximate methods allow us to perform inference using much richer models than would be the case if we were restricted to cases where closed-form inference is possible. I will show that if these assumptions underlying approximate inference are chosen in the right way, Bayesian inference can be reduced to the application of simple update rules that all have one canonical form: precision-weighted prediction errors. A prediction error is the difference between actual and predicted input, and precision is the confidence associated with the input prediction. Precision weighting of prediction errors entails that a given discrepancy between outcome and prediction means more, and leads to greater belief updates, the more confidently the prediction was made. If the mind is a model of its environment and, in order to be a good regulator, has to take recourse to Bayesian inference, which is implemented as belief updating by precision weighting of prediction error, then disorders of the mind can be described as false inference based on broken precision weighting or prediction error processing. This gives us an additional set of constraints within which to develop a scientific psychiatric nosology. To be complete, any description of a disorder in such a nosology would have to address three questions: Is this concept of a disorder able to explain the patterns seen in clinical practice? How do these patterns emerge from false inference in terms of precision-weighted prediction errors? In what way is the biological machinery underlying belief updating broken? These questions illustrate that, instead of complicating the picture, the additional constraints from systems theory simplify the task of relating clinical manifestations of disorders to underlying biological mechanisms because they serve as a conceptual bridge between them.

In what follows, I will sketch a tentative construction plan for this conceptual bridge. Its body will be formed by an elaboration of the preliminary systems theoretic reasoning given above, while the bridge ends will be formed by recent attempts to tie false inference to psychiatric symptomatology (Edwards et al. 2012; Adams et al. 2013; Lawson et al. 2014; Quattrocki and Friston 2014) and to pin down the neurobiology of belief updating (Bastos et al. 2012; Shipp et al. 2013).

Theory

The Good Regulator Theorem, Generative Models, and Inductive Reasoning

In systems theoretic terms, the mind of an organism is a system which, using its brain and other body parts, strives to survive by regulating a second system, its environment. This makes it subject to the good regulator theorem, as stated in the title of Conant and Ashby's article: "Every good regulator of a system must be a model of that system" (Conant and Ashby 1970:89). The authors explain what this means in their abstract:

> The design of a complex regulator often includes the making of a model of the system to be regulated. The making of such a model has hitherto been regarded as optional, as merely one of many possible ways.

They go on to construct a theorem which shows, under very broad conditions, that any regulator which is maximally both successful and simple must be isomorphic (i.e., of the same structure and equipped with the same properties) with the system being regulated. (The exact assumptions are given.) Making a model is thus necessary. The theorem has the interesting corollary that the living brain, so far as it is to be successful and efficient as a regulator for survival, must proceed, in learning, by forming a model (or models) of its environment.

This model of the brain's (or rather mind's) environment is a *generative model* of its sensory input, u; that is, a model of how the environment generates u. For example, u could be a slight crackling noise somebody hears while speaking on the telephone. Generative models consist of two parts. The first, called the likelihood, is the probability $p(u|x, \vartheta)$ of input u given state x of the environment and parameter ϑ. The difference between state and parameter is simply that the state changes with time while the parameter is constant. In this example, there are many possible causes that could have generated this crackling noise, some of them more plausible than others. One possible cause is that the telephone has been bugged and the crackling is caused by a listening device. The state x would then be that of the telephone, bugged or not bugged, while the parameter ϑ would govern how exactly the telephone being bugged (or not) translates into the crackling noise u. Note that both state and parameter are sets that can have many elements. The second part of the model is called the *prior distribution*, or simply *prior*. This is the probability $p(x,\vartheta)$ of state and parameter in the absence (usually before, hence *prior*) of input u. In a clinical setting, different patients can attach very different prior probabilities to telephones being bugged, to bugging leading to crackling, etc. Such a (possibly largely unconscious) view of how the environment generates sensory input is formally described by the *joint distribution* $p(u,x,\vartheta) = p(u|x,\vartheta)p(x,\vartheta)$ of input, state, and parameter; that is, by the product of likelihood and prior, which constitutes a full generative model.

Given a generative model and input u, the mind can then, in principle, calculate the posterior distribution of state and parameter by the application of Bayes's theorem:

$$p(x,\vartheta \mid u) = \frac{p(u \mid x,\vartheta)p(x,\vartheta)}{\int p(u \mid x',\vartheta')p(x',\vartheta')\mathrm{d}x'\mathrm{d}\vartheta'} . \tag{7.1}$$

This is the probability distribution of state and parameter given the input u, and the transition from $p(x,\vartheta)$ to $p(x,\vartheta|u)$ in response to u is a *belief update* in the sense that probability distributions constitute beliefs. Crucially, the update as given by Equation 7.1 is the only way to update the belief on x and ϑ that does not violate elementary requirements of inductive reasoning (Cox 1946; Jaynes 2003). Inductive reasoning is reasoning about uncertain quantities, as opposed to deductive reasoning, which deals with certain quantities. For example, if we know that every cat is an animal, we can deduce with certainty that A (an animal from the information) is a cat. However, if we at first know nothing about A and are then told it is an animal, this merely increases the probability that A is a cat without making it certain. In other words, A being a cat becomes more plausible. This increase and decrease in the plausibility of statements as a result of new information is what inductive reasoning addresses. Cox (1946) showed that the only rational way to update beliefs about the plausibility of statements is by applying the known rules of probability theory (e.g., Bayes's theorem; cf. Flagel et al., this volume). He proved this by showing that the rules of probability can be derived from three basic desiderata concerning inductive reasoning:

1. The plausibility of a statement can be represented by a real number (and the plausibilities of different statements compared this way).
2. Information that makes a statement more plausible increases the number associated with it.
3. Different ways to calculate the same plausibility should always give the same result.

This means that reasoning incompatible with the rules of probability implies a violation of Cox's three desiderata (i.e., of common sense). This may seem like a restrictive constraint, but in reality it is anything but. According to the complete class theorem (Robert 2007:411), under mild conditions, there is always a prior accounting for any given combination of posterior, likelihood, and input. This means that no conclusion that could be drawn from a given observation is impossible in the sense of violating the rules of probability. Bayesian belief updating in no way constrains the inferences the mind can make about its environment. However, it constrains the way the mind is described. A full account of a belief update has only been given once the generative model it is based upon has been fully described; that is, once a likelihood and a prior have been specified. While the good regulator theorem states that the mind will have to be a model of its environment, Bayes's theorem provides the framework within which to describe belief updating in accordance with the rules of inductive reasoning but without constraining its substance.

In what follows, I will take a fairly didactic but technical walk through some of the formal aspects of Bayesian inference in the brain. In other words, we will look at the mathematical structure of how beliefs are encoded and updated and what this tells us about neuronal processes. Although this treatment is a bit mathematical, the end point of the analysis will be something that is central to a theoretical and neurobiologically grounded understanding of false inference in psychiatry. This is the central role of gain control or neuromodulation in the brain in weighting neuronal messages that are passed from one part of the brain to the other. Neuromodulatory mechanisms are invariably implicated in both the pathophysiology and pharmacology of psychiatric conditions (e.g., implicating classical neuromodulators like dopamine and serotonin). Furthermore, this form of gain control implicates fast spiking inhibitory interneurons and synchronized neuronal activity of the sort that can be measured noninvasively using EEG and, potentially, correlated with symptoms of false inference, and response to treatment.

Sequential Updating of a Time Series Mean

Before returning to Bayesian belief updating, let us first look at ways to update the mean of a series of sequentially observed numbers. This might at first seem a distraction, but will turn out to be fundamental.

Given N observations $\{u_1, u_2, \ldots, u_N\}$, it is simple to calculate their mean $\bar{u}_n$:

$$\bar{u}_N = \frac{1}{N} \sum_{n=1}^{N} u_n. \tag{7.2}$$

However, if $\{u_1, u_2, \ldots, u_N\}$ is a time series, keeping track of the mean as new observations arise requires all observations, if the calculation is to be made according to Equation 7.2. Fortunately, there is a less memory-intensive way to achieve the same end. The following update equation can be applied sequentially:

$$\bar{u}_{n+1} = \bar{u}_n + \frac{1}{n+1}\left(u_{n+1} - \bar{u}_n\right). \tag{7.3}$$

Starting with $\bar{u}_1 = u_1$ and applying Equation 7.3 to all observations until $\bar{u}_n$ is reached, we get:

$$\bar{u}_N = \bar{u}_{N-1} + \frac{1}{N}\left(u_N - \bar{u}_{N-1}\right), \tag{7.4}$$

which gives the same result as Equation 7.2.

The sequential updating of Equation 7.3 has the advantage that it requires remembering only two numbers: the previous mean $\bar{u}_n$ and the number of previous observations n.

Since the update rule of Equation 7.3 is of fundamental importance, it is worth looking at its components. There is the previous mean $\bar{u}_n$, representing the state of belief before the new observation u_{n+1}. Since the current state of belief corresponds to the best possible prediction for any new observation, the difference $u_{n+1} - \bar{u}_n$ between the new observation and the current belief is a *prediction error*. This means that the update has the form:

$$\text{new mean} = \text{old mean} + \text{weight} \cdot \text{prediction error}, \tag{7.5}$$

where the weight of the prediction error depends on how many previous observations there have been. The more observations that have already been made, the less a new observation will be able to move the mean.

Bayesian Belief Updating

A simple example of Bayesian belief updating is the case where the likelihood $p(u\,|\,\vartheta) = \mathcal{N}\left(u; \vartheta, \pi_\varepsilon^{-1}\right)$ is Gaussian (i.e., follows a normal distribution) with known precision (i.e., inverse variance) π_ε and the prior $p(\vartheta) = \mathcal{N}\left(x; \mu_\vartheta, \pi_\vartheta^{-1}\right)$ is also Gaussian with mean μ_ϑ and precision π_ϑ. There is no time-varying state x here, the parameter ϑ is a simple scalar, and the prior hyperparameter $\{\mu_\vartheta, \pi_\vartheta\}$ (i.e., the parameter governing the prior distribution of the parameter) is taken to be known. The posterior now also turns out to be Gaussian: $p(\vartheta\,|\,u) = \mathcal{N}\left(x; \mu_{\vartheta|u}, \pi_{\vartheta|u}^{-1}\right)$, where the updated precision and mean are:

$$\pi_{\vartheta|u} = \pi_\vartheta + \pi_\varepsilon,$$

$$\mu_{\vartheta|u} = \mu_\vartheta + \frac{\pi_\varepsilon}{\pi_{\vartheta|u}}\left(u - \mu_\vartheta\right). \tag{7.6}$$

Remarkably, the update of the mean has the same structure (i.e., that of Equation 7.5) as the update of the mean of Equation 7.3. This similarity becomes even more obvious if we rearrange Equation 7.6 to read:

$$\mu_{\vartheta|u} = \mu_\vartheta + \frac{1}{\pi_\vartheta / \pi_\varepsilon + 1}(u - \mu_\vartheta). \tag{7.7}$$

This shows the correspondence between n, the number of previous observations in Equation 7.3, and the relative precision $\pi_\vartheta / \pi_\varepsilon$ of the prior with respect to the likelihood. In both cases, this represents the weight of previous evidence relative to new information. In what follows, this relative weight will be called v to emphasize that it can be any positive number, whereas n was a natural number.

This update structure is not restricted to the simple Gaussian example used above. All generative models we are ever likely to need to describe the brain (or equivalently, all generative models the brain is ever likely to need to describe its environment) will only involve exponential families of likelihoods with conjugate priors. These are families of likelihood distributions that can all be written in one canonical form, which is a generic representation of all families. A conjugate prior is one that gives rise to a posterior of the same family when combined with a given likelihood. For example, the Gaussian distribution is an exponential family, and it is its own conjugate prior. As we saw in Equation 7.6, this means that a Gaussian likelihood with a Gaussian prior leads to a Gaussian posterior. In addition to the Gaussian distribution, this includes the beta, gamma, binomial, Bernoulli, multinomial, categorical, Dirichlet, Wishart, Gaussian-gamma, log-Gaussian, multivariate Gaussian, Poisson, and exponential distributions, and many others. For all of these distributions, the Bayesian belief update has the following form:

$$\xi' = \xi + \frac{1}{v+1}(T(u) - \xi), \tag{7.8}$$

where $T(u)$ is a function of the input u called the *sufficient statistic*, ξ is the hyperparameter governing the prior, and ξ' is the updated hyperparameter, which governs the posterior.

In the case of exponential families with conjugate priors, this means that Bayesian inference reduces to tracking the mean of the sufficient statistics of observations. The weight of the prior in this update is determined by the positive number v, which can be interpreted as the number of observations preceding u, whose weight is 1. In light of Equation 7.7, v is the precision of the prior relative to that of the observation. Bayesian belief updating thus takes place by precision-weighting prediction errors on the sufficient statistics of observations.

To conclude this discussion, I will give a few examples of a more technical nature. (Less technically inclined readers can skip this without missing anything essential.) In the case of a Gaussian model with unknown mean and known precision (as in Equation 7.6), $T(u) = u$; if both mean and precision are unknown, $T(u) = (u, u^2)^T$. This generalizes to $T(\mathbf{u}) = (\mathbf{u}, \mathbf{u}\mathbf{u}^T)^T$ for a multivariate Gaussian. In the case of a gamma model, $T(u) = (\ln u, u)^T$. In the beta case, $T(u) = (\ln u, \ln(1-u))^T$, and in the categorical case, $T(\mathbf{u}) = \mathbf{u}$. Between these models, situations are addressed where observations are on an unbounded continuum, on a continuum bounded on one side or on both sides, all multivariate generalizations of these, and situations where observations are categorical. In all of these models, and in many more cases, Bayesian inference reduces to Equation 7.8 (i.e., to mean updating).

Discussion

The Bridge Ends: Clinical Phenomena and Neurobiology

If the mind is necessarily a model of the environment it regulates, and if using a model to regulate the environment entails updating beliefs according to the laws of probability (i.e., according to the rules of Bayesian inference), and if Bayesian inference entails precision weighting of prediction errors, then disorders of the mind will have to be interpretable in terms of precision weighting of prediction errors. Further, if the brain is the organ of the mind, then the brain's physiology will also have to be interpretable in terms of precision weighting of prediction errors. This is how systems theory can serve as a bridge between clinical manifestations of disorders of the mind and the disordered biological mechanisms underlying them, connecting them in a way that allows us to make sense of both.

Turning first to the side of clinical manifestations, there have been many recent attempts to understand disorders of the mind in terms of precision weighting of prediction errors. Relating to psychosis, Adams et al. (2013) give a broad overview and explain many of the manifestations of psychosis, such as hallucinations, delusions, catatonia, and sensory attenuation deficits, as the result of aberrant precision weighting of exteroceptive sensory input. For example, patients with schizophrenia show abnormalities in smooth pursuit eye movements (Thaker et al. 1999). When following a dot as it moves smoothly back and forth, right to left, they are less able to predict where it will reappear after it has been occluded by a vertical bar for a short while. When the dot disappears, patients slow down their eye movement more than healthy controls, forcing them to accelerate more to catch up with the dot once it reappears. Conversely, when the dot makes unexpected jerky movements, patients with schizophrenia are better able to follow it than healthy controls (Hong et al. 2005). Aberrant precision weighting of prediction errors can explain this apparent paradox. The observed effects are predicted by models where healthy controls rely more on top-down predictions from an internal model of dot motion to follow the dot, while patients rely more on the immediate bottom-up input. This relative disregard by healthy controls of sensory input in favor of model-based predictions is called sensory attenuation, and the patients' behavior is explained by a failure of sensory attenuation. Specifically, when the precision weight on the prediction error regarding sensory input is high relative to that regarding the model-based prediction of dot position, then the eye will remain as if glued to the stimulus, enabling quick reactions to unexpected jerks, but losing its bearings whenever the dot disappears. Failure of sensory attenuation can also explain other abnormalities in patients with schizophrenia, for example resistance to the hollow-mask illusion and to the force-matching illusion. Furthermore, aberrant precision weighting leading to a failure of sensory attenuation can explain the emergence of delusions and hallucinations (cf. Adams et al. 2013). This is because action (e.g., moving a hand or an eyeball) is impossible without sensory attenuation. When confronted with prediction errors, a biological agent existing under the constraints of the good regulator theorem (i.e., trying to make good predictions) has two general ways to reduce them: it can either update its beliefs or act to change the environment so that sensory input matches predictions. For example, if we feel cold outside, we can go inside, thereby regulating our environment to conform to a temperature range that evolution has hardwired us to find pleasant (i.e., that minimizes prediction error with respect to an unconscious model we have of how our environment will be) because it makes our survival and reproduction most likely. In this example, simply updating our beliefs about which kind of environment we will encounter would lead us to stay out in the cold, which would make our reproductive success less likely. However, starting to walk inside is also associated with prediction errors. If I have a correct proprioceptive model of myself standing still, then the way to minimize prediction errors about that is to keep standing still. For me to act, I need to attenuate proprioceptive prediction errors so that the prediction error about myself being in a cold environment can become dominant and trigger the action of going inside. If I am unable to attenuate my proprioception, I will either become catatonic or I will have to try to override the power of my proprioceptive prediction errors by ascribing my own intentions and predictions about sensory input to external forces. In other words, I will develop delusions or hallucinations (for details of these mechanisms, see Adams et al. 2013; Brown et al. 2013).

A similar line of reasoning is applied by Lawson et al. (2014) and Quattrocki and Friston (2014) to the interoceptive domain, which allows them to describe many of the symptoms of autism as a consequence of aberrant precision weighting. Edwards et al. (2012) are concerned with hysteria (i.e., functional motor and sensory symptoms, sometimes described as "psychogenic" or "medically unexplained"). They use the precision-weighting framework in an effort to introduce more rigor (more precision, one might say) into the discussion of a disorder whose mention has been all but banned, yet has stubbornly refused to go away.

On the neurobiological side of the conceptual bridge, formal models of a hierarchical, precision-weighted message passing in the brain have been developed (Friston 2008; Bastos et al. 2012; Shipp et al. 2013). These efforts build on work dating back to Helmholtz (1860/1962), who was the first to propose that the brain is a predictive machine that becomes active in response to input only insofar as the input is unexpected. This concept of the brain is theoretically reinforced by the good regulator theorem, which prescribes the presence of a model, and it underlies the Bayesian brain hypothesis (Dayan et al. 1995), which postulates that the brain uses Bayesian inference to make the predictions required in the Helmholtzian view. Neurobiologically, the Bayesian brain is taken to be implemented by predictive coding (Rao and Ballard 1999; Friston 2005, 2008), which postulates that bottom-up prediction errors and top-down predictions are processed in the cortical neuronal hierarchy by a message passing between different cortical layers at different hierarchical levels. Specifically, according to Bastos et al. (2012), there is a canonical cortical microcircuit (cf. Douglas and Martin 1991; Haeusler and Maass 2007) which receives forward connections into cortical layer 4, conveying precision-weighted prediction errors from lower levels of a message-passing hierarchy embodied in the hierarchical neuronal anatomy of the brain (Figure 7.1). These prediction errors are used to adjust predictions at the level in question and sent backward (i.e., down the hierarchy) from the deep cortical layers. In the superficial layers, backward connections from higher areas are received and compared to predictions. The resulting prediction errors are precision weighted and passed forward (i.e., up the hierarchy), where the same information processing occurs in a higher region.

Limitations

The perspective laid out here has its limitations. We pay a price for reducing Bayesian inference to the tracking of sufficient statistics by foregoing the use of (a) any likelihoods that are not exponential families and (b) any but conjugate priors. These restrictions, however, are much milder than they might appear at first sight.

I address the first limitation by looking at an example by Daunizeau et al. (2010), where a likelihood from an exponential family clearly will not do. This likelihood needs to describe the probability of sensory, in this case retinal, input where the object presented in a black-and-white image is a house or a face. Applying principal component analysis to their sixteen images (eight of each kind), Daunizeau et al. (2010) recover two clusters representing houses and faces which can adequately be described by a pair of two-dimensional Gaussian distributions in the first two principal components (Figure 7.2). While this mixture is a simple and adequate likelihood for the situation at hand, it is not from an exponential family, which means that at first sight, belief updates

Canonical microcircuit for predictive coding

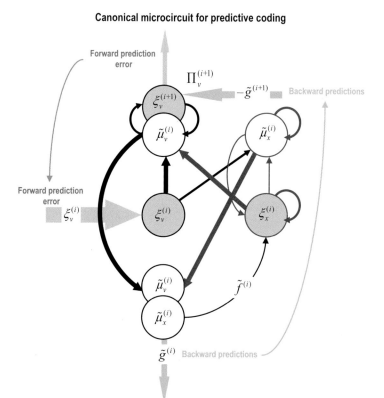

Figure 7.1 A proposed canonical microcircuit for predictive coding (reproduced with permission from Bastos et al. 2012). This is a schematic representation of a cortical column with supragranular layers at the top, infragranular layers at the bottom, and granular layers in the middle. Pink: prediction error populations. Red: inhibitory connections. Black: excitatory connections. Predictions ($\tilde{\mu}$) are encoded in supragranular excitatory and inhibitory interneurons and are passed to infragranular pyramidal cells. Prediction errors (ξ) enter granular layers from regions situated lower in the hierarchy. Prediction errors that are passed on to the next higher hierarchical level are computed in supragranular pyramidal cells. Crucially, they are weighted by the precision (Π) of the prediction ($\tilde{g}$) received from the higher level.

based on this model cannot be formulated as precision-weighted prediction errors. There is a way around this, however. The key is to formulate the problem hierarchically, with a prior on the probability of houses and faces, respectively. This allowed Daunizeau et al. (2010) to use variational Bayesian methods[1] to calculate belief updates by separating the levels of the hierarchy using a mean field approximation. Using a mean field approximation in this context means

---

[1] Briefly, variational Bayes is a method of model estimation that uses variational calculus to find the posterior distribution of parameters by maximizing the model evidence instead of calculating the posterior directly (these terms are explained in Flagel et al., this volume).

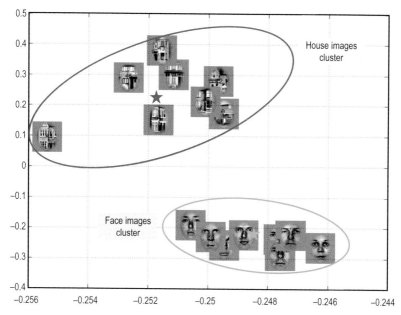

Figure 7.2　Projection of eight face and eight house stimuli onto their first two principal eigenvectors. Faces and houses form distinct clusters, and each cluster can be described by a two-dimensional Gaussian. Stars represent the means of the Gaussians, ellipses their covariances. Reproduced with permission from Daunizeau et al. (2010).

optimizing the parameters by iterating through separate subsets of them while the distributions of those not currently being optimized are assumed known (for details, see Friston et al. 2007). Building on this, we were later able to show that the belief updates for this model—and much more complicated ones—can be reduced to precision-weighted prediction errors (Mathys et al. 2011). The principles that we used are entirely general and have since been applied to many more models (e.g., Iglesias et al. 2013; Diaconescu et al. 2014; Hauser et al. 2014; Vossel et al. 2014). The procedure is straightforward: formulate the model hierarchically using mixtures of exponential family distributions and use a mean field approximation to separate the levels of the hierarchy, which then allows you to derive precision-weighted prediction error updates at each of the levels separately.

Apart from multimodality, there is another property which, at first sight, exponential families seem unable to deliver. It is sometimes desirable to have a distribution with "fat tails" (or more formally, positive excess kurtosis). Fat tails imply an increased probability of extreme values compared to a Gaussian of the same mean and variance. Distributions with fat tails are popular because the predictions they imply are conservative in the sense that they guard against underestimating the probability of extreme events. Examples of such distributions are Student's t-distribution and the Cauchy distribution, neither of which is an exponential family. The second limitation is the requirement of conjugate priors, and again the solution is a hierarchical approach. As the example from Daunizeau et al. (2010) shows, multimodal distributions, including priors, can be approximated using mixtures of distributions from exponential families. This allows for precision-weighted prediction error updating just as in the case of multimodal likelihoods.

Taken together, these limitations are not severe. All of them can be overcome by using a hierarchical approach. It might, therefore, be that the brain has evolved its hierarchical organization to take advantage of the effective and efficient predictive power of hierarchical models that are updated by precision weighting of prediction errors.

Nosologies Based on Aberrant Precision Weighting

Reducing the mind to precision-weighted belief updating implies nosologies based on false inference, owing to maladaptive weighting of prediction errors. In these terms, each nosological entity has two sides: a clinical manifestation of a particular precision-weighting disorder and a neurobiological mechanism underlying it. Steps in this direction have already been taken: Adams et al. (2013) traced out a computational anatomy of psychosis, and Lawson et al. (2014) and Quattrocki and Friston (2014) did the same with respect to autism. To identify new targets for treatments, these efforts will have to be expanded and refined with the goal of going beyond traditional diagnoses of, say, schizophrenia and autism, which lump together many disparate clinical phenomena and, we may suppose, pathophysiological mechanisms. The precision-weighting framework will help accomplish this because it tells us which questions to ask for each clinical phenomenon: Where in the inferential hierarchy is precision weighting going awry to produce this? What neurophysiological mechanism underpins the disordered precision weighting? Possible nosologies could be based on widespread aberrations in precision weighting originating in the neuromodulator systems of the brainstem and midbrain, equally widespread aberrations originating in the thalamus, or more localized aberrations originating in particular regions of the cortex or the basal ganglia, etc.

The best example of such an approach to date has been Adams et al. (2013), where many of the symptoms of psychosis are explained by a failure of sensory attenuation. Other pathologies will not be at the level of sensory input, but at other levels of the inferential hierarchy. For example, some symptoms of posttraumatic stress disorder (PTSD) could be a result of aberrant precision weighting when inferring the different possible causes of events in the environment. A loud bang is a prediction error for all of us, but while most will assign little precision to any one of the many possible explanations, enabling us to wait for more information before reacting, a PTSD patient might have very high precision on a prediction of being under fire. If this reaches such an extent that the patient is—unconsciously—constantly slightly surprised not to be under fire, a loud bang will be an opportunity for him to reduce this prediction error. Of course, this comes at the cost of a rather large prediction error about being in a safe environment. However, it depends on their relative precision, which of these contradictory predictions ("I am under fire" vs. "I am in a safe environment") dominates inference. Now that we have formulated the clinical side of the symptom in terms of precision weighting, albeit still in a very cursory and informal way, this enables us to look at the neurobiological side and know which questions to ask and how to interpret what we see. Specifically, when we investigate which neural systems are activated in PTSD patients in response to nonspecific stimuli that are over- or misinterpreted as threats, we can interpret what we see in terms of precision weighting. This could then give us a handle on manipulating precision weighting pharmacologically or psychotherapeutically, while monitoring progress neurobiologically as well as clinically. Crucially, this could enable us to transfer our interventions to other domains and—because we know the general *conceptual* mechanism in terms of which to interpret the underlying biology—allow us to make predictions about the clinical changes we expect in other domains.

Summary

In summary, I argue that the mind can only exist as a successful regulator of its environment if it continually updates model-based predictions about its interactions with that environment based on precision-weighted prediction errors. This is because the optimal way to make predictions is Bayesian inference, which can be reduced to tracking of sufficient statistics of observations under certain conditions. These conditions are that the likelihood be from an exponential family and that the prior be conjugate. These conditions are not restrictive because multimodal and fat-tailed (or otherwise nonstandard) distributions can be built hierarchically from exponential family distributions and inverted level by level by means of a mean field approximation. This amounts to a radical reduction of the mind to belief updating by means of precision-weighted prediction errors. The advantage of this reduction is that it provides terms in which both clinical phenomena and their underlying neurobiology can be understood. This enables it to serve as a bridge between the two fields and allows for the interpretation of one field's findings in terms of those of the other.

Nosology

8

Current State of
Psychiatric Nosology

Michael B. First

Abstract

Psychiatric classifications categorize how patients present to mental healthcare professionals and are necessarily utilitarian. From the clinician's perspective, the most important goal of a psychiatric classification is to assist them in managing their patients' psychiatric conditions by facilitating the selection of effective interventions and predicting management needs and outcomes. Due to the field's lack of understanding of the neurobiological mechanisms underlying the psychiatric disorders in both the Diagnostic and Statistical Manual of Mental Disorders (DSM) and the International Classification of Diseases (ICD), diagnosis and treatment are only loosely related, thus limiting clinical utility. Both DSM and the chapter on mental and behavioral disorders in ICD adopted a descriptive atheoretical categorical approach that defines mental disorders according to syndromal patterns of presenting symptoms. This chapter discusses the fundamental challenges that underlie this decision. It then reviews the Research Domain Criteria (RDoC) project, a research framework established by the U.S. National Institute of Mental Health (NIMH) to assist researchers in relating the fundamental domains of behavioral functioning to their underlying neurobiological components. Designed to support the acquisition of knowledge of causal mechanisms underlying mental disorders, RDoC may facilitate a future paradigm shift in the classification of mental disorder.

Introduction

Nosology (from the ancient Greek "noso," meaning *disease*, and "logia," meaning *study of*) is a branch of medicine that address the classification of diseases. The need to establish a classification of diseases reflects the natural human predilection to categorize for the purpose of simplifying and organizing the wide range of observable phenomena and experiences that one is confronted with so as to facilitate both their understanding and their predictability. Psychiatric nosology, with its focus on the presentations of mental and behavioral symptoms, dates back to antiquity. The first recorded depiction of mental illness dates to 3000 BC Egypt, with a description of the syndrome senile dementia attributed to Prince Ptah-hotep (Mack et al. 1994).

While some classification systems, like the periodic table, are a direct reflection of natural objective phenomena that clearly exist in nature, psychiatric classifications classify the ways in which patients present to mental healthcare professionals. They are necessarily utilitarian, and their success depends on how well they fulfill practical needs. Psychiatric classifications are used in a variety of contexts and settings (e.g., clinical, research, administrative, and educational) and thus must fulfill a variety of practical needs, including helping clinicians diagnose and treat patients, assisting researchers in selecting populations for study, facilitating administrators in their collection of health statistics, and teaching students how to recognize presentations of mental disorders. Although the original purpose of psychiatric classifications was for the collection of statistical information about institutionalized patients, the primary purpose of modern-day classifications, such as the American Psychiatric Association's Diagnostic and Statistical Manual of Mental Disorders (DSM) and the Mental Disorders Chapter of the International Classification of Diseases (ICD), is to assist mental health professionals in providing clinical care for their patients.

For the clinician, a psychiatric classification system needs to assist them in managing their patients' psychiatric conditions: it needs to facilitate the selection of effective interventions and be able to predict management needs and outcomes. For this, a clear understanding is needed of the neurobiological mechanisms that underpin psychiatric disorders; otherwise, diagnosis and treatment will only loosely be related in psychiatry and psychology, as is the case in DSM and ICD. Most currently available treatments have been found to be helpful in managing a wide range of disorders. Selective serotonin reuptake inhibitors, for example, have been shown to be useful in the treatment of depressive disorders, panic disorder, social anxiety disorder, obsessive-compulsive disorder (OCD), generalized anxiety disorder, posttraumatic stress disorder (PTSD), gambling disorder, early ejaculation, bulimia nervosa, and borderline personality disorder (Vaswani et al. 2003; Tang and Helmeste 2008). Similarly, for each diagnostic category, a wide range of treatments, both psychopharmacological and psychotherapeutic, have demonstrated efficacy (APA 2000).

Nonetheless, determining the patient's psychiatric diagnosis does provide the clinician some assistance in treatment selection and determining prognosis. Consider, for example, a patient with no premorbid psychiatric history, who presents with the very recent onset of a severe depression. Determining that the depressive symptoms arose in the context of that person having recently stopped his regular use of cocaine (which would be diagnosed as a cocaine-induced depressive disorder in DSM) has profound implications for the prognosis and management of that individual. Similarly, determining whether the symptomatic presentation of a patient with recurrent episodes of grandiose delusions and accompanying hallucinations meets the definitional requirements of bipolar disorder versus schizophrenia can be very important in clarifying both the potential role of lithium in the management of the patient (useful in bipolar disorder but not in schizophrenia) and future course.

Both DSM and the chapter on mental and behavioral disorders in ICD use essentially the same fundamental approach: they provide descriptive atheoretical categorical classifications that define mental disorders according to patterns of presenting symptoms. There are, of course, other possible approaches to implementing a psychiatric classification, namely having the classification based on etiological and pathophysiological factors as opposed to being descriptive, or adopting a dimensional approach instead of one that is categorical. In this chapter, I discuss the fundamental challenges that underlie the decision for the DSM and ICD to adopt a descriptive atheoretical approach and then describe the Research Domain Criteria (RDoC) system, created by the U.S. National Institute of Mental Health (NIMH), which is not actually a nosology or classification per se but is instead best viewed as a framework for conducting research in terms of fundamental circuit-based behavioral dimensions that cut across traditional diagnostic categories.

Approaches to Classification

Etiological versus Descriptive

Ideally, the organization and definition of conditions in any medical classification should be based on an understanding of the underlying etiology and pathophysiology, given that a classification system based on etiology and pathophysiology is most likely to be useful in helping clinicians determine disease prognosis and select the optimal treatment. For example, given that the type of infectious agent is essential for selecting the proper treatment, the diagnostic groupings in the infectious diseases chapter in ICD reflect the underlying infectious agents, starting with the intestinal infectious diseases, which are subdivided into the different bacterial, amoebic, protozoan, and viral agents that cause intestinal diseases, then the various forms of tuberculosis infections (e.g., respiratory, neurologic, etc.), and so on.

While most of the chapters of the ICD strive to follow this organization principle (Table 8.1), there are some areas of medicine (e.g., rheumatological conditions, various forms of headache, psychiatric disorders) for which a deep knowledge of the underlying etiology and pathophysiology remains elusive. For these areas, the main classificatory strategy, sometimes referred to as a "descriptive approach," is to define disease entities in terms of "syndromes"; that is, groups or patterns of symptoms which appear together temporally. The symptoms comprising a syndrome are assumed to cluster together because

Table 8.1 Organizational principles of ICD by chapter.

Disease type:	Classified by:
Infectious	Anatomic location (e.g., intestinal infectious diseases) Type of organism (e.g., protozoal diseases) Mode of transmission (e.g., infections with a predominantly viral mode of transmission)
Neoplasms	Nature of neoplasm (e.g., malignant neoplasms, *in situ* neoplasms, benign neoplasms) Malignant neoplasms subdivided by anatomic location (e.g., malignant neoplasms of the lop, oral cavity, and pharynx)
Blood and blood-forming organs; certain disorders involving immune mechanism	Etiology (e.g., nutritional anemias) Pathophysiology (e.g., coagulation defects)
Endocrine, nutritional, and metabolic	Anatomic location (e.g., disorders of thyroid gland) Clinical presentation (e.g., diabetes mellitus, malnutrition) Etiology (e.g., metabolic disorders subclassified by cause such as disorders of aromatic amino acid metabolism)
Nervous system	Pathophysiology (e.g., inflammatory diseases of the central nervous system) Clinical presentation (e.g., extrapyramidal and movement disorders, headaches) Anatomic location (e.g., nerve, nerve root, and plexus disorders)
Eye and adnexa	Anatomic location (e.g., disorders of eyelid, lacrimal system, and orbit) Pathophysiology (e.g., glaucoma) Clinical presentation (e.g., visual disturbances and blindness)
Ear and mastoid process	Anatomic location (e.g., diseases of middle ear and mastoid)
Circulatory system	Pathophysiology (e.g., ischemic heart diseases) Anatomic location (e.g., diseases of arteries, arterioles and capillaries)
Respiratory system	Pathophysiology (e.g., influenza and pneumonia) Anatomic location (e.g., other diseases of the upper respiratory tract)
Digestive system	Anatomic location (e.g., diseases of esophagus, stomach, and duodenum) Pathophysiology (e.g., noninfective enteritis and colitis)

Table 8.1 (continued)

Disease type:	Classified by:
Skin and subcutaneous tissue	Etiology (e.g., infections of skin and subcutaneous tissues) Clinical presentation (e.g., dermatitis and eczema) Anatomic location (e.g., disorders of skin appendages)
Musculoskeletal system and connective tissue	Pathophysiology (e.g., infectious arthropathies) Clinical presentation (e.g., systemic connective tissue disorders) Anatomic location (e.g., disorders of muscles)
Genitourinary system	Anatomic location reflecting pathophysiological location (e.g., glomerular diseases) Anatomic location in general (e.g., diseases of male genital organs), by clinical presentation (e.g., renal failure)

they are associated in some clinically meaningful way, presumably reflecting a common etiological process, course, or treatment response.

It is important to understand that these two classificatory strategies are not in opposition. Classifying disorders based on etiology and pathophysiology is universally regarded as the preferred approach, for the reason stated above. The syndromal "descriptive" approach is viewed as a clinically useful but temporary way station on the road to a future etiologically and pathophysiologically based classification.

The intuitive appeal of etiologically based classification systems is evidenced by their historical predominance, which goes back to Hippocrates and his classification of personality types based on whether there is excess or deficiencies in the four "humors" (blood, yellow bile, black bile, and phlegm). The problem, of course, with basing a classification on etiological principles is that their ultimate value is constrained by lack of validity of the hypothesized etiological factors. In the sixteenth century, for example, the Swiss physician Paracelsus developed a classification system in which he divided psychotic presentations into three types of disorders based on their presumed etiology. The first category, vesania, was for disorders caused by poisons and is analogous to current-day substance-induced disorders. The second, insanity, for diseases caused by heredity, is analogous to modern disorders such as schizophrenia and bipolar disorder, which appear to have a strong familial component. His third category, lunacy, described a periodic disturbance influenced by the phases of the moon. This lack of validity of the notion that the phases of the moon directly cause psychopathology severely compromises the utility of such a classificatory scheme.

In recognition of the problems inherent in basing a psychiatric classification system on unproven causal theories, an alternative approach which concentrated on careful observation of symptomatic presentations was proposed by Emil Kraepelin in the 1880s, who provided exceedingly detailed descriptions of disorders seen in inpatient settings at the turn of the century. While Kraepelin's classification viewed mental illness in terms of disease entities akin to medical disorders, his nosology was firmly based on the methods of descriptive psychiatry. Kraepelin strongly advocated that "psychiatrists should avoid postulating etiologies to make a diagnosis and should stick to the course of the illness, attend to the final state, and do follow-up studies where possible" (Decker 2007:340).

Both DSM and ICD have followed these Kraepelinian principles in their adoption of a descriptive "atheoretical" approach in which disorders are defined according to their symptomatic presentation, rather than according to unproven theories regarding underlying etiology. As noted in the introduction to DSM-III (the first edition of the DSM to adopt such an approach):

> The approach taken in DSM-III is atheoretical with regard to etiology or pathophysiological process except for those disorders for which this is well established and they are included in the definition of the disorder....The major justification...is that the inclusion of etiological theories would be an obstacle to use of the manual by clinicians of varying theoretical orientations (APA 1980:6–7).

It is important to understand that the decision to adopt a descriptive approach was not motivated by some aversion to having a classification system organized around etiology per se, but rather an aversion toward defining disorders according to unproven and potentially invalid etiological hypotheses. Thus, those psychiatric disorders in DSM for which the etiology is known (or presumed) are defined according to etiology. Such disorders include the substance-induced mental disorders and disorders due to a general medical condition, which are, by definition caused by the direct physiological effects of substance use or general medical conditions on the central nervous system, as well as disorders included in the DSM-5 diagnostic grouping Trauma and Stress-Related Disorders, each of which include exposure to a traumatic or stressful event as a required diagnostic criterion. Furthermore, given that one of the main goals of a psychiatric classification is to facilitate communication among mental health clinicians, defining disorders according to one particular theory would hinder its utility for clinicians who do not subscribe to that theory.

Although the descriptive DSM approach has been widely lauded because it established a common diagnostic language and improved diagnostic reliability, both researchers (Clark et al. 1995; Goldberg 1996; Cloninger 1998; Parker 2005; Widiger and Samuel 2005; Mellsop et al. 2007) and clinicians (McHugh 2005) have expressed great frustration with the approach taken by the DSM. As noted above, descriptive classification systems define disorders in terms of syndromes that reflected years of clinical observations regarding common cross-sectional symptom presentations and longitudinal courses. The presumption had been that, as in general medicine, the phenomenon of symptom covariation could be explained by a common underlying etiology and pathophysiology. Although based largely on expert consensus, there was a general understanding that the DSM-III syndromal definitions would be continually revised in subsequent editions of the DSM with the goal of improving diagnostic validity based on new research findings, ultimately culminating in the identification of the underlying disease processes. The process by which the validity of the DSM and ICD diagnostic categories would be iteratively refined was proposed by Robins and Guze in 1970 and entailed using five types of validity studies: studies that established clinical description, laboratory studies, studies that established differentiation from other disorders, follow-up studies, and family studies (Robins and Guze 1970). Such studies relied on the assumption that there was a one-to-one mapping between the syndromes in the DSM and ICD and their underlying disease processes.

Unfortunately, in the more than three decades that have elapsed since the publication of DSM-III, it has become increasingly clear that the DSM and ICD categories are not "carving nature at the joints" and do not represent true disease entities (Hyman 2010). Despite the discovery of many promising candidates over the years, not one single laboratory marker has been shown to be diagnostically useful for making any DSM diagnosis (Charney et al. 2002). Epidemiological and clinical studies have demonstrated extremely high rates of comorbidities among the disorders, undermining hypotheses that the DSM-defined syndromes have distinct etiologies. Twin studies have also contradicted many of DSM's assumptions that separate syndromes which have a distinct underlying genetic basis; evidence suggests, for example, that major depressive disorder and generalized anxiety disorder have the same genetic risk factors (Kendler 1996).

To reflect the observed symptomatic heterogeneity which characterizes the way patients present to clinical care, virtually all of the categories in DSM and ICD are defined polythetically; the diagnosis is made by choosing among different combinations of specific operationalized criteria defining a disorder (e.g., five out of a list of nine symptoms are required for a diagnosis of a major depressive episode). While clinically sensible, this approach has led to tremendous diagnostic heterogeneity, both in terms of symptomatic presentation (i.e., two patients with the same diagnostic label may have only one or two clinical features in common) and prediction of treatment response and prognosis. For example, the prognostic power of the diagnosis of schizophrenia for a specific patient is limited by the wide range of observed functional outcomes associated, which can range from relatively superior functioning (e.g., John Nash, winner of a Nobel Prize in mathematics) to extremely poor functioning (e.g., an individual who requires lifelong institutionalization). This heterogeneity is almost certainly a consequence of the fact that the DSM and ICD diagnostic labels, like schizophrenia, include a number of distinct diseases with different etiologies and pathophysiological mechanisms under a single diagnostic rubric.

These numerous limitations with the DSM and ICD descriptive approach sparked aspirations that DSM-5 would be able to abandon the DSM-IV descriptive approach and replace it with an etiologically and pathophysiologically based diagnostic system. Consequently, in 1999, the American Psychiatric Association initiated a DSM-V[1] research planning process, under joint sponsorship with NIMH, to focus on establishing a research agenda that would allow the DSM to move beyond the descriptive approach. Indeed, a stated goal of the DSM-V Research Agenda was "to transcend the limitations of the current DSM paradigm and to encourage a research agenda that goes beyond our current ways of thinking" (Kupfer et al. 2002:xix), with the ultimate goal of adopting an "etiologically and pathophysiologically based diagnostic system" (Charney et al. 2002:35).

The hope that neurobiological findings could play an important role in the development of the DSM-5 definition of disorders is exemplified by the inclusion within most of the DSM-V research planning conferences of at least one presentation (if not several) that explored whether neurobiological or genetic findings might be incorporated into the DSM-V diagnostic criteria. For example, The Stress-Induced and Fear Circuitry Disorders Conference included a presentation titled "The role of neurochemical and neuroendocrine markers of fear in the classification of anxiety disorders" (Andrews et al. 2009). In every case, however, the presentations concluded that the diagnostic utility of such tests remains too limited to be of use in making a psychiatric diagnosis in an individual patient. Indeed, Hyman, in a 2007 commentary that raised the question of whether neuroscience can be integrated into the descriptive DSM, concluded that "it is probably premature to bring neurobiology into the classification of mental disorders that will form the core of DSM-V" (Hyman 2007:731).

Ultimately, the quest to make DSM-5 more reflective of our current understanding of neuroscience and genetics is only evident in changes made to the DSM-5 "metastructure" (i.e., the grouping of diagnostic categories in the classification). In DSM-IV, diagnostic groupings were largely based on superficial descriptive symptomatology, with disorders sharing common presenting symptoms included in the same diagnostic grouping. For example, in DSM-IV the anxiety disorders grouping included panic disorder, the phobias, generalized anxiety disorder, OCD, and PTSD, reflecting the fact that patients with these disorders typically present with anxiety. Although not enough is known about the underlying causes of mental disorders to base their definitions on etiology and pathophysiology, enough is known about the underlying neurocircuitry, familial inheritance, risk factors, comorbidity patterns, and treatment response of OCD and PTSD to move them into their own separate groupings:

---

[1]   The designation of the revised edition changed from DSM-V to DSM-5 during the revision process to accommodate future plans for the implementation of a continuous revision model, so that the next edition could be called DSM-5.1.

obsessive-compulsive and related disorders; trauma and stressor-related disorders. The entire structure of the DSM-5 classification was thus reorganized along these lines, grouping disorders together that share putative common underlying factors (e.g., internalizing versus externalizing) and underlying vulnerabilities.

Categorical versus Dimensional

Another long running debate in nosology is whether psychiatric illness is best conceptualized as categorical versus dimensional constructs. Classification systems such as DSM and ICD define disorders categorically; that is, diagnostic definitions are provided that indicate whether an individual's clinical presentation either meets, or does not meet, the definitional requirements for a particular disorder. This method of classification is similar to what is used in the rest of medicine (i.e., a patient either has or does not have pneumonia, mitral valve prolapse, etc.). This tendency to define illnesses in terms of categories undoubtedly reflects basic human thought processes, embodied by the use of nouns in everyday speech to indicate categories of "things" (e.g., chairs, tables, dogs, cats).

In principle, however, variation in symptomatology can be represented by a set of dimensions rather than by categories. Take blood pressure, for example, which is measured along a continuum from low to high. It only becomes a categorical construct when the diagnostic label "hypertension" is applied to indicate that a patient has a significant elevation in blood pressure above a defined cut-off point that puts him or her at risk for developing serious illness in the future.

Dimensional approaches to representing psychiatric symptomatology have been proposed as well. For example, Wittenborn, Holzberg, and Simon (1953) developed a multidimensional representation of the phenomena of psychotic illness over sixty years ago, and since then others have developed dimensional models to portray the symptomatology of depressive and anxiety disorders, schizophrenia, personality disorders, and even the entire range of psychopathology (Mineka et al. 1998; Peralta and Cuesta 2000; Peralta et al. 2002; Clark 2005; Krueger et al. 2005; Watson 2005). While a categorical approach to psychiatric classification has important heuristic appeal, it may not represent the true state of things. Implicit in the categorical approach is an assumption that mental disorders are discrete entities, separated from one another, and from normality, either by recognizably distinct combinations of symptoms or by demonstrably distinct etiologies. While this has been shown to be the case for a small number of conditions (e.g., Down syndrome, fragile X syndrome, phenylketonuria, Alzheimer and Huntington diseases, and Jacob–Creutzfeldt disease), there is little evidence supporting the applicability of this model for most other psychiatric disorders. Indeed, in the last 35 years, the validity of the categorical approach has been increasingly questioned as evidence has accumulated that so-called categorical disorders, like major depressive disorder and anxiety disorders as well as schizophrenia and bipolar disorder, seem to merge imperceptibly both into one another and into normality with no demonstrable natural boundaries (Goldberg 1996; Widiger and Samuel 2005).

Dimensional approaches have some clear advantages. First, the commonly observed phenomena of excessive comorbidity (i.e., an individual receiving multiple simultaneous DSM diagnoses) is arguably a direct result of having a categorical system with more than 250 narrowly defined discrete categories. (First 2005b). A dimensional representation might characterize an individual's psychopathology by indicating the extent of his or her psychiatric symptomatology across a number of dimensions, virtually eliminating artifactual comorbidity. For example, consider an individual who presents with depression, anxiety, and social avoidance. Using the DSM-5 categorical system, criteria might be met for three diagnoses (i.e., major depressive disorder, social anxiety disorder, and generalized anxiety disorder) and thus give the appearance that the individual has three separate diagnosis. A dimensional approach, on the other hand, would simply indicate that the person has elevated values on the depression, anxiety, and social avoidance dimensions. Another advantage of the dimensional approach is that it avoids setting particular (and inevitably arbitrary) thresholds for distinguishing between pathology and normality. Thus, rather than categorically saying that an individual has major depressive disorder only if the threshold of five depressive symptoms is met or exceeded, a dimensional approach might simply say that the person is high on the depression dimension.

Dimensional approaches have other advantages as well. Research studies using dimensional scales as end points have much greater power to detect differences in groups than do studies which focus on changes in dichotomous categories (Cohen 1983; Kraemer et al. 2004). Furthermore, continuous dimensions more closely model the lack of sharp boundaries between disorders, as well as between disorder and normality, and can be developed using empirical methods that would facilitate research into the underlying etiology and pathophysiology of mental disorders (Goldberg 1996; Smoller and Tsuang 1998). Finally, dimensions can be helpful in indicating the severity of the disorder, which is relevant to making treatment decisions. For many disorders in the DSM-5, the range of appropriate treatments is related to the severity of the disorder (Andrews et al. 2007). For example, although either cognitive therapy alone or antidepressant medication alone are both reasonable options for the treatment of mild to moderate major depressive disorder, cognitive therapy by itself would not be an appropriate option for the treatment of severe forms of major depressive disorder. In such cases, treatment options would include one or more antidepressant medications or electroconvulsive therapy.

There are significant practical problems, however, with the use of a purely dimensional approach to classification (First 2005a). First, clinicians are accustomed to thinking in terms of diagnostic categories, and the existing knowledge base about the presentation, etiology, epidemiology, course, prognosis, and treatment is based on these categories. Furthermore, decisions about the management of individual patients (e.g., whether to treat, what type of treatment) are also much easier to make if the patient is thought of as having a particular disorder (with its associated prognostic and treatment implications) rather than as a profile of scores across a series of dimensions. Finally, although clinicians certainly appreciate the dimensional nature of psychiatric disorders in terms of their variable severity, the value of dimensions in terms of communicating information from one clinician to another is likely to be quite limited. As Phillips noted in his review of "The Conceptual Evolution of DSM-5" (Regier et al. 2011):

> [O]ne clinician communicates with another by saying something like, this is a bad case of depression and so far intractable to treatment, not by saying, on dimensional scales x, y, and z the patient has such and such scores (Phillips 2013:829).

Indeed, one of the greatest challenges for psychiatric classification is to craft dimensional approaches that have sufficient clinical utility to warrant the increased complexity (First and Westen 2007). In recognition of the limitations of the categorical aspects of the DSM-IV, a major emphasis of the DSM-5 revision process has been on the introduction of a dimensional component to DSM-5. During the revision process, the DSM-5 workgroups were asked to develop severity measures (clinician-administered or self-report) or to suggest existing severity measures for each DSM-5 disorder (First 2013). In addition, a disability measure, the World Health Organization Disability Assessment Schedule (WHO 2012), and a modification and enhancement of the psychiatric symptom measures from the National Institute of Health's Patient Reported Outcome Measurement Information System initiative (Irwin et al. 2010; Pilkonis et al. 2011) were proposed for inclusion in DSM-5 and tested in the DSM-5 field trials. However, because of concerns about their clinical utility, reliability, and validity, the published DSM-5 ended up relegating all of these dimensional measures to Section III, the section for proposed elements of the DSM for which "the scientific evidence is not yet available to support widespread clinical use" (APA 2013:24).

An Alternative to the Descriptive Categorical Approach: The Research Domain Criteria Project

When syndromal definitions for the various mental disorders were first introduced into DSM-III in 1980, it was widely assumed that it was only a matter of time before researchers, using the DSM definitions to select patient populations for study, would elucidate their underlying neurobiological mechanisms and pathophysiology. Although the lack of progress in understanding the causes of mental disorders stems mostly from the fact that the problem of trying to understand the underlying etiology and pathophysiology of mental disorders has turned out to be much more complex and challenging than originally anticipated, it is likely that the categorical descriptive DSM system itself is at least partly to blame. Scientists attempting to discover the neurobiological or genetic underpinnings of psychiatric illnesses have all too often treated the man-made psychiatric constructs in DSM as if they were "natural kinds," looking for the gene for schizophrenia or the neurocircuitry underlying major depression as if they were real disease entities (Hyman 2003, 2007, 2010). Perhaps whatever specificity there is between biological findings and behavioral correlates is being obscured by employing the DSM categories as if they were phenotypes, rather than focusing on more fundamental behavioral elements that cut across the various extant DSM categories.

The intent of the NIMH-sponsored RDoC project is to establish "a framework for creating research classifications that reflect functional dimensions stemming from translational research on genes, circuits, and behavior" (Insel and Cuthbert 2009:989). The RDoC project is a direct consequence of one of the aims of the NIMH 2008 strategic plan; namely, to "develop, for research purposes, new ways of classifying mental disorders based on dimensions of observable behavior and neurobiological measures" (National Institute of Mental Health 2008). Using DSM categories as the basis for selecting research subjects invites researchers to seek a one-to-one relationship between putative mechanisms and clinically defined disorder categories. The goal of the RDoC project, instead, is to shift researchers toward a focus on dysregulated neurobiological systems as the organizing principle for selecting study populations.

The initial stage of the RDoC project is to specify basic dimensions of psychological functioning and their implementing brain circuits, which have been the focus of neuroscience research over the past several decades.

Since the ultimate goal of the RDoC project is to link dysfunctions in neurocircuitry with clinically relevant psychiatric conditions, a priority in the selection of domains is that they can be related to problem behaviors found in the symptom lists of conventional disorder categories (Sanislow et al. 2010). The RDoC matrix[2] focuses on five major domains of functioning, each containing multiple, more specific constructs: *negative valence systems*, which includes constructs for fear, distress, and aggression; *positive valence*, which includes reward seeking and learning and habit formation constructs; *cognitive systems*, which includes constructs for attention, perception, working memory/executive function, long-term memory and cognitive control; *systems for social processes*, including separation fear, facial expression regulation, behavioral inhibition, and emotional regulation constructs; and *arousal/regulatory systems*, which include systems involved in sleep and wakefulness.

---

[2]   http://www.nimh.nih.gov/research-priorities/rdoc/constructs/rdoc-matrix.shtml
    (accessed July 8, 2016).

It is important to understand that the RDoC project is not intended to function as a diagnostic classification system in the way that DSM and ICD do. Unlike the DSM, ICD, and other medical classifications, which are designed to exhaustively describe and delineate the different ways that psychiatric patients might present symptomatically in terms of conceptually high-level concepts such as disease or disorder, the RDoC project is primarily a research framework to assist researchers in relating the fundamental domains of behavioral functioning to their underlying neurobiological components. As such, for each of the constructs in the RDoC matrix, the current state-of-the-art measurements/elements at several different units of analysis are to be listed, including genes, molecules, cells, circuits, behavior and self-report (Cuthbert and Insel 2010). Thus, in concrete terms, the RDoC framework is being implemented as a matrix, with the constructs forming the rows and the various units of analysis forming the columns.

The RDoC approach represents a true paradigm shift in the classification of mental disorders. It moves away from defining disorders based on descriptive phenomenology and focuses instead on disruptions in neural circuitry as the fundamental classificatory principle. Whether RDoC ultimately bears fruit in terms of eventually improving clinicians' ability to predict prognosis or treatment response will depend on how well this new approach performs for research (Insel et al. 2010), something that will takes years or even decades to realize fully.

Conclusion

From both a clinical and research perspective, the most useful nosologies in medicine are constructed around an understanding of the underlying disease mechanisms. Those areas of medicine in which the disease mechanisms remain unknown, like psychiatry, must necessarily define disorders according to symptomatic presentation. It should be noted that despite the shortcomings of the current DSM and ICD categorical descriptive systems, the major disorders which make up these classifications (e.g., schizophrenia, bipolar disorder, autism, major depressive disorder, and obsessive-compulsive disorder) pick out highly replicable features of psychopathology (Hyman 2010). Many disorders have been shown, for example, to have a high degree of familial aggregation (Kendler et al. 1997), with symptom clusters cohering both within and across generations. Twin studies (Kendler 2001) and adoption studies (Kety et al. 1971) suggest that much of this familial aggregation is explained by heredity. If DSM-defined major disorders were simply arbitrarily defined constructs created by expert consensus, the high levels of familial aggregation and heritability would be difficult to explain.

Ultimately, an improved understanding of the underlying disease mechanisms of mental disorders will result in the development of a classification system that will be more valid and more useful. Hopefully, paradigm-shifting research frameworks, such as the RDoC project, will push the research effort onto the right track and yield significant breakthroughs in our understanding of disease mechanisms. In the meantime, incremental efforts to improve the current categorical descriptive systems, as is being done with the DSM and ICD, should help refine these classifications by incorporating clinically relevant empirical data.

9

The Computation of Collapse

Can Reliability Engineering Shed Light on Mental Illness?

Angus W. MacDonald III, Jennifer L. Zick,
Theoden I. Netoff, and Matthew V. Chafee

Abstract

Computational modeling in psychiatry has generally followed from efforts to understand cognitive processes (McClelland and Rumelhart 1986) or the nervous system (Hodgkin and Huxley 1952). This stands to reason: psychiatric disorders are disorders of thought and central nervous system activity. Although there are few contributions to psychiatry from probability theorists and engineers (Shewhart 1938; Miner 1945; Lusser 1958), the tools developed for quality control of metal fatigue and failed rockets may point to a useful approach for thinking about mental illness. This chapter argues that the *computational science of collapse*, which describes the manner and likelihood of failures in complex systems, provides a framework in which to use computational modeling for relating mechanisms to behavioral outcomes. This science, known as reliability engineering, is a branch of applied probability theory that has now been used for almost a century to help understand and predict how inorganic, complex systems break down. The idea of a fault tree analysis is introduced, a tool developed in reliability engineering which may be able to incorporate and provide a broader structure for more traditional computational models. Finally, Some of the current challenges of psychiatric classification are unpacked, and discussion follows on how this framework might be adapted to provide a unifying framework for classification and etiology.

Toward a Reliability Engineering Framework for the Central Nervous System

The reliability engineering framework provides a fresh perspective on the way in which we ask questions of, and report, our data. Historically, reliability engineering developed over the twentieth century as mechanical devices became increasingly complicated. In the 1920s, for example, Bell Labs faced the problem that many of the telephone amplifiers, which they produced, failed after being buried underground. To address this, Walter A. Shewhart pioneered the field of statistical quality control, bringing together the fields of probability theory and electrical engineering. In the 1960s, at the height of the Cold War, the same company was employed to use this approach to address problems in missile launch control systems. A principle of quality control engineering is that the reliability of an entire system declines as the number of interacting components increases (Lusser 1958). A careful description of the various ways in which interacting components could lead to a system failure highlighted weak points in the device and resulted in *fault tree analysis* (FTA).

Fault Tree Analysis on the Brain

FTA is a deductive failure analysis; the fault tree identifies how faults of individual components interact with other components resulting in overall failure.[1] To generate a fault tree, the different components of the device must be identified, as well as a description of how these component failures—called faults— interact and combine into failure modes. A fault occurs when a component is unable to perform its required function, such as a mutation in an ion channel or a neurotransmitter receptor that critically impairs synaptic communication. In combination, such faults can cause a cascade resulting in a general failure mode, such as a loss of information-processing capacity in cortical networks. In psychiatry, a symptomatic expression of this cortical network failure mode might be an impairment in some aspect of cognition, emotion, or behavior. Consistent with other authors (e.g., Redish 2013), we will propose that neuropsychiatric syndromes may be thought of as failure modes of the central nervous system.

Fortunately, the failure of a single component rarely results in a general system failure due to built-in redundancy and plasticity of the brain; this enables most people with many forms of insult to function normally in the world. In this way, a FTA makes affordances for causes that are neither necessary nor sufficient for dysfunction in and of themselves. To generate a fault tree, one identifies the different components that contribute to the failure. Consider a simple circuit with a main bulb and a back-up bulb, a power generator and a back-up battery, and a controller switch. The FTA in Figure 9.1a illustrates

---

[1] FTA is only one of a number of tools that may find application for understanding neuroscientific questions. Strictly speaking, FTA is a deductive, top-down method for organizing ideas, recording probabilities, and evaluating likelihoods, whereas failure mode and effects analysis is a bottom-up approach that focuses on how a fault in a single component propagates through a system. Together these two analyses constitute a failure mode effects summary, which is commonly done after, for example, airplane crashes. Both approaches may have potential for neuroscience, and at times the best tool for a question may be adapted from still elsewhere in reliability engineering's armamentarium.

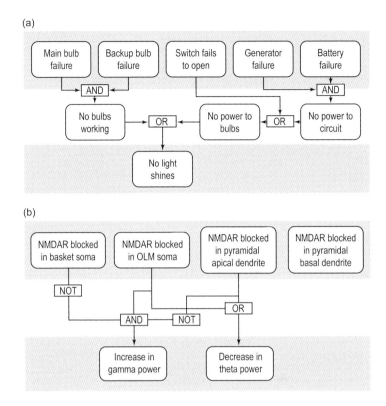

Figure 9.1 Examples of fault trees. (a) Typical fault tree of a simple mechanical circuit (after Rae and Lindsay 2004). (b) A fault tree generated from results of Neymotin et al. (2011). They used a computational model to investigate the conditions under which the power of theta frequency oscillations (3–12 Hz) decrease while the power of gamma frequency oscillations (30–100 Hz) increase, as seen in animals and human patients after ketamine administration. In their computational model, a decrease in theta power resulted when NMDA receptors (NMDARs) were blocked in either the somas of oriens-lacunosum moleculare (OLM) cells or in the apical dendrites of pyramidal cells, regardless of the function of NMDARs in other cell types. In the same model, an increase in gamma power occurred only when NMDARs were blocked in the somas of OLM cells and NMDARs were *not* blocked in basket cells or the apical dendrites of pyramidal cells. Thus the only combinations that generated both an increase in gamma power and a decrease in theta power involved blocking NMDARs in OLM soma with intact NMDARs in basket cells and pyramidal apical dendrites; the state of NMDARs in the pyramidal basal dendrites did not affect these results and can thus be said to be irrelevant in this case. Generation of a fault tree from these results allows one to visualize the roles that each factor plays in two effects, gamma and theta power.

how failures in these components might interact to cause a failure mode. It does this using Boolean logic "gates," describing how faults relate to other faults through AND/OR operators. Where the rate of such component failures is known (probability of failure over a timescale, e.g., per cycle or per day), the likelihood of a failure mode can additionally be calculated (or the range of likelihoods, using additional Markov chain expansions in which a distribution of possible outcomes is sampled many times).

Figure 9.1b is an example of an FTA reconceptualized from a biophysically realistic computational model of theta and gamma oscillation generation. In this case, Neymotin et al. (2011) looked at the functioning of different hippocampal cell types to examine the impact of ketamine administration. Ketamine, an NMDAR antagonist, is known to induce a schizomimetic state that has been traced to a contemporaneous decrease in theta and an increase in gamma brain waves, particularly in the hippocampus. The researchers developed a Hodgkin–Huxley style network model of hippocampal neurons consisting of 200 each of basket and oriens-lacunosum moleculare (OLM) interneurons and 800 pyramidal cells. All these cell types had NMDA receptors on the soma, with pyramidal cells additionally having them on apical dendrites. The models showed that blocking all NMDA receptors decreased *both* theta and gamma, inconsistent with experimental findings at schizomimetic doses. They reasoned that *differences in sensitivity* to ketamine of NMDA receptors on the different cell types must be the source of these preanesthetic effects. Therefore, they independently manipulated four types of insults—blocking NMDA receptors on somas of (a) basket, (b) OLM, (c) pyramidal neurons, and (d) apical dendrites of pyramidal neurons. This resulted in 16 binary combinations: 2 (normal/off) raised to the 4th (types of insults). The model found that pyramidal somatic NMDA receptors were largely irrelevant to theta and gamma power, whereas turning down pyramidal apical receptors alone was enough to decrease *both* theta and gamma power. In fact, the only condition in which they observed decreased theta and increased gamma was when OLM NMDA receptors were off, while the basket interneurons and apical pyramidal NMDA receptors remained functional. This result has been translated into the Boolean logic illustrated in Figure 9.1b.

The point may go without saying, but the purpose of the examples in Figure 9.1 is to illustrate the mechanisms underlying FTA and make explicit the parallels between mechanical and biological circuits. Several elements are omitted from these examples for the purpose of simplicity. First, the power of invoking probability theory is not illustrated. The probability of the co-occurrence of two independent events (pA AND pB = pA*pB) or the occurrence of either of two such events (pA OR pB = pA + pB – (pA*pB)) are familiar from introductory statistics courses. The impact of NOT, or inhibition, is straightforward. By simple extension, the inclusion of additional forms of logic (X-OR, 3 OUT OF 4, etc.) can also be readily incorporated. Furthermore, the effect of earlier event probabilities can then propagate through the FTA to examine the rate at which a general failure mode should occur. Second, many features of biological systems are not dichotomous and thus do not fall into simple categories. In such an event, classical probability theory can be augmented with a Markov chain and other methods, which involve resampling probability distributions.[2] This complicates the statistics but allows more dimensional variations of these concepts to be captured. Third, the events within the system need not be all within the organism. For example, stressors or treatments from outside the system can be modeled as externally controlled variables and can thereby alter the outcome (the rate of a specific fault and therefore the rate of general failure modes).

With these considerations in place, we come closer to imagining how FTA might provide a way of thinking about neuropsychiatric diseases and how different risk factors and treatments might interact. In this regard, the reliability-engineering framework is similar to other computational neuroscience approaches, insofar as the goal reflects a dissatisfaction with the correlative relationships that undergird much of what is known about mental illness. Like other computational approaches, the tool reflects specific causal hypotheses. Also, like other computational approaches, the tool risks becoming an armchair exercise, unless it generates hypotheses that are testable, either in patients or animal models. For example, the models can be generative—by pointing out domains and connections about which too little is known—and they can be tested and refined by assessing the extent to which known risks predict the characteristics of faults within a neural system and rates of failure modes (diseases) within a population.

Reliability Engineering on the Brain

Why have not reliability-engineering approaches been embraced more in psychiatry? One answer may be that biology in general, and the nervous system in particular, have failure rates and causes that are hard to quantify. The vast majority of the literature involving reliability engineering in the biological and medical sciences involves the traditional reliability of various medical devices. In the brain, reliability-engineering approaches are most evident in those places where engineers have had to share space with neuroscientists, such as in the study and manufacture of computer–brain interfaces (e.g., Polikov et al. 2005; Yousefi et al. 2015).

While reliability engineers may not tread into neuroscientific territory, there is a useful precedent in the work of the psychologist and neuroscientist Robert Glassman (1987). Glassman drew on Lusser's work in missiles and rocketry to speculate that component faults and failure modes were likely evolutionary constraints that led to redundancy and parallel processing in the brain (Lusser 1958). Lusser's law states that the reliability of components in series is equal to the product of the reliability of its component subsystems (an observation that

---

[2] Rather than a deterministic model, such simulation tools allow for probabilities to be assigned to several outcomes. For example, Markov chains examine a sequence of events using random draws to determine how the sequence proceeds. This then produces the distribution of probabilities informed by the internal structure of the sequences.

led Lusser in the 1950s to dismiss the possibility of reaching the Moon because of the complexity—and therefore the low reliability—of the rockets required). Building on the theorems underlying Lusser's observations, Glassman derived the observation that the brain's apparent series-parallel networking operations had evolved to overcome inevitable faults in particular neurons or neural systems over the course of a lifetime using built-in redundancies, such as large neural populations firing in concert. Glassman also saw in Lusser's law the principles underlying *diaschisis*, which is the alteration of functions in brain regions far removed from a damaged area. Diaschisis might then reflect the brain's manner for overcoming such failures such that at first it is unable to produce the given behavior at all but, with recovery, it can again produce the given behavior only by accepting a lower level of precision (a higher fault rate) from some components.

We cannot speak to the extent that these ideas informed Glassman's subsequent work. It is clear that they did not lead a stampede of neuroscientists to seek out training in reliability engineering. However, in such matters, the selection of a target problem can make all the difference. In the next section we will try again, this time by applying the framework of FTA to the challenge of integrating classification and etiology in the study of mental disorders.

Comparing Frameworks for Classification and Etiology of Mental Disorders

Although testing, or even proposing, a formal FTA for a specific mental illness is beyond the scope of this chapter, we believe that it will be a useful complement to other computational approaches in the future. Even in the absence of a realized FTA model, the reliability-engineering approach provides a framework that contrasts with the two frameworks for thinking about psychiatric and personality disorders currently prominent in the field. *Frameworks* are the premises and concepts that tacitly guide our research. For example, the number of angels that can dance on the head of a pin is now a byword for a pointless debate, but it was once a subject of serious discourse; we have long since retired the framework that led to those arguments. Are there parts of our framework for studying psychiatric disorders similarly ready for retirement?

To address this issue we will unpack the two prominent frameworks in psychiatry, which we term the neo-Kraepelinian and the reverse-engineering frameworks, and then contrast them with the reliability-engineering framework. In particular, we will examine how these frameworks affect the way in which we link the causes of a disorder to its symptoms. Since classification and cause are so central to psychiatric research, these are domains where an incorrect framing of the questions could lead us hopelessly astray. We will argue that both the neo-Kraepelinian and the reverse-engineering frameworks are misaligned with the nature of psychiatric disorders. Our aim will then be to use the tools of reliability engineering to make this discrepancy more explicit.

The Neo-Kraepelinian Framework

The founder of modern psychiatry, Emil Kraepelin (Kraepelin and Diefendorf 1907; Kraepelin 1919), was the original proponent of the quasi-medical framework used most widely in psychiatry today. This framework came to prominence in the 1970s, supplanting the psychoanalytic framework used in the second edition of the Diagnostic and Statistical Manual (DSM-II). This neo-Kraepelinian framework posited distinct categories of illness that could be assigned to someone who had a sufficient number of observable symptoms. The framework was codified in the Feighner criteria (Feighner et al. 1972), the Research Diagnostic Criteria (RDC; Spitzer et al. 1975), and eventually the third edition of DSM (DSM-III). These codes suggested that symptoms were useful for determining whether someone fulfilled the necessary and sufficient conditions for diagnosis. However, it was also acceptable for patients to share a diagnosis without sharing any symptoms. It was hoped that a formal diagnostic framework would decrease idiosyncratic noise, increase the reliability of diagnoses, and harmonize practice across laboratories and clinics (for a critique of these aspirations, see Markon et al. 2011). This served to reify the search for natural categories with distinct etiological and pathophysiological characteristics (Hyman 2010). Within this framework, theories about how the neural functions of, for example, schizophrenia patients may be distinct from the neural functions of depressed, alcoholic, or obsessive-compulsive patients were immediately salient and substantive.

Forty years on, the premises and concepts of the neo-Kraepelinian framework are hampering progress toward the grand challenges of psychiatric research (Persons 1986; Van Os et al. 1999; Krueger and MacDonald 2005; Markon et al. 2005; Hyman 2010). There is increasing evidence that upstream genetic, cellular, and neural system impairments are shared across distinct disorders, even between categorically distinct disorders. To follow up on our example, schizophrenia has at times been thought of as a categorically distinct psychiatric disorder. It is somewhat surprising, then, that 50% of people with schizophrenia also fulfill criteria for comorbid substance abuse at some point, and 50% fulfill criteria for depression (Buckley et al. 2009). People with schizophrenia are also at a 12-fold greater risk for obsessive-compulsive disorder (Pokos and Castle 2006), whereas those with obsessive-compulsive disorder are at a fourfold greater risk for schizophrenia (Tien and Eaton 1992). The levels of comorbidity between other mental disorders can be equally as high. In any case, psychopharmacology and psychotherapy frequently use the same medications and techniques in practice across different diagnoses (for additional critique, see MacDonald 2013). At some point, the neo-Kraepelinian medical framework became more useful to insurance adjusters and lawyers than to patients, clinicians, or even researchers. In the words of Thomas Insel, director of the National Institute of Mental Health (NIMH) when DSM-5 was released: "Patients with mental disorders deserve better."[3] For these reasons, some scientists are moving away from the neo-Kraepelinian framework toward something new, which we refer to as informal reverse engineering.

The Informal Reverse-Engineering Framework

Reverse engineering involves analyzing a complex system related to a function to determine the mechanisms underlying that function. This framework is already implicit in much neuroscience research, while NIMH's Research Domain Criteria (RDoC) perspective (explicitly named in recognition of the RDC framework it replaces) is the most codified version at this time. "The mandate for RDoC is to consider psychopathology in terms of maladaptive extremes along a continuum of normal functioning, to promote a translational emphasis" (Ford et al. 2014:S296). At the core of RDoC is a matrix with rows consisting of functional dimensions organized into five broad categories (positive valence systems, negative valence systems, cognition, social processes, and arousal). The columns of the matrix are levels, or units, of analysis ranging downward to genes and upward to behavior and symptoms.[4] Thus, the framework strives to organize extant knowledge about a multitude of cognitive and affective processes with research findings about brain networks, neurons, neurotransmitters, proteins, and genes (Insel and Cuthbert 2009; Stanislow et al. 2010; Cuthbert and Kozak 2013; Ford et al. 2014). The principle motivating RDoC is that patients who are sorted according to some shared functional deficits (e.g., in working memory, attention, executive control) will have more in common in terms of brain functioning than do patients placed in the same diagnostic groups according to neo-Kraepelinian schemes. The hope is that this new framework should facilitate the discovery of the underlying neural mechanisms that cause neuropsychiatric disease. RDoC is based on several reasonable, but untested, assumptions:

- A focus on functional deficits will direct research toward causal biological mechanisms more rapidly than a focus on clinical symptoms.
- Patients grouped based on functional deficits will be more homogenous with respect to underlying biological mechanisms than grouping based on clinical symptoms.

---

[3] April 2013 blog post, available at http://www.nimh.nih.gov/about/director/index.shtml. For partial retraction, see http://www.nimh.nih.gov/news/science-news/2013/dsm-5-and-rdoc-shared-interests.shtml (accessed July 7, 2016).

[4] http://www.nimh.nih.gov/research-priorities/rdoc/constructs/rdoc-matrix.shtml (accessed July 8, 2016).

- Clinical symptoms and functional deficits derive from a common set of biological mechanisms, so that studying one will provide insight into the other (which is necessary if treatments that improve functional deficits are also to improve clinical symptoms).

If patients with similar functional deficits do not end up sharing more in terms of a common set of underlying neurobiological deficits, it would suggest that either there are no meaningful categories of neuropsychiatric disease with unique neural signatures, or that new functional axes closer to neural functioning need to be identified. In spite of the uncertainty of these propositions at this stage, it seems that shifting toward functional deficits and away from clinical symptoms will enable a tighter link between biology and behavior in neuropsychiatric research. For this reason, RDoC holds enormous potential for accelerating discovery.

We refer to the RDoC approach as an instance of "informal reverse engineering" in the present context because, although it seeks to identify the biological causes underlying behavioral deficits in neuropsychiatric disease, the RDoC framework does not attempt to provide a quantitatively rigorous or unifying framework for achieving this. A potential limitation of the informal reverse engineering in general, and RDoC as a manifestation of it, is that by isolating psychological constructs from each other (rows in the RDoC grid), attention is drawn away from the sources of *structure* in psychopathology. It is this structure of covariation in cognitive and affective dysfunctions within and across patients that initially led to the delineation of diagnostic entities early in the twentieth century. Even when statistical relationships are discerned, it is not easy to append these into the cumulative science of mental illness. In particular, RDoC does not make affordances for how multiple causal factors interact, nor does it provide a basis for predicting how a complex set of interacting neural systems that collectively malfunction as a result of the disease will respond to interventions intended to normalize brain function. In short, it is not clear that RDoC in its current form will identify treatments. Alternatively, it is possible that with perfect knowledge, functional deficits in working memory, attention, or executive control may turn out to result from diverse biological causes. Under such circumstances any new grouping (or dimension) would not reduce heterogeneity within groups at all, which would limit its usefulness as well.

We argue for a more quantitatively rigorous framework that simultaneously affords an account of how causes that are neither necessary nor sufficient in and of themselves can result in a disorder, and how within-diagnosis heterogeneity and between-diagnosis comorbidity arise. With that said, there are many components of reverse engineering, even informal reverse engineering, that are a salubrious and necessary part of any FTA of a biological system. First among these are computational models.

Fault Tree Analysis Framework for Syndromes
and Computational Models in Psychiatry

If you don't know where you're going, you don't have very much control over where you arrive. The framework offered by reliability engineering allows us to envision what a description of psychopathology might look like and provides a set of tools to move forward. This possibility goes beyond the classification of people into disease categories or the measurement of people along cognitive dimensions, and begins to reintegrate causes into our conceptualization of mental disorder. Further, the framework provides a means for describing both how a single therapy might affect several different disorders and how different therapies can all reduce the same symptom.

As is well established elsewhere in this volume, computational models are mathematical formalizations of hypotheses. As noted above from the work of Neymotin et al. (2011; see also Figure 9.1b), integrating computational modeling with FTA in the context of psychiatric disease allows us to relate different risk factors to disease pathophysiology forming the outcome of a model. Such computational models can enable a mechanistic understanding of the linkages between faults that may occur in the tree; they can also relate mechanisms understood at one scale of the fault tree to outcomes at another scale.

The schematized illustration of FTA in Figure 9.2 demonstrates a number of properties of an expanded FTA framework that make it desirable for understanding the etiological and classification problems in psychiatry. Levels of analysis are illustrated in a series of gray bands, and the relationships between those gray bands is explicitly illustrated with simple logic gates. Two symptoms (A and B) represent two failure modes of the system. The co-occurrence (or comorbidity) between them is shown to be the result of the relationships among four cognitive/affective processes. Because of the dual role played by cognitive/affective process C in the two symptoms, the likelihood of co-occurrence is greater than chance (and could be explicitly calculated and compared to empirical rates), depending on the presence of other processes. Cognitive/affective process C is not the only source of comorbidity, as cell process B plays a role in three of the cognitive processes, rendering those, in turn, nonindependent.

Suppose symptoms A and B are symptoms of a given disorder. In a number of psychiatric disorders, two patients can share a diagnosis without sharing symptoms. Similarly, each patient can in turn closely resemble someone who does not fulfill the criteria. How might this be explained by FTA? In our schematic, a patient with an impairment in cognitive/affective processes A, B, C, and D will express symptom A but not B; whereas another with impairments in processes C and D will only show symptom B. This is troubling from a neo-Kraepelinian perspective, where a disease is recognized by fulfilling a series of necessary and sufficient conditions because they have a specific etiology. On the other hand, a FTA framework provides a tool for thinking about

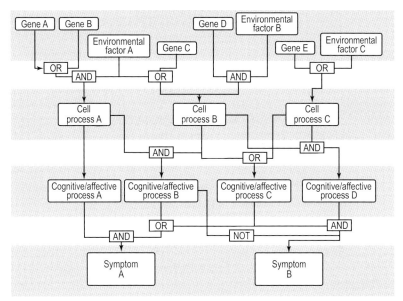

Figure 9.2 Schematic of FTA analysis illustrating the role of logic gates to integrate causes across levels of analysis and to generate nonindependence across symptoms.

syndromes. Syndromes are constructs that organize and label symptoms and other measurable signs that are often seen together. Syndromal descriptions are helpful in differential diagnosis because they prompt investigation of other features of the syndrome when the first signs are observed. Syndromes provide guidelines, but do not have rigorous necessary and sufficient conditions.

The classical psychiatric disorders are syndromes identified by the relative co-occurrence of these various failure modes, or symptoms. Our FTA example in Figure 9.2 illustrates how disorders all arise from the contributing causes of upstream faults: in some cases, several such faults will be required for a downstream failure (e.g., gene D and environmental factor B can lead to cell process B). In other cases, two alternative faults can lead to the same failure (e.g., gene E or environmental factor C causes cell process C). The result will be a statistical co-occurrence of failures in cognitive and affective processes linked to symptoms in the absence of clearly defined diagnostic boundaries. Uncovering the fuzzy logic of syndromes will be difficult to do if the focus remains on diagnostic group differences. Instead, research will add value by linking the likelihood of symptoms to the likelihood of cognitive and affective failures, and in turn by linking those failures to the likelihood of upstream faults in cellular systems, genetic polymorphisms, and environmental stressors. While this also resembles a research agenda based in informal reverse engineering, it is distinct in (a) prioritizing the importance of observing the rates of various faults, (b) making explicit their multiple causes and their various outcomes, and (c) providing potential for application of a rigorous, quantitative prediction.

Honing the relationships present in the fault tree can proceed in parallel with, and in turn complement, the efforts of computational modeling to unpack the mechanisms across different levels of analysis. For example, genomic studies have identified many mutations related to schizophrenia and depressive disorder, all of which have only weak correlations to the disorder. However, identification of changes in the genetic code does not directly link the etiology to the behavioral outcome, primarily because it is unclear how genetic mutations that change the properties of an ion channel or neurotransmitter system change the function properties of neural circuits to alter how they process information. Computational models can be used to relate what we may know at the molecular and cellular scale to the observable changes in neural function, circuit dynamics, and ultimately behavior. For example, Hodgkin–Huxley neuronal models simulate ion channel and synaptic conductances to predict cellular dynamics. Channel mutations can be modeled by changing parameters and measuring the resulting changes in excitability and/or spiking patterns in artificial neural networks. These networks can be trained to "perform" behavioral tasks (e.g., process stimuli, select between available behavioral responses) that measure specific cognitive impairment in patients. Therefore, these models are useful for linking changes at the protein scale to changes in neural function and behavior. At another scale, mean-field models simulate the average firing rates of populations of neurons in brain regions. These models can be used to relate changes in excitability or connection strengths to the emergence of synchrony and population oscillations that may be measured in system-level biomarkers such as changes in fMRI, EEG, and even to cognitive deficits.

An advantage of computational models is that they can assess how pharmacological, electrical, or optogenetic therapies could potentially modulate neural dynamics in networks to normalize information processing and behavior. By testing these predictions from the models, we are inherently testing our underlying hypothesis of the physiological mechanisms resulting in the disease state. Still, to be tractable, computational models necessarily focus on a small set of empirical observations and struggle to capture various syndromal and epidemiological aspects of psychiatry. For example, Voon et al. (2015) found that a bias toward model-free learning was more prevalent in people with binge eating disorder, methamphetamine addiction, and obsessive-compulsive disorder. Thus, diverse disorders of compulsivity are accompanied by an excessive tendency toward model-free learning. While a compelling mechanistic account of compulsive symptoms, it does not yet account for an important complication: the low level of comorbidity between these compulsivity disorders if they were indeed caused by a single fault. Thus, while the computation model is of itself mechanistic (and possibly correct), a full understanding of binge eating disorder, methamphetamine addiction, and obsessive-compulsive disorder will involve understanding why the same fault results in addiction, in one case, and obsessive-compulsive disorder, in another. Circumstances like this, where there is not a one-to-one mapping between a modeled fault and a particular symptom, are likely to be the norm. In this regard, we see FTA and computational modeling as complementary. The extent to which the predictions made by any given model correspond to observed probabilities of various outcomes provides useful validation of both the FTA and the computational model.

A Reliability Engineering-Aligned Research Agenda

While FTA has many attractive features for helping to systematize the etiology and classification of mental disorders, there are clear limitations to the application. First, generating a fault tree that spans genes to diseases is an immense task. Is it practical to complete an FTA for any given disorder, or is this really just a framework in which to think about disease diagnosis? We propose that the FTA may be practical for relating some specific etiologies to outcomes. In this sense, it provides a new way to summarize information and structure reviews of a given domain. While it may not yet be practical to generate a globally encompassing fault tree that relates all diseases into a single framework (i.e., an FTA of the brain), there are statistical tools that make the task less daunting. For example, probabilistic graphical models are an increasingly popular method for detecting nonlinear (including Boolean) relationships between observed variables (Praveen and Fröhlich 2013). Second, the FTA framework implies unidirectional causality: genes are responsible for cellular physiology, and physiology is responsible for behavior. Feedback loops are not, to our knowledge, accommodated in a simple way. Clearly, in biological systems there is feedback between every level, and this feedback can be incorporated into computational models. While feedback is complicated for the proposed approach, steady-state effects of such circuits and fluctuations in those steady states may be incorporated into expanded versions of FTA. Third, coupling between nodes within the FTA may be fit to data to best describe general outcomes, but may not represent any particular patient's connections. Therefore, an FTA alone may not be sufficient to provide useful guidance for selecting a given patient's therapy (cf. footnote 1 regarding failure mode and effects analysis, and failure mode effects summaries, which may, in time, provide precisely this type of guidance). Fourth, strictly speaking, FTA builds toward a single general failure mode. However, one of the features that may prove particularly useful for biological systems, such as the brain, is the way in which two FTAs can overlap. Thus it will be necessary to expand these models to exploit more fully the ways in which shared causes of two or more failure modes can be illustrated and calculated.

At present, there are only fragmentary parts of fault trees for complex mental disorders. FTAs are assembled from data collected from very disparate sources to relate causal relationships between different elements at adjacent scales and correlational relationships between more distant elements. The adoption of the FTA framework, perhaps in concert with probabilistic graphical modeling as needed, could therefore play a useful role in directing future research, for example, quantifying statistical relationships between key disease variables in existing datasets, and identifying high-value data to test computational models. There are some data sources that relate specific connections within the tree (e.g., genes to cell physiology, or cell physiology to network oscillations), but others relate scales that are not directly coupled (e.g., correlations between genes and disease prevalence). Mechanistic connections can be tested in computational models and further validated in animal models. These models can be used to identify the mechanisms or symptoms that relate the different components of the genomics and physiology into a biometric. The goal is to generate a single model that both explains the direct mechanistic connections and is also consistent with the indirect correlations. This will involve going beyond the statistical probabilities and effect sizes to which we are accustomed.

We have brought some attention to a field that has been working with quirks of nonbiological systems for almost a century. Many of the challenges to which this field of reliability engineering has struggled have parallels with biological systems. One tool from reliability engineering, FTA, provides an overarching framework for thinking about how complex systems, such as the brain, can break down. The reliability engineers' path that linked underground telephone amplifiers to Moon-bound rockets was filled with happy and some not-so-happy accidents. The path ahead for the computation of collapse may be much closer to home. We suggest that going on this journey begins with a mental shift, from the traditional medical neo-Kraepelinian framework to that of reliability engineering.

10

A Novel Framework for Improving Psychiatric Diagnostic Nosology

Shelly B. Flagel, Daniel S. Pine, Susanne E. Ahmari,
Michael B. First, Karl J. Friston,
Christoph Mathys, A. David Redish,
Katharina Schmack, Jordan W. Smoller,
and Anita Thapar

Abstract

This chapter proposes a new framework for diagnostic nosology based on Bayesian principles. This novel integrative framework builds upon and improves the current diagnostic system in psychiatry. Instead of starting from the assumption that a diagnosis describes a specific unitary dysfunction that causes a set of symptoms, it is assumed that the underlying disease causes the clinician to make a diagnosis. Thus, unlike the current diagnostic system, this framework treats both symptoms and diagnostic classification as consequences of the underlying pathophysiology. Comorbidities are therefore easily incorporated into the framework and inform, rather than hinder, the diagnostic process. Further, the proposed framework provides a bridge—which did not previously exist—that links putative constructs related to pathophysiology (e.g., RDoC domains) and clinical diagnoses (e.g., DSM categories) related to signs and symptoms. The model is flexible; it is expandable and collapsible, and can integrate a diverse array of data at multiple levels. Crucially, this novel framework explicitly provides an iterative approach, updating and selecting the best model, based on the highest-quality available evidence at any point. In fact, the scheme can, in principle, automatically ignore data that is not relevant or informative to the diagnostic trajectory. Finally, the proposed framework can account for and incorporate the longitudinal course of an illness. This chapter details the theoretical basis for this framework and provides clinical examples to illustrate its utility and application. Multiple iterations of this framework will be required based on available information. It is hoped that, with time, the framework will enhance our understanding of individual differences in brain function and behavior and ultimately improve treatment outcomes in psychiatry.

Introduction

Nosology is defined as the branch of medicine that addresses the classification of disease (see First, this volume). In psychiatry, such classifications capture the ways in which patients present to clinicians and are meant to assist mental health professionals in providing optimal clinical care. The current fundamental approach to nosology in psychiatry is to classify disorders as categorical syndromic clusters of signs, symptoms, and potentially laboratory findings, as outlined in the Diagnostic and Statistical Manual of Mental Disorders (DSM). Clearly, this approach has had a major positive impact on clinical practice and has led to substantial improvements in research and improved understanding of some neural mechanisms underlying dysfunction. Nevertheless, the clinician's reliance on the DSM categorical descriptive approach to diagnosis has had significant limitations in terms of assisting clinicians in treatment selection and prediction of prognosis. Although it was clearly hoped that the DSM system would reflect the etiology or the pathophysiological processes that underlie the disorders, it has become equally clear that this is not the case (Hyman 2010). Although relatively little was known in this regard when DSM-III emerged more than thirty years ago (APA 1980), there is little to no mention of neurobiological processes, even in the more recent version, DSM-5 (APA 2013), despite great advances in psychiatric neuroscience. As a result, consensus has emerged regarding the need to go beyond the categorical descriptive approach of DSM, with the hope of improving outcome prediction and treatment response for individual patients.

The Research Domain Criteria (RDoC), sponsored by the U.S. National Institute of Mental Health (NIMH), were established to provide "a framework for creating research classifications that reflect functional dimensions stemming from translational research on genes, circuits, and behavior" (Insel and Cuthbert 2009:989). The goal of RDoC was to shift researchers toward a focus on dysregulated neurobiological systems, as the organizing principle for delineating dysfunction. For example, constructs in RDoC, such as "cognitive control," emerge from neuroscience research that links functions in neural circuits to measures of information processing obtained in the laboratory (e.g., Morris and Cuthbert 2012). Although RDoC has not, to date, led to great improvements in diagnostic nosology, the foundation it provides for linking nosology and neuroscience is critical for advancing the field in both research and clinical domains.

Clearly, both dimensional and categorical approaches such as RDoC and DSM offer complementary advantages, and both approaches to nosology are likely to be invaluable for many years to come. The problem, however, is that there is currently no link between the neurobiological systems that form the basis of the fundamental RDoC domains and the symptoms that form the basis for the syndromes that comprise the DSM. As a result, neuroscience remains far removed from clinical decision making. Although there are advantageous features of the DSM and RDoC which, if combined, could potentially improve and advance clinical treatment and research related to mental illness, there is currently no means to bridge these two systems. That is, there is no framework that will allow clinicians or researchers to link domains of psychological or neurobiological function to existing DSM diagnostic categories. Further, there is no mechanism in place to relate RDoC domains or DSM constructs to measures of clinical utility (e.g., prognosis, effective treatment, cost). Current approaches to categorization are also deficient in their ability to account for comorbid diagnoses and the longitudinal evolution or trajectory of disorders. Here, we introduce an integrative framework, based on Bayesian principles, that builds upon the current systems. It integrates the process of clinical decision making and the process by which individual differences in brain function give rise to individual differences in behavior. The ultimate goal of this framework is to improve diagnoses and treatment outcomes in psychiatry.

Utilizing an Integrative Framework to Improve the Diagnostic Process

Computational (or theoretical) neuroscience likely has much to offer in terms of developing a mechanistically informed nosological structure; however, it may also have a useful role in providing a formal probabilistic structure to the ensuing diagnostic process in and of itself. In short, our approach treats symptoms, signs, and clinical decisions about a patient as observable consequences of latent constructs (such as deficits in "cognitive control"), which emerge from hidden physiological states (see Figure 10.1). Such a framework can be used progressively to reduce uncertainty about clinical decision making by guiding the clinician to the most informative questions to ask or diagnostic tests to perform. Further, inherent in this framework is the ability to compare models of pathophysiological and psychopathological dynamics, and their causes and consequences, to quantify the relative confidence in differential diagnoses. This, in principle, could include the clinicians' more intuitive (expert) inferences, which could recorded as diagnostic scores (see Friston, this volume), much like scores from clinical questionnaires. Prior to presenting the technical components that constitute this framework, we first introduce some key concepts and terminology.

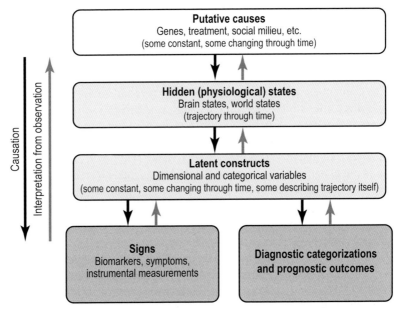

Figure 10.1 Conceptual drawing of the Bayesian Integrative Framework. In this novel integrated nosological framework, diagnoses are placed with symptoms and other observations that arise from underlying pathophysiological causes. Specific causes in the subject or the world create physical and physiological states which are understood through latent constructs that produce observations observable in the world. Thus, causation is assumed to flow from top to bottom in this figure. Causes can be constant (e.g., the genome of a subject) or changeable over time (e.g., time since an externally driven trauma). The underlying physical and physiological states are assumed to be unobservable and to be changing through time. These states are understood through dimensional and categorical latent constructs, such as cognitive control or negative affect. These latent constructs produce observable signs, such as symptoms, measures on instruments, or detectable neural signals. Diagnostic categorizations are additional observations based on clinical expertise. Similarly, prognostic outcomes are observable categorizations (e.g., recovery or relapse).

Normative versus Process Models

When thinking about how the brain works, there is an important distinction between models in computational science that is indicated by the adjectives "normative" and "process." This distinction is important both conceptually and practically. In brief, *normative models* describe how an "optimal" system would work given the goals; in other words, normative models describe "what" the brain is trying to do. In contrast, *process models* are fundamentally about the mechanisms, thus describing "how" it is done. Although these two types of models are often placed in contrast to each other, they actually represent two sides of the same coin. Briefly, every process model implies the existence of a normative model, asking what the optimal solution is given the underlying limitations of that process. Similarly, every normative model has implications for the potential processes that can produce that optimal solution. For example, many normative decision models do not take into account the time it takes to compute a predicted expected outcome, nor do they take into account the limited perceptual abilities humans have. However, once one has specified the processes for computing expected outcomes or the limitations of one's perception, one can ask about the normative model, given those limitations.

More technically, normative models rest on the assumption that the behavior at hand can be cast as an optimization process, where the states or parameters of the normative model optimize a well-defined function. In contrast, a process model specifies the algorithmic and implementational details of how the objective is attained. Process models can be formulated in terms of putative neuronal processes. Indeed, the utility of process models is that they can be used as observations or statistical models of observed responses such that, when fitted to data, their parameters associate biological processes with a functional role (Boly et al. 2011). Thus, process models have the useful property of characterizing neurophysiological responses in terms of well-defined computations.

The standard process model implicit in most nosologies assumes that a particular entity exists, as indicated by a diagnostic term, which causes symptoms. For example, "schizophrenia" is viewed as an external reality in a patient, which causes hallucinations and other manifestations of the illness. Here, we adopt a different view of diagnosis, considering it an observation based on the judgment of the clinician. From this perspective, continuously distributed latent variables cause a diagnosis. As described below, a latent variable is an unobservable state or parameter that is inferred or hypothesized based on the observable states or parameters. For example, a neuronal failure of perceptual inference causes hallucinations, delusions, and a concomitant diagnosis of schizophrenia. When a clinician arrives at a diagnosis, this represents one important piece of observational data that can be placed in the context of other observations. Thus, the diagnosis does not cause symptoms, but rather reflects the impact of multiple underlying processes. Using Bayesian inference, probabilistic statements can be made about the nature of these underlying processes. Moreover, as information accumulates, these statements can be progressively updated in a process that, over time, may improve clinical prognosis. This Bayesian process model is based on the notion that a diagnosis does not cause a disease process—the disease process causes the diagnosis—and its accompanying symptoms and signs (Figure 10.1).

Use of a Bayesian Model to Improve Diagnostic Nosology

Bayesian inference allows reasoning about uncertain quantities (i.e., latent variables) according to the rules of probability theory. In other words, Bayesian inference is the extension of deductive logic (i.e., reasoning about quantities which are certain) to inductive reasoning (reasoning about quantities which are probabilistic). One can show (Cox 1946) that any violation of the rules of probability (or, equivalently, of inductive reasoning or Bayesian inference, which are different names for the same thing) entails a violation of common sense. We can therefore use Bayesian models to relate the cause-effect inferences above.

Latent Variables/Latent Constructs

Latent or hidden variables are states or parameters that are not directly observed, but can be inferred by inverting a model of how observations depend on them. To give a simple example, consider the interpretation of a diagnostic test of HIV—a single observation in this system. We interpret the relationship between the presence of HIV infection (A) and the results of the test (B) based on Bayes's theorem:

$$p(A\mid B) = \frac{p(B\mid A)\cdot p(A)}{p(B\mid A)\cdot p(A) + p(B\mid \neg A)\cdot p(\neg A)}, \qquad (10.1)$$

where A and B refer to statements that can be true or false, and $\neg A$ is the negation of A. Applied to clinical data, A could be, "The patient is HIV positive," and B could be "The patient's HIV test is positive." $p(A)$ denotes the probability that statement A is true, and $p(A \mid B)$ denotes the probability that A is true given that B is true. Assuming that the infection rate for the demographic profile of the patient is 1% (i.e., $p(A) = 0.01$ and $p(\neg A) = 0.99$), that the true positive rate of the test is 95% (i.e., $p(B \mid A) = 0.95$), and that the false positive rate is 2% (i.e., $p(B \mid \neg A) = 0.02$), Bayes's theorem tells us that, given an initial positive test, the probability that the patient is HIV positive is 32%:

$$\frac{0.95 \cdot 0.01}{0.95 \cdot 0.01 + 0.02 \cdot 0.99} = 0.32. \qquad (10.2)$$

The usefulness of Bayes's theorem derives from the fact that it allows us to quantify the effect on the probability that A is true, after establishing that B is true. Moreover, as observations accrue, we also are able to incorporate each emerging observation into this accrual process. Before taking any initial observation on B into account, based on the infection rate in the relevant population, $p(A)$ is 0.01 and $p(\neg A)$ is 0.99. This is called the *prior probability distribution*, often referred to simply as the *prior*, because it is the prior belief before taking any additional observations into account. After observing B, Bayes's theorem then gives us $p(A \mid B) = 0.32$ and $p(\neg A \mid B) = 0.68$, which is the *posterior probability distribution*, or simply *posterior*, because it is the (updated) distribution after taking the observation into account.

The latent variable here is the presence of HIV infection in the blood of the patient (A or $\neg A$). It cannot be observed directly; rather, it has to be inferred by inverting (i.e., applying Bayes's theorem to) a probabilistic model of how HIV status leads to test outcomes. Evidence regarding the latent variable can be accumulated by repeated observation. Since a probability of 32% for positive HIV status is a poor basis for a treatment decision, we can choose to apply the test again. Assuming the test comes back positive again, we now have a probability of 96%:

$$\frac{0.95 \cdot 0.32}{0.95 \cdot 0.32 + 0.02 \cdot 0.68} = 0.96. \tag{10.3}$$

In this calculation, we simply had to replace 0.01, our original estimate based only on demographics, with 0.32, our estimate that followed from our initial observation of B. Moreover, in this calculation, we replace 0.99 with 0.68, similarly reflecting the influence of the first test. The probability of the patient being HIV positive had risen from 0.01 to 0.32, and conversely, the probability of his or her being HIV negative shrank from 0.99 to 0.68. In Bayesian terms, the *posterior* after the first observation is the new *prior* before the second observation.

Information about latent variables cannot only be accumulated by repeated observations of the same kind, but also by integrating information from different sources. If we assume that after the first test was positive, a second, perhaps more expensive, test with true positive rate 99% and false negative rate 0.5% was used and came back positive, then the probability of HIV positive status after the second test would have been 99%:

$$\frac{0.99 \cdot 0.32}{0.99 \cdot 0.32 + 0.005 \cdot 0.68} = 0.99. \tag{10.4}$$

These examples illustrate that if we have models of how latent variables lead to observations, we can use Bayesian belief updating to infer the status of latent variables and base treatment decisions and prognoses on those inferences. Indeed, in the proposed integrative framework, we are using Bayesian principles to do just this for psychiatry.

Bayesian Model Comparison

Bayesian inference can also be used to score the goodness of formal generative models. For example, one model might describe the evolution of a particular psychopathology as cyclical, leading to a prediction of periodically fluctuating clinical observations, whereas another might describe the evolution of the same psychopathology as linear, leading to a prediction of linearly progressing clinical observations. Given actual observations, the goodness of each model can be scored in terms of its expected predictive power, which has two aspects: the ability to explain existing data and the ability to generalize to new data. These two requirements are jointly quantified by one number: the model evidence. When applied to a given dataset, each competing model has a certain amount of evidence for it. Formally, the model evidence is calculated by marginalizing (i.e., taking a probability-weighted sum of) the model likelihood over all possible latent variable values:

$$\text{Mode evidence} = p(\text{Observations} \mid \text{Model})$$
$$= \int p(\text{Observations} \mid \text{Variables}) \, p(\text{Variables}) \, d\text{Variables}, \quad (10.5)$$

where dVariables denotes integration (i.e., summation) over the whole variable space, p(Observations | Variables) is the likelihood, and p(Variables) weights the likelihood by the prior probability of the variables taking a particular value. While the resulting number is not interpretable in isolation, the *ratio* of two model evidences is the *Bayes's factor* which indicates the *relative* quality of two models. This is because the Bayes's factor is what relates the prior odds (i.e., the ratio of the probability of one model to the probability of the other before making any observations) to the posterior odds (i.e., the same ratio after making observations):

$$\text{Posterior odds} = \text{Bayes's factor} \cdot \text{prior odds}, \quad (10.6)$$

or more formally:

$$\frac{p(\text{Model 1} \mid \text{Observations})}{p(\text{Model 2} \mid \text{Observations})} = \frac{p(\text{Observations} \mid \text{Model 1})}{p(\text{Observations} \mid \text{Model 2})} \cdot \frac{p(\text{Model 1})}{p(\text{Model 2})}. \quad (10.7)$$

A Bayes's factor greater than 1 indicates more evidence for Model 1 than for Model 2, while a Bayes's factor of less than 1 indicates more evidence for Model 2 than for Model 1.

It is important to note that Bayesian model comparison does not decide which model is correct; it quantifies the evidence supporting each of the candidate models. In doing this, it automatically accounts for both the accuracy and complexity afforded by each model. One can show (Penny et al. 2004) that:

$$\text{Model evidence} = \text{Accuracy} - \text{Complexity}, \quad (10.8)$$

where accuracy and complexity both have formal definitions. Introducing additional latent variables increases the complexity of a model, but these additions might yet improve model evidence if the complexity increase is outweighed by an increase in accuracy. In general, there will be a peak in model evidence for a certain number of latent variables, after which adding more complexity is no longer warranted.[1]

Bayesian Integrative Framework

Although we can only observe (e.g., symptoms, measurements), we assume that these observations arise from an underlying (unobservable) reality. This underlying reality includes many diverse factors (e.g., the social milieu, the brain state of the subject, epigenetics) that are too complex to measure directly. We define each of these variables to be a dimension within a space. A given subject occupies some point in this very high-dimensional space.[2] Since these variables are assumed to be unobservable, we assume that there are a set of *latent variables* or *constructs* that capture the most important aspects of this underlying reality. A given subject at a given moment in time occupies some position within this space of latent variables that is defined by dimensional constructs. Over time, the subject traces a trajectory through that multidimensional space. Because we do not actually observe the latent variables directly, we use Bayesian analysis methods (see below) to derive a probability distribution over a position at a given time, and over the trajectories a subject is taking through that space. An example of trajectories through this dimensional space is given by Friston (this volume), who provides a simulation of how such a framework could work.

These latent variables reflect dimensional constructs that arise from our understanding of neuroscience, psychology, and other sciences. For example, one might wish to quantify the attention abilities of a subject, how the subject arbitrates between deliberation and habit-based decision making, whether the subject's behavior reflects problems with emotions or impulse control, how reactive a subject's amygdala is to emotional stimuli, etc. These latent variables are, by definition, dimensional, incomplete, and mutable with new discoveries. Furthermore, in this framework, some subjects at particular locations within this state-space of latent variables manifest clinical observations that would lead a clinician to place the subjects into one or more diagnostic categories at any time. Indeed, generating a categorical diagnosis from latent variables that correspond to the dimensions allows us to model comorbidity in terms of underlying pathophysiological or psychopathological dimensions, and to assign a unique mapping to diagnostic categories. Because diagnostic categories reflect (potentially overlapping) areas of state-space, a single point in state-space may correspond to several diagnostic categories. Of course, the actual location of the subject in state-space is unknown, but a probabilistic distribution over

---

[1]   Popular model scoring measures such as the Akaike information criterion or the Bayesian information criterion are approximations to the model evidence which use the exact accuracy term but approximate the complexity term because this term is harder to compute.

[2]   The state-space is the set of all possible values the latent variables can take; given the large number of potential variables, the state-space is described here as "high dimensional."

points in state-space can be inferred from the symptoms shown by the subject. Because the multiple diagnostic categories overlap in some, but not other, areas of state-space, comorbidity can increase the accuracy of this state-space prediction.

Of note, there had been hope that the diagnoses in the DSM and its ilk would reflect specific latent variables—that all of the subjects diagnosed with a disorder, such as obsessive-compulsive disorder (OCD), would have a similar underlying neuropsychological dysfunction. However, decades of research have established this hypothesis to be incorrect. Importantly, and perhaps unfortunately, a diagnosis made using taxonomies such as the DSM is a single measurement, while the reality is much more complex. For example, the relationship between a diagnosis and neuropsychological dysfunction shows both equifinality (a particular symptom can arise from multiple dysfunctions) and multifinality (a particular dysfunction can generate multiple symptoms; see below for further discussion). The Bayesian Integrative Framework we propose captures this complexity by separating the diagnoses from the underlying latent variables/constructs. For this reason, we treat the diagnosis as an observation reflecting underlying latent variables, with a diagnosis being a clinician's measurement of the patient within a scheme such as DSM or the International Classification of Diseases (ICD).

The levels of this framework are connected by probabilistic (Bayesian) reasoning (Figure 10.2). We can measure the probability of a symptom S_i arising from each latent variable (or combination of latent variables) LV_j as $P(S_i | LV_j)$ or $P(S_i | LV_j, LV_k, ...)$. Using Bayes's rule (see above and Mathys, this volume), we can invert these probabilities to infer the probability of a latent variable having a specific value, given the observations $P(LV_j | S_1, S_2, ... S_n)$. What this means is that we can use observations to predict which values we expect the latent variables to take. For example, a diagnosis of OCD may only be partially predictive of a dysfunction in the balance between goal-directed and habit-based decision-making systems (Gillan et al. 2011). Thus, if a clinician diagnoses a patient as having OCD, we can use this framework to infer the probability of having a dysfunction in the balance between goal-directed and habit-based decision-making systems.

To be thorough, we should describe the probability of each observation S as arising from a trajectory through the space of latent variables $P(S_i | \text{path of } LV_j, \text{path of } LV_k, ...)$, and again, we can invert this probabilistic model to describe the likelihood of following a path through the space of latent variables from the observations $P(\text{path of } LV_j | S_1, S_2, ...)$. Thus, for example, many patients presenting to a clinician with a diagnosis of major depressive disorder will have suffered from a series of relapsing depressed episodes. The number and spacing of these episodes may be more informative for the likelihood of treatment successes than the particular constellation of symptoms that any given patient expresses when seen by the clinician (e.g., Tundo et al. 2015). The relapsing and remitting nature of these episodes forms a trajectory through an

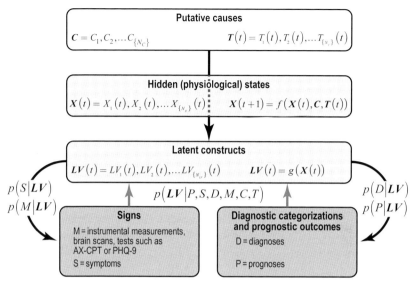

Figure 10.2 Mathematical formulation of the Bayesian Integrative Framework. As in Figure 10.1, the variables are connected through Bayesian inference, and causation is assumed to flow from top to bottom. Thus, putative causes probabilistically cause hidden physiological states, which probabilistically cause latent constructs, which probabilistically produce signs and outcomes. Because we cannot directly observe the hidden physical/physiological states, we will directly use the probability of obtaining a latent construct given the putative causes. Because we are using a Bayesian framework that can be inverted using Bayes's rule, we can calculate the probability of obtaining a latent construct given the putative causes and the observed signs and diagnoses. Similarly, we can calculate the probability of observing a given sign or diagnosis from the derived latent constructs. Thus, for example, we could measure signs (such as PHQ-9 scores or brain scans), calculate how that changes the probability distribution across latent constructs, and then calculate how that changes the probability of a given diagnosis or prognosis. AX-CPT: continuous performance task; PHQ-9: patient health questionnaire; *LV*: latent variable; *X*: hidden physiological states; *C*: putative causes; *p*(…): probability function; *t*: time point; *T*: treatment.

underlying space of both symptoms (such as reports of dysphoria) and latent variables (such as negative cognitive schema). In addition, we can incorporate the probabilities arising from static underlying causes such as genetics.

The interaction of probabilities arising from this scheme means that latent variables that are informative provide tight probabilities to observations, and latent variables that are not informative do not. Thus, we do not need to commit to specific latent variables at any particular point in time; rather we can allow the latent variables to change with basic science discoveries and the development of new conceptual frameworks in the coming years.

The following important points need to be considered in this framework:

1. The position of the patient within the latent variable space is probabilistic. That is, we do not say that the patient has followed a specific

trajectory or will follow a specific trajectory through that latent variable space, but rather that there is a probability that any trajectory is the real one.

2. There are multiple levels of understanding and data included. The recognition that the diagnosis of the clinician is actually a measurement, not a latent variable in itself, is important because it allows the incorporation of information from neuroscience that derives from dimensional conceptual frameworks such as RDoC. Importantly, there is no limit to the set of conceptual frameworks that can be so incorporated.

3. This framework allows one to move both up (from observations to constructs) and down (from constructs to observations).

Because the framework is fundamentally dynamic, it includes mechanisms to incorporate new variables and to remove variables that are not informative. Thus, when developing the framework, the following points are important:

1. We do not need to rebuild this framework from scratch. Because the framework is modifiable, we can use the current DSM categories and insights available from the RDoC project (or elsewhere) as starting points and modify them as needed. In the Bayesian terminology, these are our priors.

2. This framework can incorporate new measurements, whether they are new tasks, new symptomologies, or new diagnostic schemata. It can be updated based on new evidence to enhance its explanatory value.

3. This framework allows one to use Bayesian model comparisons to make decisions about which models and latent variables to include.

In this framework, the patient is described as taking a trajectory through the multidimensional space of latent variables. What those latent variables are will change as new findings from basic science emerge. It might be useful to think of this as a blurry set of possible trajectories: some of them may be very sharp and clear, whereas others less so.

In this framework, one could take a diagnosis made by a clinician at a given time, use that to predict a probability distribution across latent constructs, and then use that probability distribution across latent constructs to predict future outcomes, including symptoms, task-related observations, and diagnoses. In general, the strength of this proposal is that it allows a set of observations (e.g., task performance, measurements) to be used to predict a distribution across constructs which can then be used to predict a future diagnosis. In this scheme, *prediction* arises from the future trajectory. The trajectory describes how subjects typically pass through the paths in the space of latent variables. This feature will allow a diagnostician to better infer future outcomes on a case by case basis. *Treatment*, in this case, is about bending the curve; it is about changing the probabilities along the future trajectories, which changes the latent variables, which changes the observations. An example of how treatment can bend trajectories can be found in the simulations in Friston (this volume).

Because trajectories are very high-dimensional (i.e., they describe the time course of many latent constructs) and complex, we can also view future trajectories as *prognoses*. Prognosis is really a set of predicted observations (diagnoses or other outcomes) at some future time. Therefore, we could capture the concept of prognosis by including both present and predicted future outcomes in our analyses. Thus, it is not necessary to specify a prognosis explicitly within this framework. However, it may be useful to do so. Within this framework, a prognosis is an outcome/observation. One could, for example, create a new outcome/symptom/observation of interest (e.g., "will relapse/will not relapse"), which could then be predicted from the distribution across constructs. The decision to treat or not could then be determined by examining how the prognosis is conditional on treatment.

Phases of Application

Some of the pragmatic advantages of the Bayesian Integrative Framework can be appreciated in terms of the phases of its application:

Phase I: Construction of the Framework. This step could begin with an expert consensus that provides the clinician with the optimized model in a computerized form. As part of this phase, all of the empirical observations (e.g., signs, symptoms, treatment history and diagnoses at various time points) would be used to build a generative model that evolves over time and describes a trajectory. Of course, the process of generating such a model is quite important, and the details of this process would need to be explicated. However, the process of refining the model is more important than the process of generating the initial model. As a result, our discussion here devotes space primarily to discussing refinement of the model. Broadly conceptualized, to create this initial model, the number of latent variables, causal inputs, and associated functions that best describe outcomes in terms of symptom profiles and other measurements, such as the clinician's diagnosis, would be optimized using standard Bayesian inversion schemes and Bayesian model selection (see above). Thus, this phase uses existing data, or priors, to optimize the model per se. It should also be noted, as described below (see Phase Ia), that theories specific to a particular disorder or cluster of symptoms can be tested with this framework and used to further refine and enhance the model.

Phase II: Application. This step is the application for the clinician. After having selected the best model, the posterior relative to the parameters (i.e., weights) of that model can be used as priors to estimate the posterior constructs and hidden states that provide the best explanation for a patient's symptom profile and associated signs (e.g., neuroimaging, or clinical) and measures. Thus the clinician would use all available data to make such estimates. This might only include data from a mental-status examination and history. Alternatively, such clinical data could be accompanied by data from neuropsychological tests and brain-imaging experiments. Regardless of the content, the process of estimation relies on the same procedures; the posterior distribution of latent causes of the patient's symptoms can now be used to create a posterior predictive density over differential diagnoses (e.g., DSM) at the current time and, crucially, in the future. Furthermore, one could simulate probabilistic responses to therapeutic interventions in terms of (probabilistic) trajectories over future (diagnostic or therapeutic) outcomes. Taken together, the knowledge gathered in Phase I can be integrated into an application in Phase II for a clinician to use to get from observations (symptoms, diagnoses) to predicted trajectories of future outcomes (prognoses). The predicted trajectories would be reported as probabilities over the multiple possible future outcomes.

Phase III: Refinement.   This step would again emerge from a consensus of experts. The generative model could be refined by using all the clinician's observations made during the application of the model. The accumulated data from application could be assimilated using Bayesian belief updating (as described above) to improve the parameter estimates and model selection. This recursive procedure could be iterated indefinitely, providing an increasingly efficient description of "good diagnostic practice" and therapeutic outcomes. In addition, one would have the opportunity to include (or eliminate) constructs and hidden states using Bayesian model selection. Ultimately, the trajectories of hidden states underlying disease progression (or its resolution) may acquire increasingly mechanistic details as our understanding of pathophysiology accumulates.

Examples of Application of the Bayesian Integrative Framework

To illustrate how the proposed integrative nosological framework could enable the integration of findings from psychiatric research into the clinician's diagnostic act in a well-defined and quantifiable manner, we present three specific examples. Using the phases of the application described above, we explain how complex empirical results could inform the clinician's prognostic predictions and treatment decisions for individual patients presenting with attenuated psychotic symptoms, posttraumatic stress disorder (PTSD), or OCD.

Attenuated Psychotic Symptoms

Attenuated psychotic symptoms can, in some cases, herald late onset schizophrenia. A diagnosis for "attenuated psychosis syndrome" is now included in the DSM-5 (Research Appendix) as a condition warranting further investigation. Epidemiological studies suggest, however, that less than 40% of the individuals with attenuated psychotic symptoms will develop full-blown schizophrenia within five years after diagnosis (Fusar-Poli et al. 2012a, 2013a). Despite accumulating evidence about risk and protective factors for the transition from attenuated psychotic symptoms into schizophrenia (Fusar-Poli et al. 2013b), prognostic predictions in clinical practice are still very imprecise.

Phase I: Construction of the Framework

Empirical observations from various time points are used to construct a generative model of the underlying static and dynamic unobservable causes, states, and constructs (Figure 10.3). These empirical observations can be specific *symptoms* (e.g., ideas of reference or delusions), *functional outcomes* (e.g., social or occupational functioning), or *diagnoses* (e.g., attenuated psychosis syndrome or schizophrenia) assessed at various time points by longitudinal studies. These empirical observations can further comprise any *sign* or *biomarker* that has been associated with the transition from attenuated psychotic symptoms into psychotic disorder. For instance, a positive family history for psychosis is a strong predictor for transition (Seidman et al. 2010; Thompson et al. 2011), indicating that genetic risk might be a putative cause of the development of schizophrenia.

Moreover, the trajectory from attenuated psychotic symptoms into psychotic disorder has been associated with reduced cortical volume in prefrontal, cingulate, and insular regions (Smieskova et al. 2010), as well as with a range of neurocognitive deficits and impaired social perception (Fusar-Poli et al. 2012b). This points to a role for constructs, such as "cognitive control" or "perception and understanding of others," in the development of schizophrenia. The generative model can also accommodate observed influences of *treatment* on the trajectory toward schizophrenia, such as the reduction of transition rates into schizophrenia by psychosocial interventions (Preti and Cella 2010; van der Gaag et al. 2013).

Phase Ia: Refinement and Testing of the Hypothesis

After comparing models with different observed and unobservable variables, the best generative model is then selected. To illustrate the potential power of model selection, let us assume that the selected generative model for the trajectory of attenuated psychotic symptoms into schizophrenia includes both the constructs "perception and understanding of others" and "cognitive control." From a pragmatic perspective, the use of this refined model would yield more precise predictions regarding the trajectory of an individual patient (see Phase II: Application). Most importantly, however, the models derived in the integrative framework are not agnostic with respect to the mechanisms underlying

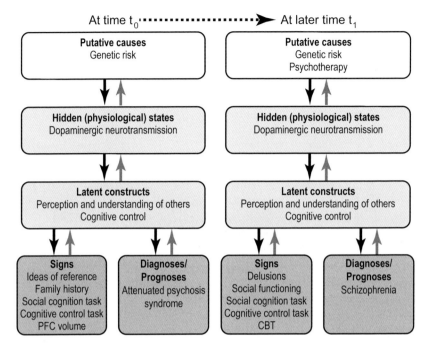

Figure 10.3   Generative model describing the trajectory from attenuated psychotic symptoms to schizophrenia. Putative causes can be either constant (genetic risk) or time-varying (psychotherapeutic interventions) and are assumed to generate neurophysiological states (altered dopaminergic neurotransmission). These neurophysiological states are linked to constructs of mental function (perception and understanding of others and cognitive control). The unobservable causes, neurophysiological states, and latent constructs $(LV(t))$ generate time-varying observations in the form of symptoms (e.g., ideas of reference), signs (e.g., reported family history of psychosis), and biomarkers (e.g., performance in a social cognition task). By model inversion, the observations can be used to infer the unobservable causes, states, and constructs to enable clinical predictions: Will this patient develop schizophrenia? How should we treat this patient? (See also Figure 10.4.) They can also provide mechanistic insight into the pathophysiology of psychotic symptoms: What is psychosis? PFC: prefrontal cortex; CBT: cognitive behavioral therapy.

the trajectory to schizophrenia. For instance, selecting between models which include either "perception and understanding of others" or "cognitive control," or both, would provide formal tests of hypotheses regarding the neuropsychology of psychotic symptoms.

Phase II: Application

Let us imagine that a 20-year-old patient presents with intermittent ideas of reference with mostly preserved reality testing. The clinician can now take the patient's history and use the reported symptoms to make a probabilistic

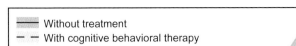

Patient, 20 years old, intermittent ideas of reference

Figure 10.4 Example of a probabilistic trajectory for an individual patient as predicted by the generative model. In the application phase, the clinician can enter clinical observations (e.g., severity of ideas of reference, family history of psychosis) into the refined and estimated generative model, and get time courses of posterior distributions over diagnoses (e.g., schizophrenia) or symptoms (e.g., social functioning). The lines represent the mean of these posterior distributions; the shaded areas depict confidence intervals. Entering additional clinical observations (e.g., task scores in a social cognition task or prefrontal cortical volumes measured by an MRI scan) would result in a narrowing of the confidence intervals equivalent to an increased precision of the predicted probabilistic time courses. Further, the model could simulate probabilistic time courses in response to a therapeutic intervention (e.g., cognitive behavioral therapy).

prediction about the patients' trajectory into full-blown schizophrenia. Alternatively, it would also be possible to make a prediction about the trajectory of any sign, symptom, or functional outcome (e.g., social functioning). Moreover, a probabilistic prediction for a trajectory without treatment can be compared to the probabilistic prediction for a trajectory with treatment (Figure 10.4). The clinician can also decide to obtain further information for the patient (e.g., a structural MRI scan to measure prefrontal cortical volume or a neurocognitive test to quantify cognitive control). This information can then be used along with the reported symptoms to increase the precision of the probabilistic predictions about the trajectory.

Phase III: Refinement

Given that the proposed integrative framework can incorporate different types and amounts of data, the generative model can be continuously refined using new empirical evidence. For instance, a recent large-scale genome-wide association study (GWAS) established an association between schizophrenia and several polymorphisms in genes related to immune function (Schizophrenia Working Group of the Psychiatric Genomics Consortium 2014). To accommodate this finding, the generative model could be extended by the observed genotypes and the neurophysiological state "immune function." The refined model would enable more precise predictions about the trajectory of attenuated psychotic symptoms in individual patients that have been genotyped for the respective genetic variants. Most notably, however, by concurrently considering the new neurophysiological state ("immune function") together with other neurophysiological states ("dopaminergic neurotransmission"), novel mechanistic hypothesis regarding the etiology of psychotic symptoms could be generated and tested.

Posttraumatic Stress Disorder

PTSD represents another especially complex disorder (Shay 1994; Kessler et al. 1995; Cantor 2005). It has interesting temporal components that are more easily accommodated within this framework than in standard practice.

Phase I: Construction of the Framework

First, we need to take what is known about PTSD and build a framework based on these known interactions. Within our integrative framework, we could include measurements of symptoms or signs (e.g., number of recalls, emotionality of recalls, where the recalls occurred, lack of sleep) as well as the DSM categorizations as observations. We would also want to include the time since the occurrence of the traumatic event as an observation. Importantly, this integrative framework is able to add in other factors, which may or may not be related to one another. For example, we could factor in a preexisting cause based on hippocampal size. Data from twin studies show that soldiers with PTSD as well as their non-trauma-exposed twin (who does not have PTSD) have smaller hippocampi than soldiers without PTSD and their twins (Gilbertson et al. 2002). This can be factored in by including an additional constant, Hipp, reflecting hippocampal size (Figure 10.5), allowing us to ask whether the addition of this constant (hippocampal size) changes the predicted probabilities of different latent variables.

Phase Ia: Refinement and Testing Theories

There are three classes of theories regarding dysfunction in PTSD: it entails (a) an *encoding* error (Brown and Kulik 1977), (b) a *decoding* error (Nadel and Jacobs 1996), and (c) a *recovery* error (Redish 2013). The latter can be separated into a lack of normal posttraumatic recovery or a worsening of symptoms.

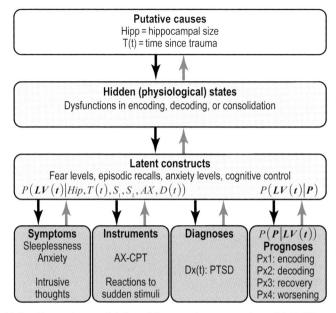

Figure 10.5 Generative model describing a patient presenting with PTSD symptoms. A set of hypothesized and explanatory latent constructs ($LV(t)$) are assumed to vary over time and can be predicted from putative causes, including constant causes, such as hippocampal size (Hipp), and temporally changing causes, such as time since trauma ($T(t)$) as well as from observations such as symptoms (S), measures on instruments (AX-CPT: continuous performance task), and clinical diagnoses (D). The latent constructs will show a progression through time, likely following one of the four hypothesized theories: an encoding deficit, a decoding deficit, a lack of recovery deficit, or a worsening (anti-recovery) trajectory. Prognoses are effectively a categorization of these trajectories. The integrative framework permits theories to be tested and prognoses to be predicted, allowing it to be used in fundamental science (e.g., causes of PTSD) as well as clinical science (e.g., PTSD treatments).

All of these hypotheses suggest differences across the patient's trajectory after the trauma. The *encoding* hypothesis suggests that the trauma is encoded differently in the patients who develop PTSD compared to those who do not. For example, a patient who develops PTSD after a bus explosion may have encoded this event with more emotional valence than one who does not develop PTSD, and this is perhaps related to more stress-related cortisol in their brain at the time of the trauma (LeDoux 1996; Jacobs and Nadel 1998; Shors 2004). The *decoding* hypothesis suggests differences in how the patient recalls the traumatic event later. That is, the patient may have a generalization deficit, in which they do not successfully identify the circumstances that indicate danger (Nadel and Jacobs 1996; Jovanovic and Ressler 2010). For example, a soldier may overreact to a surprising touch on the shoulder or may be unable to sit in a crowded restaurant after returning home from a combat zone because they incorrectly retrieve danger signals (Shay 1994). The *recovery* hypothesis suggests that trauma is encoded similarly between people who develop PTSD and those who do not; however, people who do not develop PTSD recover from their trauma differently than people who do (with PTSD patients possibly getting worse over time). For example, in "normal" trajectories of memory function, memories are initially encoded episodically with a strong "you-are-there" component, which creates a mental time travel component during re-call. Subsequently, with time, storytelling, and sleep, those memories become semantic narratives and are decoupled from the episodic mental time travel (Nadel and Moscovitch 1997; Squire 2004; Redish 2013). For PTSD patients, however, the memory continues to be stored episodically.

These three theories can be differentiated through prospective research that charts the trajectory of the PTSD symptoms and their associated effect on functioning over time. Imagine comparing changes in symptoms over time after trauma between two groups of individuals, one that eventually develops PTSD and another that does not. The encoding hypothesis suggests that there would be large differences in reactions to the trauma and associated symptoms virtually immediately after the trauma. In contrast, the decoding hypothesis suggests that the differences in symptoms would not manifest immediately after the trauma, but would emerge soon thereafter and continue over time, but with little change. Finally, the recovery hypothesis suggests that non-PTSD subjects would change more over time than PTSD subjects. We could test all three of these hypotheses using the proposed integrative framework.

Phase II: Application

The advantage of this framework is that it does not assume that PTSD is a single phenomenon represented by one of those three theories—it is possible that any given patient may have an encoding error, a decoding error, or a recovery error. The Bayesian Integrative Framework provides probabilities of each of these underlying dysfunctions from the set of observed symptoms.

Presumably, each of these dysfunctions will require different treatments. As the scientific community is refining and testing the theories (Phase Ia), it will also be necessary to determine how future outcomes (symptoms, diagnoses, prognoses) are affected by different treatments. From the probabilities of each of these trajectories, it should be possible to identify which treatments would be best suited to which patient on an individualized basis.

Phase III: Refinement

As stated above, one of the major advantages of this integrated framework is its inherent flexibility. As new neurophysiological and neuropsychological measures become available, it is easy to incorporate those new observations into the Bayesian equations. For example, we could add another neurobiological variable of functional connectivity. Georgopoulos et al. (2010) and his team have reported connectivity differences as detected with magnetoencephalography (MEG) in patients categorized with PTSD relative to those who do not have the disorder. These measurements could be integrated as new instruments in Figure 10.5. With our integrative framework, we could ask whether or not repeated measures of these connectivity differences allow clearer identification of the latent variables or of the patient's trajectory through that space of latent variables. They have, for example, also found that these measurements change over time, with more changes in controls than in veterans with PTSD (Anders et al. 2015). This example is illustrative of how advances in psychiatric neuroscience research will continuously inform and improve the results of this model in years to come.

Obsessive-Compulsive Disorder

For several reasons, OCD provides another useful test case for our proposed integrative framework. First, compared to other neuropsychiatric disorders, OCD has excellent convergence in results from neuroimaging studies, demonstrating consistency in identified brain regions with abnormal structure and function (orbitofrontal cortex, anterior cingulate cortex, striatum, and anterior thalamus) across imaging modalities (e.g., structural MRI, fMRI, PET, DTI) and research sites (Baxter et al. 1988; Swedo et al. 1989; Rauch et al. 1994; Alptekin et al. 2001; Menzies et al. 2008). This contributes to multiple types of reliable observations that can be plugged into the model, including neuroimaging findings as well as symptoms and DSM diagnoses. This, in turn, may lead to greater power due to the generation of smaller confidence intervals. In addition, with OCD we have a clear theoretical hypothesis regarding a potential set of latent variables that may have importance in driving OCD symptoms; this will be described further below in Phase Ia.

Phase I: Construction of the Framework

In Phase I, as described in the previous two examples, the proposed Bayesian Integrative Framework can be used to assemble empirical observations about OCD from both clinical experience (e.g., DSM-based diagnoses) and the clinical literature (e.g., genetic risk factors, performance on neurocognitive tasks, treatment outcomes) to build a generative model. As delineated in Figure 10.6, these observations could include a broad array of data such as:

- clinical measurement of symptom presence (obsessions, compulsions, anxiety levels, tics),
- symptom types (e.g., contamination obsessions/compulsions, doubt obsessions/checking compulsions),
- symptom levels (as measured by reliable instruments such as YBOCS, HAM-A, HAM-D; see Figure 10.6),

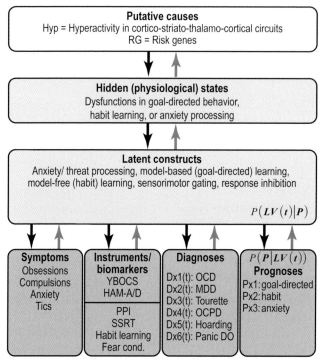

Figure 10.6 Generative model describing a patient presenting with OCD symptoms. As for PTSD, a set of hypothesized and explanatory latent constructs (*LV(t)*) are assumed to vary over time and can be predicted from putative causes—including constant causes, such as risk genes (RG), and temporally changing causes, such as hyperactivity in cortico-striato-thalamo-cortical circuits (Hyp)—and from observations (e.g., symptoms, measures on instruments, biomarkers, and clinical diagnoses). Latent constructs will show a progression through time, likely following one of the three hypothesized theories: deficits in goal-directed behavior, habit learning, or anxiety processing/expression. Prognoses are effectively a categorization of these trajectories. The integrative framework allows theories to be tested and prognoses to be predicted, making it useful to address basic issues in fundamental science (e.g., causes of OCD) as well as clinical science (e.g., effective treatments for OCD). YBOCS: Yale-Brown obsessive-compulsive scale; HAM-A: Hamilton anxiety rating scale; HAM-D: Hamilton depression scale; PPI: prepulse inhibition; SSRT: stop signal reaction time; Fear cond.: fear conditioning; MDD: major depressive disorder; OCPD: obsessive-compulsive personality disorder; Panic DO: panic disorder.

- level of insight,
- presence of comorbidities (e.g., major depressive disorder, tic disorder/Tourette syndrome, obsessive-compulsive personality disorder, hoarding disorder),
- response to both pharmacotherapy and psychotherapy treatments,
- performance on measures of neurocognitive functions implicated in OCD (e.g., habit-learning tasks, prepulse inhibition),

- fear conditioning,
- stop signal reaction time tasks, and
- putative causes (e.g., hyperactivity in cortico-striato-thalamo-cortical circuits that can be observed with both PET and fMRI; risk genes).

Importantly, these variables can be weighted according to the strength of the available evidence so that consistently replicated findings would make a greater contribution to the model. For example, despite the fact that no GWAS study to date has identified strong genome-wide candidates for OCD risk genes (Stewart et al. 2013; Mattheisen et al. 2015), genetic association studies have implicated an association between the glutamate transporter, SLC1A1, and OCD (Bloch and Pittenger 2010; Wu et al. 2012). Having an *SLC1A1* risk allele would therefore impact the model. In addition, as previously discussed, variables can be readily added or subtracted from the generative model as the literature evolves, and the effect of these changes on the probabilistic outcome (e.g., prognosis, illness trajectory, likely response to treatment) could then be assessed.

Phase Ia: Refinement and Testing Theories

In part because there is a solid foundation of evidence pointing to the likely role of both cortico-striato-thalamo-cortical circuits and anxiety/fear-related circuits in the pathophysiology of OCD, the field has likewise converged on several theories regarding the evolution of OCD symptoms. For purposes of this example, we will focus on three main theories which suggest that OCD results from (a) dysfunction in goal-directed behavior systems, (b) overactive/ dysfunctional habit systems, and/or (c) dysregulation of threat processing and/ or anxiety expression.

Recent work has suggested that OCD symptoms can result from an imbalance in the systems guiding action selection: the goal-directed "model-based" and the habitual "model-free" systems. Initial evidence has pointed mostly to dysfunction in the model-free, or habit, system (Gillan et al. 2011, 2014, 2015), with OCD patients being more prone to forming habits both in neutral conditions and "in avoidance" (i.e., to avoid a perceived threatening stimulus, such as a shock). This would suggest that excessive "model-free" action selection could be used as a latent construct reflecting a hidden physiologic state in our nosological framework of OCD. However, other evidence from both clinical deep-brain stimulation studies (Greenberg et al. 2006, 2010; Goodman et al. 2010; de Koning et al. 2011; Figee et al. 2014; Mantione et al. 2015), imaging studies (Baxter et al. 1988; Swedo et al. 1989; Rauch et al. 1994; Alptekin et al. 2001; Mataix-Cols et al. 2004), and preclinical studies in mice (Ahmari et al. 2013) suggests that dysfunction in medial orbitofrontal and ventral striatal regions linked to goal-directed systems can lead to abnormal compulsive behaviors. Thus it would be useful to be able to determine how changing the balance between the model-based and model-free systems would affect symptom presentation, diagnosis, and other observable entities. Using our model, this could be accomplished using different latent variables for model-based, model-free, or arbitrators between model-based and model-free systems (Dayan and Balleine 2002; Redish et al. 2008; Dezfouli and Balleine 2012; Wunderlich et al. 2012a; Dolan and Dayan 2013; Dayan and Berridge 2014; Lee et al. 2014).

Finally, there has been debate, in part heightened by the recent separation of OCD from other anxiety disorders in DSM-5, about whether abnormalities in anxiety regulation or threat processing play a pathologic role in OCD. A potential role for anxiety dysregulation in the pathogenesis of OCD is supported by factors including its prominence as a clinical symptom in OCD patients, observations of fear-conditioning abnormalities (Milad et al. 2013), and enhanced avoidance habits in OCD (Gillan et al. 2014), and reports of trauma-induced OCD (Dykshoorn 2014). Because the observable entities of symptoms (including anxiety), brain metabolic state (functional imaging), and neurocognitive task performance are all known in the recent studies of habit formation in OCD, these data could potentially be used to validate the efficacy of the model in one defined case.

Phases II and III: Application and Refinement

As described above, the potential power of this Bayesian framework lies in the fact that it is agnostic about the latent constructs and theories which make up a particular mental illness. It therefore easily incorporates the possibility that a disease currently classified as a single entity in DSM-5 could be broken down into several different theory-based categories with different disease and treatment trajectories. In fact, in an ideal case, the framework would actually help to identify valid patient subgroups through an iterative process of including progressively higher quality data in the model as the field evolves, and examining treatment outcomes in subpopulations of patients. Ultimately, delineating whether valid patient subcategories exist, or identifying which (if any) of the theories described in Phase Ia underlies pathophysiology for a particular patient, could be extremely important for guiding treatment. For example, a patient with excessive habits leading to an increased propensity to develop compulsions might benefit most from habit-reversal training, whereas a patient with high levels of anxiety might benefit more from treatment with selective serotonin reuptake inhibitors (SSRIs) and classic exposure therapy with response prevention.

There are, of course, potential limitations in using OCD as a test case. For example, first, unlike several other psychiatric disorders, there is limited data available on longitudinal trajectories in OCD throughout the life span, due to a lack of large-scale multisite longitudinal studies. Second, though similar brain regions are consistently highlighted across multiple imaging studies, behavioral studies of neurocognitive functions such as set shifting, response inhibition, and reversal learning are more variable. Although these factors may currently limit the richness of the data that can be incorporated into the model, they do not affect utility. Indeed, it is important to emphasize that "bad data" or data not relevant to the probable outcomes will automatically be removed from the integrative framework.

Special Considerations: Comorbidity

OCD is a particularly useful case example for exploring the issue of how our Bayesian framework can integrate comorbidities. OCD is highly comorbid with several other DSM-5 diagnoses, including major depressive disorder, panic disorder, Tourette syndrome, hoarding disorder, and obsessive-compulsive personality disorder (Murphy et al. 2013). In fact, hoarding disorder was only separated from OCD in the most recent edition of DSM based on findings from neuroimaging studies which identified distinct neurobiological substrates and differences in treatment response. This has raised the question of whether OCD patients with these comorbid disorders should be considered as belonging to separate diagnostic categories. This is an important consideration because there is some evidence that OCD patients with different comorbidities may have distinct responses to treatments. For example, it is commonly known that hoarding, which used to be considered an OCD symptom, is more resistant to both pharmacotherapeutic and psychotherapeutic treatment than classic OCD symptoms (Bloch et al. 2014; Mataix-Cols 2014). In addition, a recent study suggested that augmentation with atypical antipsychotic medications may be most useful in OCD patients who have comorbid tics (Bloch et al. 2006). Within the context of our proposed framework, comorbid conditions can be easily included in the generative model as further empirical observations (see Figure 10.6). By combining this with other observations, including neuroimaging data, genetic information, and neurocognitive task performance, we can determine the impact of comorbidities on OCD illness trajectories and potentially glean information about the most effective treatment interventions through the iterative process described above.

Discussion

The Bayesian Integrative Framework that we propose inverts the standard model of nosology. That is, rather than subscribing to the notion that particular entities that are classified by diagnostic categories cause symptoms, we are proposing that diagnostic classification and symptoms are a consequence of latent variables (or constructs) which themselves are caused by evolving but hidden pathological states. We consider the problem of nosology as modeling the diagnostic (and prognostic) process, where diagnosis is an observation or an outcome as opposed to a cause. Implicit in this framework is the mapping from the hidden or unobservable causes of psychopathology to observable symptoms and signs, and "good practice" diagnostic outcomes. This framework, therefore, accommodates risk assessment and the resolution of ambiguous nosological problems (e.g., comorbidity). Crucially, it accommodates, but is not limited to, currently accepted diagnostic categories (e.g., DSM). Further, it accounts for developmental and longitudinal trajectories of mental illness. Finally, it furnishes a formal link between putative constructs (e.g., RDoC constructs) and clinical categories, thereby harnessing the complementary perspectives afforded by dimensional and categorical approaches. Below we will further discuss the value of this model over the current diagnostic process and practical considerations for the implementation of this framework.

How Do We Define Success of a Diagnostic System?

Given that diagnostic systems such as nosologies exist primarily to assist clinicians in the management of their patients, the measure of the success of a diagnostic system can be framed as the degree to which it achieves that goal. There are several ways in which a diagnostic system can assist clinicians, all of which are covered under the broad rubric "clinical utility."

The first such use, the one in which categorical classifications like the DSM have been the most successful, is facilitating communication among clinicians, between clinicians and patients/families, and between clinicians and administrators. It will be important to retain this strength in any newly developed scheme.

A second use is to help clinicians select the optimal treatment for a patient. Ideally, making a diagnosis would be an initial critical step in guiding the choice of treatment (i.e., if the clinician makes a diagnosis of X, he or she can be confident that treatment Y is very likely to be effective). In actuality, however, there is an uncertain relationship between the current DSM diagnostic categories and treatment. Most psychiatric treatments are at least somewhat effective for a variety of categories that cut across the various DSM categories: SSRIs work for depressive disorders, OCD, PTSD, premenstrual dysphoric disorders, and others (Wagstaff et al. 2002; Saxena et al. 2007; Rapkin and Winer 2008). Moreover, the effectiveness of a particular intervention in treating a particular diagnosis has been disappointing for some patient subgroups within a diagnostic category. In treating major depressive disorder, the likelihood of any significant clinical benefit from antidepressant treatment is no better than 70%, and the likelihood of full remission is far lower (e.g., Rush et al. 2006b; Khin et al. 2011).

A third use is to help clinicians predict the future course and outcome of a psychiatric presentation (e.g., to inform the patient how likely it is for the symptoms to remit, or get worse, over time, as well as what environmental or other factors are likely to make the symptoms worse, or better). As with predicting treatment response, because of the range of potential course and outcome trajectories associated with each disorder, meeting criteria for a DSM category provides limited information in terms of predicting a future course. For both treatment and prognostic prediction, much of the problem stems from the fact that the diagnostic categories are essentially "black boxes" which obscure crucial mechanistic differences between individual patients included in a particular category. This suggests that in order for a diagnostic system to be successful, it must incorporate mechanistic processes in the diagnostic formulation, which is exactly what the proposed framework will do.

One example of a framework that incorporates the kind of categorical diagnostic decisions that are essential to clinical practice, while grounding these decisions in a biologically meaningful process, is the "harmful dysfunction" model (Wakefield 1992a, b, 2007). The harmful dysfunction model defines psychiatric disorders as requiring both a value judgment (harm, negative consequences) and a dysfunction judgment (when the condition represents dysfunction of a naturally selected mechanism or trait). Importantly, harm is not a scientific question but rather a value question. Harm is defined as something that causes distress or is socially disvalued, thus "harm" inevitably involves a value judgment. The harmful dysfunction analysis specifies that neither harm without dysfunction (e.g., procrastination, illiteracy, grief) nor dysfunction without harm (e.g., synesthesia) is a disorder. While the harmful dysfunction analysis provides an illustrative example of incorporating mechanisms into categorical diagnoses, this model has not been widely accepted due to the limited knowledge about the nature of "natural functions." Regardless, nosologies that reflect the harmful dysfunction combinatorial approach possess major advantages. For progress to occur, such nosologies should utilize the knowledge we have about the brain to inform diagnostic decisions, as we have proposed to do with the Bayesian Integrative Framework.

Problems with the Current Diagnostic Process

Creating Categories out of Continuous Data

Research demonstrates the continuous nature of many psychiatric problems, without a natural breakpoint between health and disease (Fergusson and Horwood 1995; Kendler and Gardner 1998; Fergusson et al. 2005). As a result, a major problem faced by diagnosticians arises from the need to impose a categorical structure on information that is largely continuous in nature (Van Os et al. 1999). For example, when we examine overall levels of multiple symptoms related to anxiety, we find that we can arrange individuals in a continuous distribution, from few to many symptoms. Moreover, when we examine the predictive relationship between symptom number and various external validators, such as outcome or treatment response, we see no natural break point in these relationships. More symptoms predict worse outcome or treatment response in a continuous fashion.

Another example of continuous data that can be problematic relates to the risk of a false-positive or false-negative diagnosis. When an individual receives a psychiatric diagnosis in error, this carries risk, both due to delivery of treatment and its resulting harm as well as to the stigma associated with many psychiatric diagnoses. Again, this gives a set of continuous relationships, such that the more extreme the treatment, the greater the risk. Other areas of medicine provide guidance for the clinician in the circumstances that currently confront the psychiatrists. Clinicians can derive categories based on the point on a continuous scale where the benefits associated with treatment outweigh the risks of false positives. Such an approach has been, for example, used in obstetrics. In some consensus guidelines, age 35 had been considered a breakpoint when considering the appropriate age for amniocentesis. This age was a point where the rate of a positive diagnosis of a genetic anomaly became greater than the rate of miscarriage associated with the procedure. However, this crossing of risks assumes that the parent's judgment of negative consequences is equivalent between genetic anomaly and miscarriage. Clearly, in current practice, other factors influence decisions about amniocentesis. Discussing the risks separately allows the parents to make decisions based on their own value judgments of risk. Similar computations apply for the treatment of hypertension, where treatment is initiated when the benefits, in terms of reducing risk, outweigh the risks associated with side effects.

Psychiatry currently faces a few problems in the applications of this approach. Importantly, we need more data to precisely quantify the nature of risks associated with various levels of symptoms, various treatments, and false positive diagnoses. In addition, we need a process for identifying thresholds at which risks from declaring a positive diagnosis outweigh the risks from a false positive diagnosis. The proposed Bayesian Integrative Framework addresses these problems and provides guidance for better predicting outcomes and treatments in a way that maintains the continuous nature of risk and thus minimizes the adverse consequences of false-positive diagnoses.

Accounting for Trajectories

Most psychiatric disorders have a longitudinal evolution, with underlying risk factors and symptoms changing over time. Still, current diagnostic systems incompletely incorporate a developmental perspective and do not fully utilize repeated, longitudinal observations. This is important because clinical decisions are often based on estimating the probability of future trajectories for a patient that can only be determined by repeated observations. For example, the first episode of a mild disorder in adolescence (e.g., subthreshold depression) might spontaneously remit, progress to a full major depressive episode, or herald future bipolar disorder (Rutter et al. 2006). Many psychiatric disorders are preceded by earlier difficulties in childhood (Kim-Cohen et al. 2003; Pine and Fox 2015). For example, mood disorders are commonly preceded by anxiety and behavioral problems (conduct disorder). Schizophrenia can be preceded by earlier childhood symptoms and neurodevelopmental impairments involving anxiety, inattention, as well as cognitive, motor, language, and social communication impairments (Kim-Cohen et al. 2003; Rutter et al. 2006; Dickson et al. 2012). Most individuals with childhood anxiety or conduct disorder, however, do not go on to develop serious forms of mood disorder (e.g., bipolar disorder), and most children with neurodevelopmental impairments do not develop schizophrenia.

Importantly, these variable trajectories pose problems for clinical decision making. Clinical symptoms, such as depression, observed at a particular point in development, can emerge through multiple earlier developmental trajectories—a process often referred to as *equifinality* (see Cicchetti and Rogosch 1996). The idea of equifinality can be applied to multiple factors beyond development. Thus, risk factors or brain dysfunction can also be viewed from this perspective. Two distinct risk factors or two different types of brain dysfunction can predict the same symptomatic presentation. Heterogeneity from equifinality arises when different pathophysiological processes give rise to the presentation of similar symptoms. Another form of heterogeneity arises from a process termed *multifinality*. This means that one risk factor, such as a traumatic life event, can give rise to multiple different outcomes. Like equifinality, the idea of multifinality can be applied to development, whereby one developmental profile has many different outcomes (as discussed above). This can also be applied to risk factors (e.g., genetic variants) and brain function, whereby a single risk factor or a single type of brain dysfunction gives rise to many different outcomes.

The proposed integrative framework will provide a means to incorporate and appropriately weight all of the available observations (e.g., clinical variables, cognitive variables, family history of bipolar disorder), and, in turn, inform clinicians as to the probability of the trajectory their patient is most likely to follow. This will be important in influencing decisions about whether or not to intervene, balancing the risks versus side effects of diagnostic labeling or treatment, choosing an appropriate intervention, and selecting the intensity of treatment and follow up—ultimately improving diagnostic and treatment outcomes.

Knowing if the Model Is Right or Wrong

We have cast the problem of optimizing a Bayesian Integrative Framework among generative models of diagnostic/prognostic outcomes and their associated symptoms and signs. So how do we know if a model generated from this framework is correct or incorrect? Strictly speaking we can dissolve this question by noting that all models are wrong but some have much more evidence than others (recall that model evidence scores the goodness of a model in terms of the right balance between the accuracy of fitting some data and the complexity of the model; see above).

The notion of right and wrong presupposes that there are only two models—a correct and an incorrect model. Formally, the idea of right and wrong models is implicitly assumed in classical statistics and corresponds to the null and alternative hypotheses, respectively. We reject the null model as wrong if there is enough evidence to make one chosen statistic sufficiently large. However, in Bayesian model comparison the number of competing models or hypotheses can be much greater than two, and each model has its own evidence. Bayesian model comparison then reduces to selecting those models that have the greatest evidence. This is a simple thing to do because the difference in log evidence between one model and another corresponds to the log of their relative (marginal) likelihood. For example, if the best model has a log evidence of three or more, relative to the next best model, then the best model is twenty $(\exp(3) = 20)$ times more likely than all of its competitors. This, however, is a relative statement; it only pertains to the set or space of models considered. In this sense, there is no right or wrong model—only models which are better or worse at explaining any given data in a parsimonious fashion. In this regard, the current integrative framework will allow us to identify the "best" fit model on a case by case basis and will, undoubtedly, provide a foundation to improve upon the current diagnostic system in psychiatry.

Potential Problems

The implementation of this framework faces technical and community acceptance challenges—challenges that will need to be addressed with outreach, training, and continuous communication between the fields of computation, psychiatry, and neuroscience. A number of advances have already developed in this regard, with the emergence of new training programs and funding opportunities and new journals (e.g., *Computational Psychiatry*). In addition, a Transcontinental Computational Psychiatry Workgroup has recently emerged out of this Ernst Strüngmann Forum. This group consists of scientists from the fields of computation, neuroscience, and psychiatry who have begun to convene on a regular basis to discuss and advance the field of computational psychiatry. In turn, new conferences are being developed and the field is gaining increasing recognition. Moreover, we are aware that there will be financial and political barriers that will need to be overcome in adopting this framework. Finally, it is important to note (as outlined in the discussion of the *Phases* above) that multiple iterations of this model will be required for continual updates and improvements based on the information that we have at any given time.

Summary

In this chapter we have proposed and described the Bayesian Integrative Framework—a novel integrative framework intended to integrate neuroscience and clinical psychiatry research, enhance the current diagnostic process, and improve treatment outcomes in psychiatry. We have identified several important features of the proposed framework that will allow us to circumvent some of the challenges currently being faced in the field of psychiatry. For one, we will be able to integrate mechanistic processes in diagnostic formulation; that is, we can incorporate any knowledge we have about underlying pathophysiology to inform diagnostic decisions. Indeed, the Bayesian Integrative Framework provides the necessary bridge between putative constructs (e.g., RDoC) and clinical diagnoses, thereby linking the complementary perspectives afforded by dimensional and categorical approaches. It also allows us to incorporate many different flavors of data at multiple layers. One of the most valuable features of this framework is perhaps its ability to account for and incorporate longitudinal trajectories that may nuance diagnosis, prognosis, and treatment. This framework will yield a better understanding of individual differences and, importantly, how individual differences in brain function give rise to individual differences in behavior.

Acknowledgments

We would like to thank David Redish and Joshua Gordon for their foresight in proposing this Forum and their dedication to making it happen. We also thank the other Program Advisory Committee members, especially Julia Lupp, for organizing and hosting the Forum. We would also like to thank the staff for their administrative help with this chapter, particularly Eleanor Stephens. Finally, we would like to thank all of the Forum participants for helpful discussions and comments on early drafts of this chapter.

11

Computational Nosology and Precision Psychiatry

A Proof of Concept

Karl J. Friston

Abstract

This chapter provides an illustrative treatment of psychiatric morbidity that offers an alternative to the standard nosological model in psychiatry. It considers what would happen if we treated diagnostic categories not as putative causes of signs and symptoms, but as diagnostic consequences of psychopathology and pathophysiology. This reconstitution (of the standard model) opens the door to a more natural formulation of how patients present and their likely response to therapeutic interventions. The chapter describes a model that generates symptoms, signs, and diagnostic outcomes from latent psychopathological states. In turn, psychopathology is caused by pathophysiological processes that are perturbed by (etiological) causes, such as predisposing factors, life events, and therapeutic interventions. The key advantages of this nosological formulation include: (a) the formal integration of diagnostic (e.g., DSM) categories and latent psychopathological constructs (e.g., the dimensions of RDoC); (b) the provision of a hypothesis or model space that accommodates formal evidence-based hypothesis testing or model selection (using Bayesian model comparison); (c) the ability to predict therapeutic responses (using a posterior predictive density), as in precision medicine; and (d) a framework that allows one to test hypotheses about the interactions between pharmacological and psychotherapeutic interventions. These and other advantages are largely promissory at present: the purpose of this chapter is to show what might be possible, through the use of idealized simulations. These simulations can be regarded as a (conceptual) prospectus that motivates a computational nosology for psychiatry.

Introduction

One of the key issues addressed by our working group at this Forum (Flagel et al., this volume) was the status of psychiatric nosology and how it might be informed by advances in computational neurobiology (Redish and Johnson 2007; Montague et al. 2012; Wang and Krystal 2014). In brief, our starting point was the realization that diagnostic categories are not the causes of psychopathology—they are (diagnostic) consequences. Although rather obvious in hindsight, this was something of a revelation, largely because it disclosed the missing link between putative causes of psychiatric illness (e.g., genetic predisposition, environmental stressors, iatrogenic) and the consequences, as observed by clinicians (e.g., symptoms, signs and, crucially, diagnostic outcome). In what follows, I briefly rehearse the ideas—borrowed from computational neurobiology—that we hope might close this gap (for full discussion, see Flagel et al., this volume). This treatment provides a technical summary of our conclusions, using an illustrative (simulated) case study of nosology, diagnosis, and prognosis.

The principal contribution of a formal or computational approach to nosology rests on the notion of a *generative model*. A generative model generates consequences from causes—in our case, symptoms, signs, and diagnoses—from underlying psychopathology and pathophysiology. Generally, these models are state-space models that describe dynamics and trajectories in the space of latent (e.g., pathophysiological) states. These states are latent or hidden from direct observation and are only expressed in terms of measurable consequences, such as symptoms and signs. The utility of a generative model lies in the ability to infer latent states from observed outcomes and, possibly more importantly, assess the evidence for one model relative to others, given a set of measurements. This is known as (Bayesian) model comparison, where the evidence is simply the probability of any sequence of observations under a particular model (Stephan et al. 2009b). To assess the evidence for a particular model, it is necessary to fit the model to observed data, a procedure known as *model inversion*. This is because the mapping from causes to consequences is inverted, to infer from consequences to causes (e.g., inferring pathophysiology from symptoms). Furthermore, having inverted a model—by optimizing its parameters to maximize model evidence—one can then simulate or predict new outcomes in the future using something called the *posterior predictive density*. Heuristically, this is the technology behind weather forecasts, where the generative model is a detailed state-space model of meteorological dynamics (Young 2002). So can we conceive of an equivalent meteorology for psychiatry?

In the past decade, there have been considerable advances in using state-space models of distributed neuronal processes to understand (context-sensitive) connectivity and functional architectures in the brain. This is known as *dynamic causal modeling*, which accounts for a significant number of papers in the imaging neuroscience literature (Friston et al. 2003; Daunizeau et al. 2011a). In what follows, I apply exactly the same (computational neuroscience) principles to the problem of modeling the causes of nosological outcomes in psychiatry. Indeed, all the examples below use standard Bayesian model inversion schemes that are available in freely available academic software: the simulations described below can be reproduced by downloading the SPM software (http://www.fil.ion.ucl.ac.uk/spm/) and invoking the Matlab script *DEM_demo_ontology.m*. The dynamic causal modeling of psychopathology can, in principle, offer a number of advantages over the standard nosological model. As noted in the abstract and by Flagel et al. (this volume), these include: (a) the formal integration of diagnostic (e.g., DSM) categories and latent psychopathological constructs (e.g., the dimensions of RDoC) (Stephan and Mathys 2014), (b) the provision of a hypothesis or model space that accommodates formal evidence-based hypothesis testing (Krystal and State 2014), and (c) the ability to predict therapeutic responses (using a posterior predictive density). Crucially, by adopting a dynamic modeling approach, one can properly accommodate the personal history and trajectory of individual patients in determining (and predicting) the course of their illness.

Our overall approach to nosology (and its promissory advantages) may seem rather abstract and perhaps even grandiose. This chapter should therefore be taken as a prospectus for future discussions about nosology and the potential for individualized or precision psychiatry in the future. Its purpose is to illustrate what could be possible, in an idealized world, if we were able to use clinical data to optimize generative models of psychopathology. Whether or not this is possible with current data is an outstanding question. In short, this chapter offers a (mathematical) sketch of what a computational nosology could look like.

I begin by describing a formal model that generates symptomatic and diagnostic outcomes from latent (pathophysiological and psychopathological) causes. This model should not be taken too seriously; it is just used to illustrate the promise of such modeling initiatives and to show how a formal approach to nosology forces one to think carefully about the known and unknown variables in psychiatric processes and how they influence each other. In the next section, I consider the use of ratings of symptoms and signs (and diagnosis) to estimate or infer their latent causes. This is a necessary prelude for model comparison and is discussed briefly. Finally, prognosis and prediction are considered by using the generative model to predict the outcome of a (simulated) schizoaffective process and its response to treatment.

Generative Models for Psychiatric Morbidity

This section introduces the general form of generative (dynamic causal) models for psychiatric morbidity and a particular example that will be used to illustrate model inversion, model selection, and prediction in subsequent sections. As noted above, a generative model generates consequences from causes. The basic form assumed here starts with the (observable) causes of psychiatric illness, such as genetic or environmental predispositions, and therapeutic interventions. These factors induce pathophysiological states, such as aberrant dopamine receptor availability or glucocorticoid receptor function, to alter their trajectories (i.e., time courses or fluctuations) over a course of weeks to months. These latent pathophysiological states then determine psychopathology, cast in terms of latent cognitive, emotional, or behavioral function (e.g., low mood, psychomotor poverty, thought disorder). Psychopathological states correspond to the constructs underlying things like Research Domain Criteria (RDoC) (Kaufman et al. 2015) and clinical brain profiling (Peled 2009). Finally, psychopathological states generate measured symptoms using, for example, standardized instruments (e.g., PANSS [Kay 1990], Beck depression inventory, mini mental state) or diagnostic outcomes (e.g., schizophrenia, major affective disorder, schizoaffective disorder).

Note that in this setup, a diagnosis represents an outcome, provided by a clinician. In other words, symptoms, signs, and diagnosis have a common cause, where the diagnostic categorization provides a useful summary outcome that integrates aspects of psychopathology which may not be covered explicitly by standardized symptom ratings or particular signs (e.g., psychomotor poverty, EEG abnormalities, abnormal dexamethasone suppression).

This formulation of psychiatric nosology is largely common sense and reiterates what most people would understand about psychiatric disorders. However, can this understanding be articulated formally in a way that can be used to make quantitative predictions and test competing etiological hypotheses? This is where a formal nosology or generative model comes into play. The first step is to construct a *graphical model* of dependencies among the variables generating measurable outcomes. Figure 11.1 (left panel) shows the graphical model that summarizes the probabilistic dependencies among etiological causes $u(t)$, pathophysiological states $x(t)$, psychopathology $v(t)$, and, finally, symptoms and diagnosis $s(t)$, $\Delta(t)$. In this format, the variables in white circles correspond to latent states that are hidden from direct observation, while the observable outcomes are in the cyan circle.

Probabilistic dependencies are denoted by arrows that entail (time-invariant) parameters ($\theta = \theta^s$, θ^n, θ^p). This formulation clarifies the roles of different quantities and makes their interdependencies explicit. For example, a diagnostic classification at a particular time would be an outcome variable, whereas a patient's drug history would be an etiological cause that influenced pathophysiology. Having established the form of the graphical model, it is now necessary to specify the nature of the dependencies within and among latent variables. An example is provided in the right panel and can be described as follows.

A Generative (Dynamic Causal) Model

This example of a generative model is deliberately very simple and restricts itself to modeling a limited differential diagnosis that includes (a simulation of) schizophrenia, affective disorder, schizoaffective disorder, and a remitted state. These diagnostic outcomes are accompanied by six symptom scores that have been normalized to lie between plus and minus one. Furthermore, only a limited number of exogenous causes and latent variables are considered: specifically, a single therapeutic cause perturbs the evolution of three physiological states, from which two psychopathological states are derived. At any one time, the location in the psychopathological state-space determines a symptom profile (through a linear mixture of psychopathological states that is passed through a sigmoid function). Diagnostic outcomes are, here, encoded by a probability profile over the differential diagnosis (i.e., the relative confidence a clinician places in the differential diagnoses). In the model, the probability of any one diagnosis corresponds to the relative proximity to a particular point in psychopathological state-space. In other words, as the disorder progresses, a trajectory is traced out in a two-dimensional psychopathological state-space, where, at any one time, the prevalent diagnosis is determined by the location to which the current state is closest. Technically, this has been modeled by a softmax function of diagnostic potential, where diagnostic potential is the (negative) Euclidean distance between the current state and the locations associated with each diagnostic category (encoded by θ_i^{Δ}, the color dots in Figure 11.1).

The trajectory of psychopathology is determined by the corresponding trajectory through a pathophysiological state-space which has its own dynamics. These are encoded by *equations of motion* or flow based, in this example, on a Lorenz attractor (Lorenz 1963). The Lorenz form is an arbitrary choice and could easily be replaced by other plausible equations: Moran et al. (this volume) provide an example based on normal form stochastic dynamics and indeed optimized on the basis of Bayesian model evidence (see below). Having said this, the Lorenz attractor provides a simple model of chaotic dynamics in the physical (Poland 1993) and biological (de Boer and Perelson 1991) sciences. Interestingly, it arose in the modeling of convection dynamics, which speaks to the analogy between the current modeling proposal and weather forecasting. In this setting, the ensuing dynamics can be regarded as a canonical form for nonlinear coupled processes that might underlie pathophysiology in psychosis. It has a canonical form because, as seen in Figure 11.1 (on the right), one can regard the parameters as specifying coupling coefficients or connections that mediate the influence of one physiological state on the others. Crucially, some of these connections are state dependent. This is important because it means one can model fluctuations in pathophysiology in terms of self-organized (chaotic) dynamics that have an underlying attracting set. In other words, we have a rough model of pathophysiological dynamics that summarize slow fluctuations in neuronal (or hormonal) states that show homoeostatic or allostatic tendencies (e.g., Leyton and Vezina 2014; Misiak et al. 2014; Oglodek et al. 2014; Pettorruso et al. 2014; see also Krystal et al., this volume).

The equations in Figure 11.1 all include random fluctuations. These fluctuations render the generative model a probabilistic statement about how various

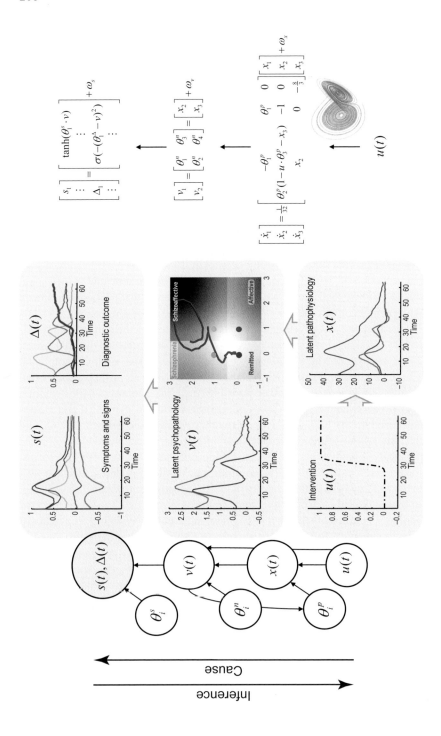

Figure 11.1 Schematic of a generative model for psychiatric morbidity. The model is shown in terms of a (probabilistic) graphical model on the left. In this format, the quantities in white circles correspond to random variables that include (unknown) parameters on the left and hidden or latent states on the right: etiological causes $u(t)$, pathophysiological states $x(t)$, psychopathology $v(t)$. The arrows denote conditional dependencies and describe the influences among latent variables that generate observations or outcomes in the cyan circle at the top. Here, the outcomes are clinical symptom scores and a differential diagnosis, measured as a probability distribution over diagnostic classifications. The only difference between these outcomes is that the diagnostic probabilities are constrained to the nonnegative and sum to one. The outcomes are generated as functions of psychopathological states which, themselves, are mixtures of pathophysiological states. Finally, the pathophysiological states are perturbed by inputs (like therapeutic interventions). An example of the form of the conditional dependencies is provided on the right, in terms of functions and random (Gaussian) fluctuations. In this example, a single therapeutic intervention enters the dynamics (or equations of motion) governing the evolution of pathophysiology states. Here, the therapeutic intervention changes the influence of the second physiological state on the first, where this coupling is itself state-dependent (and changes with the third state). These dynamics are based upon a Lorenz attractor (illustrated with the insert). The psychopathological states are generated as a linear mixture of the last two physiological states. In turn, the psychopathological variables are mapped to clinical outcomes through sigmoid functions (to generate symptom scores) and a softmax function of diagnostic potential (to generate a differential diagnosis). The diagnostic potential is based on the proximity of the psychopathological state to locations representing diagnostic categories. The middle panels illustrate a particular realization of this generative model over 64 time bins (i.e., weekly assessments). The lower panels show the therapeutic input (starting at 32 weeks) and the dynamic response of the three pathophysiological states. The ensuing psychopathology is shown as a function of time and as a trajectory in state-space in the left and right middle panels respectively. The state-space of psychopathology contains the locations associated with diagnostic categories (colored dots), which determine the diagnostic classification that tiles state-space (shaded gray regions). In this example, there are four diagnostic categories: schizophrenia, schizoaffective and affective disorder, and a state of remission. The symptoms and differential diagnosis, $s(t)$, $\Delta(t)$, generated by this trajectory are shown in the upper panels on the left and right respectively. In this and subsequent simulations, the initial (physiological) states were [8,10,32], the parameters for the symptom scores are sampled from a unit Gaussian distribution, while the remaining parameters are shown in Figure 11.4.

variables influence each other. Here, the random fluctuations can be regarded as observation noise (when assessing symptoms) and random fluctuations or perturbations to psychological or physiological processing (e.g., life events or drug misuse). In this chapter, these random fluctuations are smooth processes with a Gaussian correlation function with a correlation length of half an assessment interval (i.e., a few days).

There is no pretence that any of these states map in a simple way onto physiological variables; rather, they stand in for mixtures of physiological variables that have relatively simple dynamics. The existence of mixtures is assured by technical theorems such as the center manifold theorem and the slaving principle in physics (Carr 1981; Haken 1983; Frank 2004; Davis 2006b). Basically, these theorems say that any set of coupled dynamical systems can always be described in terms of a small number of patterns (known variously as order parameters or eigenmodes), which change slowly relative to fast and noisy fluctuations about these patterns.

Crucially, this generative model has been constructed such that the therapeutic intervention changes the state-dependent coupling between the first and second pathophysiological states (in fluid dynamics, this control parameter is known as a *Rayleigh number* and reflects the degree of turbulent flow). This means we could regard this intervention as pharmacotherapy that changes the coupling between different neuronal (or hormonal) systems, e.g., an influence of an atypical antipsychotic (Hrdlicka and Dudova 2015) on dopaminergic and serotonin receptor function responsible for monoaminergic tone in the ventral striatum and serotoninergic projections from the amygdala to the paraventricular nucleus (Wieland et al. 2015; Muzerelle et al. 2016). Furthermore, I have introduced a parameter θ_3^p that determines the sensitivity to the intervention that may be important in determining a patient's responsiveness to therapy (Brennan 2014).

The middle panel of Figure 11.1 provides an illustration of how a patient might present over time under this particular model. Imagine we wanted to model (six) symptom scores and a probabilistic differential diagnosis over four diagnoses (schizophrenia, schizoaffective, affective, and remitted), when assessing an outpatient on a weekly basis for 64 weeks. A therapeutic intervention, say an atypical antipsychotic, is introduced at 32 weeks and we want to model the response. This therapeutic input is shown in the lower left panel as a dotted line and affects the evolution of physiological states according to equations of motion on the right. These equations generate chaotic fluctuations in (three) pathological states shown on the lower right. Two of these states are then mixed to produce a trajectory in a psychopathological state-space. This trajectory is shown in the middle panel as a function of time (middle left) and as a trajectory in state-space (red line in the middle right panel). In turn, the psychopathology generates symptom scores (shown as colored lines on the upper left) and diagnostic probabilities (shown on the upper right). The relationship between the continuous (dimensional) latent space of psychopathology and the (categorical) differential diagnosis is determined by diagnostic parameters θ_i^Δ defining the characteristic location of the ith diagnosis. These locations are shown as dots in the state-space: the blue dot corresponds to a diagnosis of remission, while the green, cyan, and red locations correspond to a diagnosis of schizophrenia, schizoaffective, and affective disorder, respectively. One can see that initial oscillations between schizophrenia and schizoaffective diagnoses are subverted by the therapeutic intervention. At this point, the latent pathophysiology is drawn to its (point) attractor at zero, the most likely diagnosis becomes remission and the symptom scores regress to their normal values of zero. In short, this models a successful intervention in a pathophysiological process that shows chaotic oscillations expressed in terms of fluctuating symptoms and differential diagnosis. Later we will see that, in the absence of therapy, these chaotic oscillations would otherwise produce a relapsing-remitting progression with an ambiguous diagnosis that fluctuates between schizophrenic and schizoaffective. This intervention is formally similar to what is known anecdotally as *chaos control* (e.g., Rose 2014). This example suggests that the goal of therapy is less about countering pathological deviations and more a subtle problem of suppressing chaotic or turbulent neurohormonal processes that are equipped with many self-organizing feedback mechanisms. Heuristically, the role of a clinician becomes much more like the captain of the ship that uses prevailing winds to navigate toward calmer waters.

This particular example is not meant to be definitive or valid in any sense. It is just one of a universe of potential models (or hypotheses) about the way psychiatric morbidity is generated. (Discussion will return to procedures for comparing models in the next section.) This example does allow us, however, to make a few key points about the nature of pathology and its expression. First, in any generative model of psychopathology there is a fundamental distinction between (time-invariant) parameters and (time-sensitive) states. This distinction can be regarded as the formal homologue of the distinction between *trait* and *state* abnormalities. For example, the patient illustrated above had a particular set of parameters θ_i^p determining the family of trajectories (and their attracting sets) of pathophysiology. Simply knowing these parameters, however, does not tell us anything about the pathological state of the patient at a particular time. To determine this, one needs to infer the latent pathophysiology in terms of the current state $x(t)$ using model fitting or inversion. This presents a difficult (but solvable) problem, because we have to estimate both the parameters (traits) and states of a patient to determine their trajectory in the short term.

The second distinction this sort of model brings to the table is the distinction between parameters that are patient specific and those conserved over the population to which the model applies. In statistical terms, this corresponds to the difference between *random* and *fixed* effects, where patient-specific effects model random variations in traits that may reflect predisposing factors (e.g., genetic predisposition). Conversely, other parameters may be fixed over patients and determine the canonical form of nosology.

In the example above, this distinction is illustrated by the difference between parameters that are specific to each patient or pathophysiology θ_i^p and those that are inherent in the nosology θ_i^n. The nosological parameters define a generic mapping from pathophysiology to psychopathology that is conserved over patients. Understanding this distinction is important practically, because nosological parameters can only be estimated from group data. Examples of this are given in the next section.

Model Inversion and Selection

Here let us consider the inversion and selection of generative models based on measurable outcomes. The ultimate aim of modeling is to predict outcomes for a particular patient. The quality of these predictions rests upon a model that is both accurate and generalizes to the sorts of patients encountered. The quality of a model is scored in terms of its evidence, given some data. However, to evaluate model evidence, one needs to be able to invert or fit data. This means that we first have to ensure that models can be inverted. In other words, can we recover the unknown parameters and latent variables responsible for clinical data? In what follows, the simulated patient above is used to see whether the latent states can be recovered, given the (known) therapeutic input and clinical outcomes (symptoms and diagnosis). Thereafter I will briefly review Bayesian model comparison and discuss its crucial role in hypothesis testing and elaborating a more mechanistic nosology for psychiatry in the future.

Model Inversion and Bayesian Filtering

The problem of estimating unknown parameters and latent states from time series data is known as deconvolution or filtering in the modeling literature. Because we have to estimate both parameters and states, this presents a *dual estimation problem* that is usually accommodated by treating parameters as very slowly fluctuating states. I will illustrate Bayesian filtering using an established procedure called *dynamic expectation maximization* (DEM). DEM was originally devised to infer latent neuronal states and the connectivity parameters generating neurophysiological signals in distributed brain networks and has been applied in a number of different contexts (Friston et al. 2008). Special cases of DEM include Kalman filtering (when the states are known and the state-space model is linear).

Figure 11.2 shows the results of Bayesian filtering when applied to the symptom and diagnostic time series shown in the previous figure. The format is similar to the middle panel of Figure 11.1; however, here, the colored lines correspond not to the true values generating data but to the estimated

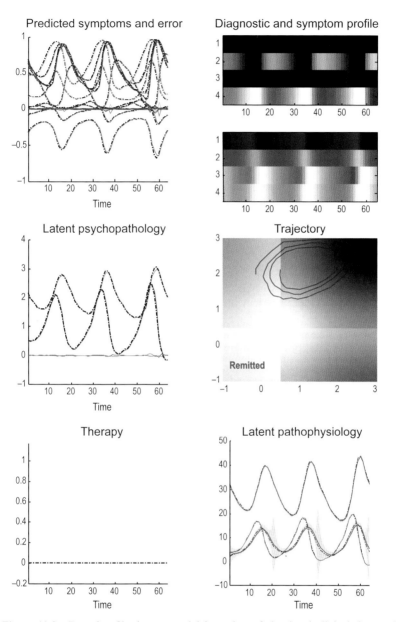

Figure 11.2 Bayesian filtering or model inversion of simulated clinical time series data: this figure reports the results of model inversion using the data generated by the model described in Figure 11.1. In this instance, the clinical data were simulated, without any therapeutic intervention, over 64 weeks and show the results using a similar format to the middle panels in Figure 11.1. Here, the solid lines represent predictions, while the broken black lines are the true values and actual outcomes being predicted. The gray areas correspond to 90% confidence intervals on hidden states. (*continued on next page*)

Figure 11.2 (continued)   The colored lines correspond to clinical outcomes (upper left panel) and latent psychopathology (middle left panel). The symptom scores and differential diagnosis are shown as functions of time (upper left) and in image format (upper right). The top image shows the changes in differential diagnosis, with the diagnosis of remission in the first row; the lower panel shows the fluctuations in the first four symptom scores. Note that, in the absence of treatment, the chaotic fluctuations between schizophrenia and schizoaffective regimes of latent psychopathology slowly increase in amplitude.

trajectories based upon Bayesian filtering (as implemented with the Matlab routine *spm_DEM.m*). In this example, I simulated clinical progression in the absence of any therapeutic intervention (as shown by the flat line in the lower left panel). In the absence of any check on pathophysiology, chaotic oscillations of slowly increasing amplitude emerge over a period of 64 weeks. These fluctuations are shown in terms of a trajectory in the state-space of psychopathology (lower right panel) and as functions of time (middle row). Here, the solid lines correspond to posterior expectations (the most likely trajectories) that are contained within 90% Bayesian confidence intervals (gray areas). The true values are shown as dotted black lines. In this example, the true and estimated values were almost identical. This is because very low levels of random fluctuations were used—with log precisions of twelve, eight, and four—to control the amplitude of random effects at the level of outcomes, psychopathology, and pathophysiology, respectively (see Figure 11.1). Precision is the inverse variance or amplitude.

The upper panels show the resulting fluctuations in symptom and diagnostic scores as a function of time in graphical format (upper left panel) and in image format (upper right panel). One can see clearly that the differential diagnosis of schizophrenia and schizoaffective disorder vacillate every few months, reflecting an unstable and ambiguous diagnostic picture. Filtering was then repeated but with a therapeutic intervention at 32 weeks. The simulated response and inferred latent states are shown in Figure 11.3. These reproduce the results of Figure 11.1 and show the success of the intervention—as indicated by the emergence of a remitted diagnosis as time progresses (solid blue line on the upper left and cyan circle on the upper right).

In these illustrations, I estimated both the unknown states generating (simulated) clinical data and the patient-specific (trait) parameters governing pathophysiological dynamics. The estimated and true parameters are shown in the upper left panel of Figure 11.4: estimated values are shown as gray bars, true values in black, and white bars show 90% confidence intervals. The accuracy of these estimates is self-evident, with a slight overconfidence that is characteristic of approximate Bayesian inference implicit in dynamic expectation maximization (MacKay 1995). Although these estimates show that, in principle, one can recover the traits and states of a particular subject at a particular time, I used the true values of the nosological parameters coupling pathophysiology to

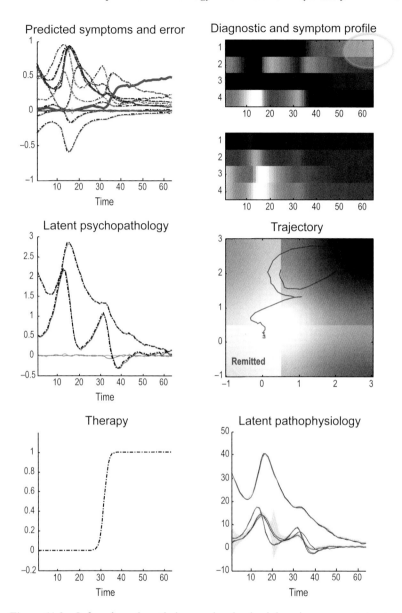

Figure 11.3   Inferred psychopathology and pathophysiology in response to treatment. This figure uses the same format as Figure 11.2. The only difference here is that a therapeutic intervention was introduced that destroyed the chaotic attractor, replacing it with a point attractor in the remitted regime of latent psychopathology. As a consequence, the pathophysiological variables flow toward zero and the symptom scores normalize. At the same time, the most probable diagnosis becomes one of remission (solid blue line in the upper left panel; see also the cyan circle in the corresponding predictions in image format).

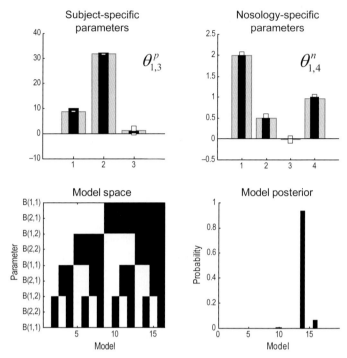

Figure 11.4 Bayesian model identification and comparison: this figure summarizes the results of model identification (parameter estimation) and model selection using Bayesian model reduction. Upper panels show the posterior estimates of subject-specific (left) and nosological (right) parameters based on single subject time series and (eight subject) group data, respectively. Gray bars correspond to the posterior mean; white bars report 90% Bayesian confidence intervals. These are superimposed on black bars that correspond to the true values used to simulate the clinical data. The lower left panel shows the combinations of nosological parameters that define 16 competing models that were compared using Bayesian model reduction. This comparison entails evaluating the evidence for each model; namely the probability of the data under each model, having marginalized over unknown parameters and states. Model evidence is also known as the marginal or integrated likelihood. Under uninformative or flat priors over models, this also corresponds to the model posterior. The posterior probability over 16 models for the group data is shown on the lower right, suggesting that a model that precludes coupling between the second physiological state and first psychopathological state (mediated by the third nosological parameter) have greater evidence than all other models.

psychopathology (and generating clinical outcomes) during model inversion. One might ask: Can these parameters also be estimated?

To illustrate this estimation, the upper right panel of Figure 11.4 reports the estimates (and 90% confidence intervals) based on any *empirical Bayesian* analysis (Kass and Steffey 1989) of eight simulated patients (using *spm_dcm_peb.m*). Again, the estimates are remarkably accurate, suggesting that, in principle, it is possible to recover parameters that are conserved over subjects.

Empirical Bayesian analysis of dynamic causal modeling estimators refers to the hierarchical modeling of within- and between-subject effects which may or may not be treated as random effects. Note that the third nosological parameter has an estimated value of zero. This is important because it leads us into the realm of Bayesian model comparison and evidence-based hypothesis testing.

Bayesian Model Comparison and Hypothesis Testing

Above, we saw that this sort of model can, in principle, be inverted such that underlying (latent) psychopathological and pathophysiological states can be inferred, in the context of (unknown) subject-specific parameters or traits. However, this does not mean that the model itself has any validity or will generalize to real clinical data. In other words, how do we know whether we have a good model?

This is a question of model comparison. In short, the best model provides an accurate explanation for the data with the minimum complexity. The model evidence reflects this, because model evidence is equal to accuracy minus complexity. The complexity term is important and ensures that models do not overfit data, and will thus generalize to new data. The model evidence is evaluated by marginalizing (averaging over) unknown parameters and states to provide the probability of some data, under a particular model. The model here is defined in terms of the number of states (and parameters) and how they depend upon each other. A simple example of model comparison is provided in Figure 11.4 (lower panels).

In our model, there are a number of ways in which the physiological states could influence psychopathology. There are two physiological states that can influence two pathophysiological states, creating four possible dependencies that may or may not exist. This leads to $16 = 2^4$ models which cover all combinations of nosological parameters (see the lower left panel of Figure 11.4). We can evaluate the evidence for each of these 16 models by inverting all 16 and evaluating the evidence or, as illustrated here, inverting the model with all four parameters in place and computing the evidence of all reduced models, with one or more parameters missing. This is known as *Bayesian model reduction*, which is an efficient way of performing Bayesian model comparison (Friston and Penny 2011). The results of this model comparison are shown in the lower right panel of Figure 11.4 and suggest that the posterior probability of model 15 is much greater than any of the others. In this model, the influence of the second pathophysiological state on the first psychopathological state has been removed. Removing this parameter reduces model complexity without any substantial loss in accuracy and therefore increases model evidence. We might have guessed that this was the case by inspecting the posterior density of the third nosological parameter mediating this model component (see the upper right panel).

This is a rather trivial example of model comparison but illustrates an important aspect of dynamic causal modeling; namely, the ability to test and compare different models or hypotheses. Although not illustrated here, one can imagine comparing models with different numbers of pathophysiological states and different forms of dynamics. One could even imagine comparing models with a different graphical structure. One interesting example here would be the modeling of psychotherapeutic interventions that might influence pathophysiology through experience-dependent plasticity. This would necessitate comparing models in which therapeutic intervention influenced psychopathology, which couples back to pathophysiology, through the parameters of its dynamics. This is illustrated by the dotted arrows in Figure 11.1.

Many other examples lend themselves to speculation: crucial examples involve an increasingly mechanistic interpretation of pathophysiology, in which pathophysiological states could be mapped onto neurotransmitter systems through careful (generative) modeling of electrophysiological and psychophysical measurements (Stephan and Mathys 2014). One could also contemplate comparing models with different sorts of inputs or causes, ranging from social or environmental perturbations (e.g., traumatic events) to genetic factors (or their proxies like family history). Questions about whether and where genetic polymorphisms affect pathophysiology are formalized by simply comparing different generative models that accommodate effects on different states or parameters. For example, do models that include genetic biases on physiological parameters have greater evidence than models that do not?

The potential importance of model comparison should not be underestimated. Here I have tried to give a flavor of its potential. It is also worth noting that this field is an area of active research; fast and improved schemes for scoring large model spaces are continually being developed (e.g., Viceconti et al. 2015). One can construe an exploration of model space as a greedy search over competing hypotheses and a formal statement of the scientific process. This may be especially relevant for psychiatry, which addresses the specific problem of integrating both physiological and psychological therapies, and points to the need for generative models that map between these two levels of description. In the final section, let us turn to the more pragmatic issue of predicting response to treatment for an individual patient.

Prediction and Personalized Psychiatry

Let us assume that we used Bayesian model comparison to optimize our generative model of psychosis and prior probability distributions over its parameters. Can we now use the model to predict the outcome of a particular intervention in a given patient? In the previous section, we saw how clinical data from a single subject could be used to estimate subject-specific parameters (traits) and states at a particular time. In fact, the parameter estimates in Figure

11.4 were based on the first 32 weeks of data *before any treatment began*. This means that we now have estimates of how this particular subject would respond from any physiological state and the physiological state at the end of the period of assessment. Given these (posterior probability) estimates, there are several ways in which we can predict clinical prognosis and responses to different treatments. The simplest way would be to sample from the posterior distribution and integrate the generative model with random fluctuations to build a probability distribution over future states. We will take a related but simpler approach and apply Bayesian filtering to null data with zero precision; in other words, data that has yet to be acquired. This finishes a predictive distribution over future trajectories based on the posterior estimates of the subject's current parameters and states.

Figure 11.5 shows the results of this predictive filtering using the same format as Figure 11.3. However, there are two crucial differences between Figures 11.5 and 11.3. First, we are starting from latent states that are posterior estimates of the subject's current state and, more importantly, the trajectories are pure predictions based upon pathophysiological dynamics. One can see that the predicted response to treatment (at 16 weeks) has a similar outcome to the actual treatment (although the trajectories are not exactly the same, when comparing the predicted and actual outcomes in Figures 11.3 and 11.5, respectively). Figure 11.6 shows the same predictions in the absence of treatment, again showing the same pattern of fluctuation between schizophrenia and schizoaffective diagnosis encountered in Figure 11.2.

The right panel of Figure 11.6 also includes trajectories with increasing levels of therapeutic intervention (ranging from 0 to 2). The final outcomes of these interventions are summarized on the lower left in terms of the probability of a diagnosis of remission at 48 weeks. This illustrates the potential for predictive modeling of this sort to provide dose-response relationships and explore different therapeutic interventions (and combinations of interventions). In this simple example, there is a small probability that the patient would remit without treatment, which dips and then recovers to levels of around 50% with increasing levels of therapy. The apparent spontaneous recovery would not, however, be long lasting, as can be imputed from the chaotic oscillations in Figure 11.1.

Conclusion

In this chapter, I have illustrated what a computational nosology could look like using simulations of clinical trajectories, under a canonical generative model. The potential of this approach to nosological constructs can be motivated from a number of perspectives. First, it resolves the dialectic between categorical diagnostic constructs (e.g., DSM) and those based on latent dimensions of psychopathology or pathophysiology (e.g., RDoC). Both constructs play an

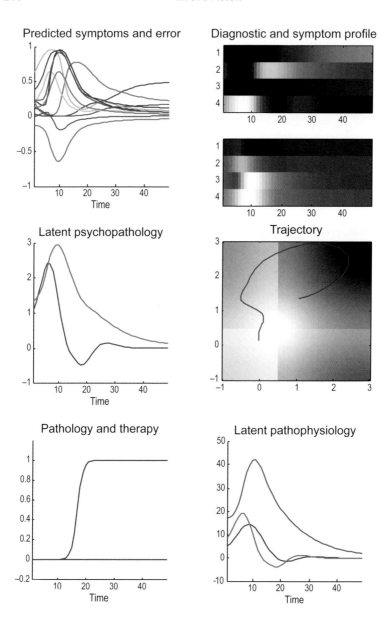

Figure 11.5 Predicted responses to treatment. This figure uses the same format as Figure 11.3; however, here, the results are purely predictive in nature. In other words, the predictions are driven entirely by pathophysiological dynamics based on subject-specific estimates of the model parameters and starting from the state last estimated on the basis of an assessment prior to therapy.

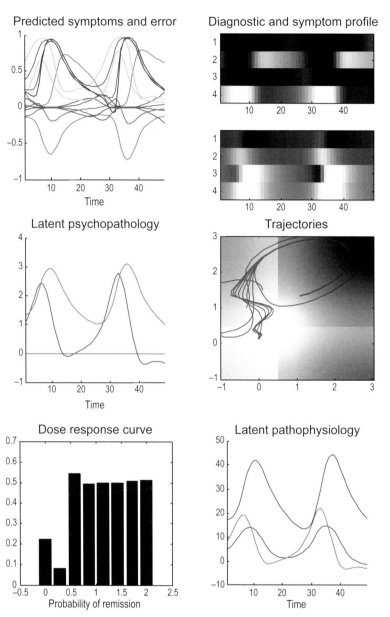

Figure 11.6 Predicted responses to different treatments. This figure uses the same format as Figure 11.5; however, here, it shows the results in the absence of any treatment. The trajectories in the middle right panel report eight simulations over increasing levels of therapeutic intervention. The endpoints of these trajectories are summarized on the lower left in terms of the probability of receiving a diagnosis of remission. This can be regarded as a predicted dose-response curve, illustrating the potential of the model to optimize treatment strategies.

essential role in a generative modeling framework, as diagnostic outcomes and latent causes, respectively. To harness their complementary strengths, it is only necessary to determine how one follows from the other, which is an inherent objective of model inversion and selection. Second, I have tried to emphasize the potential for an evidence-based approach to nosology that can operationalize mechanistic hypotheses in terms of Bayesian model comparison. This provides an integration of basic research and clinical studies which could, in principle, contribute synergistically to an evidenced-based nosology. Finally, I have illustrated the practical utility of using the predictions of (optimized) generative models for individualized or precision psychiatry (Chekroud and Krystal 2015), in terms of providing probabilistic predictions of responses to therapy.

Although I have emphasized the provisional nature of this approach, it should be acknowledged that one could analyze existing clinical data using the model described in this chapter with existing algorithms. Indeed, there are hundreds of publications using dynamic causal modeling to infer the functional coupling among hidden neuronal states in the neuroimaging literature. In other words, it would be relatively simple to apply the techniques described above to existing data at the present time. The real challenge, however, lies in searching the vast model space to find models that are sufficiently comprehensive yet parsimonious to account for the diverse range of clinical measures—in a way that generalizes from patient to patient. This challenge is not necessarily insurmountable: one might argue that if we invested the same informatics resources in psychiatry as has been invested in weather forecasting and geophysical modeling, then considerable progress could be made. Ultimately, one could imagine model-based psychiatric prognosis being received with the same confidence with which we currently accept daily weather forecasts. There are, of course, differences between psychiatric and meteorological forecasting. For example, the latter must handle the "big data problem" with a relatively small model space. Conversely, psychiatry may have to contend with a "big theory problem," with a relatively large model space but more manageable data sets.

Perhaps the more important contribution of a formal nosology is not in the pragmatic application to precision medicine (i.e., through the introduction of prognostic apps for clinicians), but in the use of Bayesian model selection to test increasingly mechanistic hypotheses and pursue a deeper understanding of pathogenesis in psychiatry. This is the way in which dynamic causal modeling has been applied in computational neuroscience and, as such, is just a formal operationalization of the scientific process.

Acknowledgments

KJF is funded by the Wellcome Trust. The author declares no conflicts of interest.

Exemplars

12

Candidate Examples for a Computational Approach to Address Practical Problems in Psychiatry

Rosalyn Moran, Klaas Enno Stephan, Matthew Botvinick,
Michael Breakspear, Cameron S. Carter, Peter W. Kalivas,
P. Read Montague, Martin P. Paulus, and Frederike Petzschner

Abstract

Scientists and clinicians can utilize a model-based framework to develop computational approaches to psychiatric practice and bring scientific discoveries to a clinical interface. This chapter describes a general modeling perspective, which complements those derived in previous chapters, and provides distinct examples to highlight the scientific and preclinical research that can evolve out of a computational framework to offer new tools for clinical practice. It begins by reviewing areas of theoretical and modeling studies that have reached a critical mass and outlines the pathophysiological insights that have been revealed. Three particular models are used to demonstrate how clinical questions, relating to understanding disease mechanisms and predicting treatment response, could be potentially addressed using an integrated computational framework. First, the phasic dopamine temporal difference model shows how neurophysiological and neuroanatomical research, incorporated into a learning circuit model, provides a constrained hypothesis testing framework, related to the likely multiple mechanisms contributing to addiction. Second, a potential application of generative models of neuroimaging measurements (dynamic causal models of EEG data) is described to predict individual treatment responses in patients with schizophrenia. The third example offers a novel approach to quantifying patient outcomes under a "recovery model" of psychiatric illness. This involves a dynamical system appraisal of allostasis, using the amygdala-HPA axis with its role in anxiety disorders and depression as a clinical target syndrome to which the model could be applied. In conclusion, consideration is given to the community efforts needed to support the validation of these and future applications.

Introduction

The promise of computational approaches to psychiatric clinical practice is evidenced by the breadth and scope of developments from computational and systems neuroscience that are targeted directly at understanding the etiology (Winterer and Weinberger 2004), pathogenesis (Kheirbek et al. 2012), and clinical course (Huys et al. 2015a) of psychiatric illnesses. We outline where these scientific points of contact are concentrated and how their development could further evolve into pragmatic tools for the practicing psychiatrist. The primary motivation for this endeavor is the lack of diagnostic technologies that could be used to probe whether the symptoms of a particular patient are more likely to be explained by one putative pathophysiological process or another. In other words, unlike cardiologists, for example, who are equipped with a mechanistic understanding of how the heart works and have access to a broad arsenal of tools for differential diagnosis, psychiatrists face a multitude of competing pathophysiological theories and lack the technology to disambiguate alternative disease mechanisms in individual patients. So far, the translation of the current corpus of neuroscience into the clinic has had limited penetration and practical value (Kapur et al. 2012; Millan et al. 2012). A fundamental goal of computational psychiatry is thus to develop a framework that situates the algorithmic properties of the brain (its information-processing capacity and the supporting neural substrates) as the basic scientific level of inquiry (Maia and Frank 2011) and to provide tools for inferring individual disease mechanisms and predicting individual clinical trajectories and treatment response (Stephan and Mathys 2014).

In this chapter, we consider a general computational approach that could be applied to produce interventions and characterizations across the history of a patient's disease (Figure 12.1). In much the same way as Flagel et al. (this volume) developed a computational model to formalize current psychiatric nosology and putative disease trajectories, our framework similarly employs models populated by discoveries from basic science for use by clinicians and patients to monitor disease risk, progression, and recovery. Specifically, we propose that unobservable biological parameters (B) affect unobservable computational parameters (C), which can be estimated using observable behavioral symptoms and signs (S), biological measurements (M), and diagnosis (D). Our model allows for treatment response prediction by incorporating

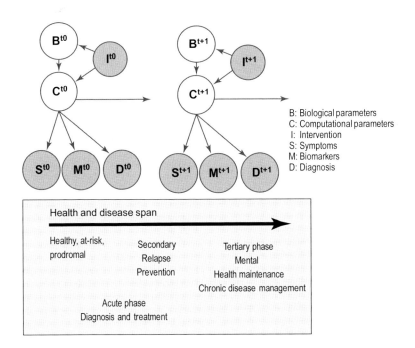

B: Biological parameters
C: Computational parameters
I: Intervention
S: Symptoms
M: Biomarkers
D: Diagnosis

Figure 12.1 General model-based framework made up of a generative model to enable inference on the biological and computational causes of illness, and a formulation where therapeutic interventions affect model parameters. Model outputs are symptoms, biological measurements such as imaging data, and diagnosis. In principle, given interventions, symptoms, measurements, and diagnosis, a model inversion could be performed to infer disease causes, progression, and intervention effects as well as their trajectories over time.

intervention states (I) that can influence the underlying biological or computational variables. We envisage that this framework offers both a platform for testing putative computational and biological substrates of disease as well as a prototypical system that could eventually bridge basic science to a clinical end user, and present concrete examples of where this type of modeling platform could prove useful.

Different instantiations of this general model should be aimed at different stages of medical interventions. For example, when considering the challenge of primary prevention in healthy, at-risk, and prodromal stages, circuit and behavioral models of normal function are essential. From here, consideration of the acute disease phase necessitates developments in diagnostic and treatment prediction models, whereas the stage of secondary prevention is more concerned with relapse and recurrence susceptibility, prognostic classification, and monitoring. Finally, tertiary prevention during an emergent chronic phase mandates recovery-based approaches aimed at assisting a patient's ability to function in daily life (e.g., learning to live with negative symptoms of schizophrenia).

Current developments span all of these stages with focused advances clustering around diagnosis and early detection. These have arisen from over a decade of work in computational neuroimaging (for review, see Stephan et al. 2015), which has produced normative descriptions of computational neural substrates in healthy populations, their developmental, and aging trajectories (Eppinger et al. 2013; Plichta and Scheres 2014; Thomas et al. 2014); the identification of unique neuronal correlates of high risk and prodromal states (Breakspear et al. 2015); and the identification of deviations from health in a range of psychiatric conditions (Frank et al. 2011; Morris et al. 2012; Robinson et al. 2012). In particular, algorithmic approaches to midbrain-striatal interactions have been used to uncover aberrancies in latent valuation and reward processing. These developments have been buttressed by biological circuit models that capture the downstream effects of these signals, helping to understand later, chronic disease-stage processes, such as the psychiatric symptoms of impulse control disorders that emerge following long-term dopaminergic treatment in Parkinson disease (Voon et al. 2011).

Ideally, the field of computational psychiatry will provide a broad armamentarium: from computational stress tests for teens at risk of schizophrenia, to a computational model that predicts pharmaceutical response in mania. In pursuit of these long-term goals, we develop specific examples of the general framework presented in Figure 12.1, to illustrate the "value added" by our burgeoning field. Overall, our objective pertains to inducing a conceptual shift in the qualification of psychiatric illness, placing computational formalism as a developing language designed to illuminate the link between the psychological and behavioral symptoms of psychiatric disorders and their neurobiological underpinnings. Our examples are designed to incorporate the translational potential of a shared mathematical brain language, whereby animal models of psychiatric disease can be compared and adjudicated in light of commonalities between theoretical and model-based constructs.

We begin by reviewing the state of the art and highlight those domains where a critical mass of knowledge has accrued to enable new scientific hypothesis testing in psychiatric illness as well as implementation and potential clinical take-up in the near term. We then proceed to develop focused prototypes based on our general model. These include computational approaches that have led to new and testable hypotheses in addiction research (the dopamine "fruit fly"), a computational model for predicting treatment responses in schizophrenia, based on a dynamic causal model (DCM) of neural circuit dynamics, and a model of allostatic regulation with relevance for chronic disease management. These candidate examples are diverse and generally intended to illustrate potential ways of moving computational models forward into the clinical application domain.

Cases of Computational and Theoretical
Neuroscience Offering Biological Insight

To define the space of models that may be fit for purpose, let us first consider the definition of "computation" or "information processing" as it is relevant for psychiatry. Marr's tri-level hypothesis (Marr 1982; Ullman and Poggio 2010) partitions information processing in the brain into computational goals, algorithmic solutions, and implementational/physical levels. (As discussed by Kurth-Nelson et al., this volume, this nomenclature may not be ideal, and an alternative would be to refer to levels of purpose, computation, and implementation.) While computational psychiatry may fit most naturally at the intermediate algorithmic level (Montague et al. 2012), the field has developed biophysical instantiations of circuits that can also link psychiatric dysfunction to neurobiological substrates directly (Cooray et al. 2015). Thus for our toolkit, we consider both algorithmic (i.e., information-processing models) *and* implementation levels (i.e., biophysical models) as two broad categories that link computation and pathophysiology (for an overview, see Kurth-Nelson et al., this volume). Here we focus on examples that have provided direct insights into pathophysiology.

Biophysical Models

Biophysical models describe the dynamic activity of neurons, neuronal circuits, and large neuronal ensembles typically using either conductance-based models based upon simplifications of the Hodgkin-Huxley equations (Hodgkin and Huxley 1952; McCulloch and Pitts 1943; Morris and Lecar 1981) or current-based neural mass models (Freeman 1975). Their utility lies primarily in characterizing or identifying (through model inversion) the biological substrates of information transmission, ion channels, synaptic weights, transmitter levels, etc. These models are typically agnostic to the type of information processing (transformation of cognitive variables) that emerges from their activity. Exceptions do exist, however, particularly in circuitry where cognitive variables have been well investigated (e.g., in the direct and indirect pathways of the basal ganglia; Frank et al. 2004). Thus, these models serve primarily to provide causal explanations of observed neurophysiological data (Friston et al. 2003).

At a microscopic level, models of synaptic dynamics consider the subcellular milieu in which information is communicated (Jaeger and Bower 1999). These efforts are important to identify molecular targets for pharmacological interventions and, more importantly, can accommodate a detailed understanding of the effectors (Luscher et al. 2000) and dynamic function (Rubinov et al. 2009) of synaptic plasticity. With relation to psychiatric pathophysiology, models at the synaptic level have provided new insights into maladaptive plasticity in neuronal circuits. For example, components of glutamatergic homeostasis have been used to explain a crucial nexus of circuit dysfunction in addiction (Kalivas et al. 2005). This has provided important translational insights in the context of new treatments for addiction. In this approach (Pendyam et al. 2009), the model was initially parameterized, based on a collation of quantities representative of anatomical and cellular physiological data acquired over years of experimental rodent work. Thereafter new disease data features were simulated by investigating the parameter space with the intention of identifying a limited set of parameters that were altered by chronic cocaine administration. The data for which the model was optimized was made up of measured microdialysis levels of extracellular glutamate. Critically, using a biophysical model of cocaine-induced cellular changes, neuronal plasticity, and larger network effects, it was possible to elucidate the influence of prefrontal inputs on the nucleus accumbens (Kalivas 2009). This example demonstrates a model-based identification of a putative target for disease treatment.

Above synaptic-level dynamics, single neuron models can be used to describe cellular input–output transformations whereas neural network models simplify the cellular processes and employ, for example, integrate and fire dynamics (Rudolph and Destexhe 2006) to represent a cell in an ensemble of connected neurons. These models can be used to describe the pathology of network connections, for instance, along the perforant pathway (from the entorhinal cortex to hippocampus), where they have been used to simulate the effects of NMDA receptor antagonism on memory impairment in schizophrenia (Siekmeier et al. 2007). At a scale above these models lie neural mass and mean field models (Deco et al. 2008). Mean field approaches engage the statistical properties (typically first- and second-order moments) to model the evolution of neuronal ensembles probabilistically. These meso- and macroscale dynamics are governed by the interaction of their statistical quantities (e.g., the mean and variance of membrane depolarizations within a cortical macrocolumn; Marreiros et al. 2009) and can be described by Fokker–Planck or path integral formulations (Knight et al. 2000). In terms of insights to pathophysiological processes, these population equations have most widely served as models for understanding seizure activity (Breakspear et al. 2006; Jirsa et al. 2014). More recently they have been proposed as component models in large connectomic analyses of neuropsychiatric disorders with the intention of developing a field of "pathoconnectomics" (Deco and Kringelbach 2014; see also (Horga et al. 2015). This class of model is also used in *dynamic causal modeling* and has been applied to a range of psychiatric disorders to test network hypotheses. For example, in the study of schizophrenia, DCMs of both fMRI and EEG data were used to study hierarchical brain connectivity associated with visual illusions, revealing a reduction of top-down effects on visual processing (Dima et al. 2009, 2010), effects mirrored by rodent electrophysiological DCMs in a ketamine model of psychosis (Moran et al. 2015). Below we highlight the potential of the DCM approach for deriving treatment predictions, in the specific context of schizophrenia.

Information-Processing Models

In contrast to biophysical models, which are designed primarily to answer the question of *how* the brain performs a particular operation, information-processing models ask *what* it is that the brain—its neural circuits, cells, and molecules—is computing. In other words, information-processing models are designed to uncover the neuronal computations that drive behavior and expose latent states, which can be used to characterize an individual patient's traits, such as how emotion affects valuation of immediate relative to delayed reward (Lempert et al. 2015). A more complete characterization of the distinction and overlap between biophysical and information-processing models is given by Kurth-Nelson et al. (this volume). Here we offer examples of where information-processing models have informed pathologies in the brain's algorithmic performance.

One area where computational modeling has already had a profound and sustained impact on understanding clinical phenomena is dementia. The syndrome of semantic dementia (SD), which is associated with a subset of neurodegenerative disorders as well as herpes encephalitis, involves a disruption of conceptual knowledge, including both knowledge concerning specific facts (e.g., Paris is the capital of France) and richer patterns of associative knowledge and inference (e.g., penguins are birds, consistent with their having wings and beaks, but also have the atypical characteristic in that they do not fly). SD patients show impairments in tasks which tap into such knowledge and yet other forms of memory, including episodic and working memory, are relatively spared. In a series of studies beginning in the 1990s, Timothy Rogers, Jay McClelland, and colleagues developed a computational account of SD, leveraging the tool of neural network modeling (Rogers et al. 1999; Rogers and McClelland 2008). Neural network models (also referred to as connectionist or deep learning models), are closely aligned with biophysical models and involve simple neuron-like units, which carry activation levels analogous to neuronal spike rates and connect to one another through idealized excitatory and inhibitory synapses. A key aspect of neural networks is that they are associated with well-developed learning algorithms which permit the strengths of the synapse-like connections in a network to be adjusted to allow the network to perform target tasks, producing desired output patterns in response to particular inputs (McClelland et al. 2010).

McClelland and Rogers (2003) have modeled semantic knowledge as involving associative relations among object properties. For example, a network might be trained to map from inputs representing *robin* and the relation *has* to the outputs *wings, beak,* and *feathers,* and from the inputs *goldfish has* to *fins* and *gills.* Following such training, if a new item *bluejay* is introduced and the network is trained to respond to *bluejay* and *has* with *beak*, the network is likely to infer that *bluejays* also have wings and feathers. Beyond reproducing such intuitive patterns of learning and inference, the Rogers–McClelland model captures detailed patterns of behavioral data related to human category knowledge and conceptual development. More germane to the present topic, however, the model reproduces and explains detailed aspects of task performance in SD. When synaptic connections in the network are weakened or removed, simulating the effect of disease, the network model displays a degradation of conceptual and category knowledge that parallels the pattern of progressive memory loss seen in SD. Furthermore, detailed analysis of the conditions under which such impairments arise in the model have given rise to novel ideas, subsequently validated, about the anatomical locus of the critical lesion in SD—pointing to the importance of highly convergent multimodal inputs into a "hub" region, which leads to a resulting focus on the temporal pole as a candidate region (Hoffman and Ralph 2011; Irish et al. 2014).

A second class of information-processing models are prediction error-based models which have served as a basis for theoretic approaches to understanding learning, inference, and decision making (Botvinick et al. 2009; Friston 2009). Starting from Pavlov's conditioned stimulus reflexes, these models have evolved to where predicted future rewards are used to estimate the value of states and actions (Montague, this volume). Temporal difference reinforcement-learning models use errors in expected future reward to update expectations (Sutton and Barto 1998). The now iconic correlate of phasic activity in VTA dopamine firing levels with temporal difference updates (Schultz et al. 1997) offered a paradigm shift in terms of the discovery of formal equivalencies between computation and neuronal activity. Moreover, the discovery highlights a key point in computational approaches to understanding (patho)physiological mechanisms; namely, raw data can be difficult to understand without models. Pathophysiological consequences of dysregulated dopaminergic reward prediction errors have been used to explain behavioral observations of aberrant learning patterns in dopamine-associated disorders, for instance when medicated Parkinsonian patients exhibit an impairment in learning from negative predictions in the presence of high tonic levels of striatal dopamine (Frank 2006). Using model-based fMRI (O'Doherty et al. 2007), pharmacological studies, genetic associations, and PET studies, different positive and negative learning signals have been linked to D1 and D2 binding, respectively (e.g., Cox et al. 2015). This dopamine-reliant signaling has also been used to identify striatal dysfunction in human neuropsychiatric conditions, including attention-deficit/hyperactivity disorder, substance abuse, and schizophrenia (Whitton et al. 2015). Expansions of these habitual reinforcement-learning models have been used to unravel the role of serotonergic dysregulation in depression, where Markov chain-transition probabilities become value dependent (Dayan and Huys 2008), thereafter permitting a reconciliation of complementary serotoninergic processes in depression; that is, where its role in predicting aversive events (Paulus and Angela 2012) can lead to a bias toward optimistic prediction and, through altering stopping policy and pruning, of action options. These findings inform rumination, low mood, and perseverative thinking which pervade subjective descriptions of depression

(Dayan and Huys 2015). These approaches are closely linked to accounts where an agent comprises a dyadic structure of models representing both value and the causal structure of the environment. This allows for further goals and neural systems to be examined by our models and can be imbued with temporal hierarchical structure (see Botvinick and Weinstein 2014). Indeed, this "promiscuity of models" may be a useful metaphor for brain function and is expanded below in the dopamine "fruit fly," where we sketch a role for "model-based" versus "model-free" brain circuits in mediating addiction.

Bayesian inference has been used formally to instantiate neuronal codes such as predictive coding under the free energy principle (Friston et al. 2006). This account posits a particular neuronal machinery that is designed to perform probabilistic reasoning, whereby the brain models, learns, infers, and acts on its world so as to minimize precision-weighted prediction errors. The neurobiological circuits required for this type of predictive coding have been shown to recapitulate key anatomical features of canonical cortical microcircuits (Bastos et al. 2012). This account is appealing to computational psychiatry as it posits prediction error updating processes throughout cortex. Indeed, precision-weighted prediction errors have been found to be reflected by fMRI signals all over the brain, even for simple sensory tasks (Iglesias et al. 2013). Moreover, predictive coding makes testable predictions about neurobiological substrates of belief updating in cortex, based on precision-weighted prediction errors. For example, glutamatergic top-down connections are supposed to mediate predictions via NMDA receptors, prediction errors are believed to be signaled via both AMPA and NMDA receptors at bottom-up connections, and their precision weighting is thought to depend on postsynaptic gain control through neuromodulatory transmitters and local GABAergic mechanisms (Corlett et al. 2011; Adams et al. 2013). This framework has been applied recently to outline a "computational anatomy of psychosis" (Adams et al. 2013) and offers testable substrates of the misheld belief structures that pervade psychiatric symptomatology.

The Value of Generative Models for Building
Clinical Application Prototypes

Developments in biophysical and information-processing models, together with advances in molecular, cellular, and systems neurobiology, are building the foundations of a basic science of computational psychiatry. They proffer deep mechanistic insights into mind–brain relationships, which might apply directly to psychiatric clinical practice, and offer an avenue to amalgamate detailed neurobiological accounts into clinically relevant process models. In other words, the models accommodate an important translational aim whereby they harness and apply findings from animal models of psychiatric illness to build better, more detailed descriptions of algorithmic and circuit breakdowns. To produce methodologies for improving patient care, prototypical examples of the modeling framework should be prefaced by a simple question: Will any of these models help diagnose, enable prognosis, tailor treatment, or prevent psychiatric illness? Independent of "big-data" analytics and disease predictions, where clinical predictions rest on black-box statistical relations among descriptive data features, the computational psychiatry approach seeks a mechanistic understanding of how treatment can be improved. While it may take longer to develop our form of model for clinical practice compared to black-box counterparts, the mechanistic approach allows for interpreting a successful prediction, in terms of the underlying biology and/or computation, and helps identify targets for new treatment approaches.

Our rationale is that data generated from a probabilistic model is most efficiently described by the parameters that generated it, in the sense that those parameters capture everything there is to say about the data that is not noise. For instance, if one were to generate noisy behavioral output from a simulation of a learning model, then the most succinct and theoretically optimal description of that data would be in terms of the parameters that were used to generate the data in the first place. This means that a good model which describes a data set well can be used to condense the data set optimally (cf. generative embedding, described below), resulting in measures that most efficiently remove the noise.

The examples we develop below are designed to illustrate:

1. *the uncovering of disease mechanisms*, where a learning model applied to dopamine signaling is used to inform testable hypotheses related to the causes of addiction;
2. *treatment prediction*, in which a computationally derived nosology may help in subcategorizations and tailored therapies for patients with schizophrenia; and
3. *a monitoring system for chronic disease management*, where the goal of reaching a patient's "new normal" is instantiated in a dynamic model applied to markers of brain activity.

Based on the overall framework of this generative probabilistic model of disease, we now present distinct examples tailored to specific clinical problems and highlight how different types of data may be used to further inform these models.

Testing Computationally Informed Theories of Addiction

Addiction has been termed "a pathology of motivation and choice" (Kalivas and Volkow 2014), with dopamine implicated in its pathophysiology given the direct effects of drugs of abuse such as cocaine in blocking dopamine transport and enhancing striatal dopamine levels. Model-based accounts of dopamine's role in learning have provided a detailed theory of how aberrant learning mechanisms could contribute to addiction. Here we examine how the temporal difference model of reinforcement learning (TDRL) (Schultz et al. 1997) offers precise computational counterparts (C) to specific neural substrates (B) (Figure 12.2). We examine how together these mechanisms inform particular aspects of addictive behaviors, specifically hypervaluation of drugs of abuse, while remaining relatively agnostic to other aspects of the illness, such as individual susceptibility to compulsive drug-seeking behavior. The fact that TDRL cannot computationally expose all of the signs, symptoms, and markers of addiction is viewed here as an opportunity for computational psychiatry and our modeling approach. In particular, it is clear that multiple neural systems, instantiating habitual, goal-based, and emotional control over behavior, play a role in this disease (Redish et al. 2008; Everitt and Robbins 2013). Thus in the future, our framework provides a formalism to consider, simulate, and test the precise interaction of these systems. Indeed, a recent review of potential decision-making vulnerabilities in addiction highlights ten system functions that could play a role in maladaptive choice. Rather than treating each vulnerability separately, the value added by a model such as that presented in Figure 12.1 lies in their joint consideration, accessible through the clarity of a mathematical description.

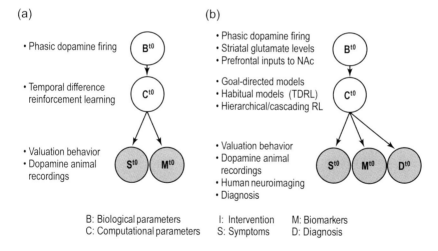

Figure 12.2 Developments in building a model for addiction. (a) The temporal difference model of reinforcement learning (TDRL) has formalized the role of phasic dopamine responses in evaluating states in the environment that predict reward and how the valuation process may be disrupted by pharmacologically enhanced dopamine. This model explains overvaluation in addiction and voltammetry findings in rodents. (b) Future accounts may use extended neurobiological and computational parameters to predict the full spectrum of symptoms associated with addiction and drug dependency.

The TDRL model (Sutton and Barto 1998) of phasic dopamine responses (Montague et al. 1996) goes beyond traditional reinforcement-learning and Pavlovian descriptions of stimulus-response associations, with two key components emerging from its particular mathematical formulation (Glimcher 2011). First, the formulation treats time as continuous discrete steps, removing the arbitrary notion of a "trial"; second, the formulation captures sequential associations, where an agent exposed to a series of conditioned stimuli, predicting the same unconditioned stimulus, can learn the history and redundancy of cues. These are both important constructs when considering addiction since real lives traverse through time not trials, and sequential operations may be bypassed by this system, given that reward predictions are transformed to the earliest patterns of predictive stimuli. Perhaps most importantly, the TDRL framework proffers a fundamental goal of a system instantiating its rules: it must carry zero prediction errors when a reward is encountered.

This mathematical goal has aided in explaining the apparent aberrancy in choices related to drug taking (Redish 2004), and in this context, addiction might be considered a valuation disease. In line with this model, prediction at synapses in the striatum is a potential neural substrate for instantiating the goal, which could be hijacked by an interacting pharmacological agent. Using a temporal difference model, the agent learns the value of each state in its world. Dopamine signals an error when things are better than expected, reduces its activity when outcomes are worse than expected, and stops modulating when rewards are predicted with zero error. However, imagine a case where reward is coupled with a pharmacologically magnified dopamine transient, via a drug: physiological assignment to predictive stimuli could reach ceiling levels and the agent would thus not reach its goal of zero error or learned state values. In other words, the values of the states which lead to drug receipt dwarf all other state trajectories, and maladaptive learning signals ensue. Based on this profound overvaluation, predictive environmental cues follow that enslave the agent to further drug-seeking behaviors.

This account has been used to explain several empirical features of addictive behavior and neurophysiology, including a decreasing elasticity to non-drug choice options in addicts over time, a resistance to blocking (learning associative redundancy) in animal models, and the concomitant dual dopamine signals observed in dopaminergic neurons in rodents (Redish 2004; Figure 12.2). As mentioned above, the model does not explain the whole range of phenomena encountered in addiction research, such as extinction (though the negative dip in firing is asymmetrically smaller than the phasic burst) and individual drug dependence. However, it stands as a canonical computational model system—a *computational fruit fly*—built directly on research which offered a model to explain neurophysiological firing patterns. Using this model as a basis, developments in formalizing additional decision systems—their parametric form and interactions (Servan-Schreiber et al. 1998)—will likely aid in building an expanded hypothesis set to uncover mechanisms of susceptibility, chronic self-administration, relapse, and prevention. The idea of addiction as a shift from goal-directed to habitual networks has already been developed in a neurobiological circuit framework that includes shifting involvement from ventral to dorsal striatum, with a crucial role for prefrontal regions in drug reinstatement (Kalivas 2004; Everitt and Robbins 2005). This neurobiological extension (Figure 12.2) has been allied by theoretical developments which treat the formal adjudication between interacting "model-based" and "model-free" networks (Gläscher et al. 2010; Wunderlich et al. 2012b). The model also forms a basis from which hypotheses related to temporal extensions of a reinforcement-learning effect though hierarchical representation (Botvinick et al. 2009) and cascading corticostrital circuits (Collins and Frank 2013) could be studied (Figure 12.2).

Developing the neurobiological substrates of addiction may also refine model unknowns, such as the sort of ceiling effects that could be reached in overvaluing states. One crucial nexus is the role of the prefrontal cortex and its glutamatergic inputs for plasticity settings in the striatum. In other words, developing the latent biological parameters of the model will coincide with developing the latent computational parameters (Figure 12.2). In rodents, important clues regarding glutamate's role in striatal dysfunction has been elucidated using biophysical models. Pendyam et al. (2009) developed a structural and dynamic model of synaptic and extrasynaptic glutamate regulation to test *in silico* the effects of chronic cocaine administration and glutamate on neuroplasticity and withdrawal symptoms in the nucleus accumbens. Specifically, a differential-diffusion equation was used to understand the complex dynamics of extracellular glutamate levels. Diffusion properties were formalized by parameters, including diffusion coefficients given by the proposed local glial geometry ("tortuoisity"), while parameters of glutamate flux rates were controlled by synaptic and nonsynaptic glutamate release and exchange, reuptake transporters, and glutamatergic autoreceptors. By fixing model parameters to known empirical values—both physiological and cocaine-induced levels in glutamate exchange and autoreceptor signaling—different models of striatal geometry allowed the model to reproduce the basal reductions in extracellular glutamate observed in rodents after chronic cocaine administration. In addition to this withdrawal effect, the model predicted that enhanced extracellular glutamate levels which occur during rodent drug-seeking behavior could result from a specific change in the model's parameter space; namely, an alteration of the astrocytic XAG transporter. Thus the model predicted a molecular cause of synaptic overflow during drug-seeking behavior, providing a mechanism for reduced effectiveness in glutamatergic synaptic transmission in the nucleus accumbens. This susceptibility to extracellular glutamate accumulation has been proposed as a pathophysiological adaptation mechanism that could impair larger-scale corticostriatal communication (Kalivas 2009) and fits comfortably in an extended model-based framework of addiction phenomena. Couching these biophysical properties as computational effectors could provide crucial links that enable a deeper understanding of the computational circuitry of addiction.

Dynamic Causal Modeling of Mismatch Negativity Circuitry for Treatment Prediction in Schizophrenia

As described by Huys (this volume), it would be intriguing to use computational models to predict which treatment should be assigned to an individual patient (e.g., in depression, psychotherapy vs. pharmacotherapy). Here we describe a potential (so far fictitious) concrete application in the domain of schizophrenia: predicting individual treatment response to a switch in pharmacotherapy. Clinically this is a highly relevant issue in the management of schizophrenia because at present there are no predictors to inform us as to which patient will respond to which drug. In clinical practice, antipsychotic drug treatment rests on trial and error processes: after several weeks of treatment, drugs (often chosen based on relevant side-effect profiles rather than predicted efficacy) are exchanged if no beneficial effect has been achieved.

Here, we consider a potential application of computational modeling to address this clinical prediction problem, using a concrete scenario. This potential application is guided by theories which highlight the pathophysiological role of NMDA receptors (Lisman et al. 2008; Gonzalez-Burgos and Lewis 2012) and, more specifically, their interaction with neuromodulatory transmitters (dopamine, acetylcholine, and serotonin) (Friston 1998; Stephan et al. 2006). This "dysconnection hypothesis" postulates that individual variability in clinical trajectories and treatment response results from individual variability in dysfunctional interactions between NMDA receptors and neuromodulators (Stephan et al. 2009a). This implies that a tool capable of inferring NMDA receptor function and its regulation by neuromodulatory effects within disease-relevant circuits should have predictive power with regard to outcome and treatment response.

Methodologically, the approach described below conforms to the notion of "generative embedding" (Brodersen et al. 2011). The general approach of "embedding" is at the heart of a computational rationale, where model parameters have been demonstrated to better capture and classify patients than raw data alone (Wiecki et al. 2015).This entails using a generative model of measured data to obtain subject-specific parameter estimates of mechanisms (with a physiological or computational interpretation) for use in unsupervised learning procedures (e.g., clustering) to detect mechanistically defined subgroups. One can then test, in a second step, whether the assignment of individual subjects to subgroups has prognostic value; that is, whether belonging to one subgroup or another predicts differential response to treatment. In the potential application described here, the idea is to use a DCM (a generative model of neuroimaging or electrophysiological; here, EEG responses) to infer values of synaptic parameters, at a circuit-level of description, that are empirically known to be sensitive to changes in NMDA receptor and muscarinic receptor status.

Concretely, the DCM considered here concerns an auditory-prefrontal circuit (with bilateral primary and secondary auditory cortex, as well as right inferior frontal gyrus) known to be involved in mismatch negativity (MMN), an event-related potential in response to unexpected or surprising auditory events (for review, see Garrido et al. 2009). This particular combination of model and task is of special interest for three reasons:

1. MMN is significantly reduced in schizophrenic patients: a meta-analysis including more than thirty studies has indicated a robust effect at the group level (Umbricht and Krljes 2005).
2. Pharmacological studies in both animals and humans indicate that this reduction can be mimicked by administering antagonists of NMDA and cholinergic receptors (e.g., Javitt et al. 1996; Umbricht et al. 2000; Pekkonen et al. 2001; Schmidt et al. 2012).
3. Several previous studies have applied different DCMs to MMN data acquired under pharmacological manipulation and have demonstrated that appropriate physiological parameters of the DCMs are sensitive to selective pharmacological interventions. For example, parameters encoding the strength of glutamatergic connections from primary to secondary auditory cortex are sensitive to ketamine, an NMDA receptor antagonist (Schmidt et al. 2012). Furthermore, parameters controlling the postsynaptic gain of supragranular pyramidal cells in primary auditory cortex reflect the level of acetylcholine under manipulation by the acetylcholinesterase inhibitor galantamine (Moran et al. 2013).

In the proposed application, different types of DCMs could be used. The simplest variant would be a current-based neural mass model predicated on the formulation of Jansen and Rit (1995). This model has been used successfully in the ketamine DCM study of MMN (Schmidt et al. 2012) but offers a relatively limited representation of physiological mechanisms. A more sophisticated alternative is a conductance-based DCM, which represents a circuit of interacting cortical modules, each characterized by a mean-field model, where the neuronal state equations describe the change in average membrane potential as a function of conductance changes in ionotropic receptors (AMPA, NMDA, GABA) with sufficiently distinct time constants (Marreiros et al. 2009; Moran et al. 2011).

These equations of hidden neuronal dynamics can be coupled to an observation model which predicts sensor-level EEG measurements as a linear superposition of sources (Kiebel et al. 2006). Under Gaussian assumptions about the observation noise and Gaussian priors on the parameters, the model can be inverted using a variety of techniques (e.g., variational Bayes or Markov chain Monte Carlo), yielding posterior parameter estimates. In other words, the model described here in brevity allows one to obtain probabilistic estimates, from conventional EEG measurements, of synaptic parameters within a circuit of interest, some of which have been previously found empirically to be sensitive to pharmacological perturbations of NMDA receptor function and acetylcholine levels (Schmidt et al. 2012; Moran et al. 2013).

In the proposed application, this model is applied to MMN data obtained from EEG measurements in schizophrenic patients who have not adequately responded to a first course treatment with the antipsychotic drug risperidone, and for whom the treating physician recommends a switch to another drug, olanzapine. We consider this particular constellation for two reasons. First, the sequence of treatments considered here represents a common clinical algorithm for schizophrenia treatment and has been investigated by other recent studies, which examined the success of a treatment switch from risperidone to olanzapine (e.g., Agid et al. 2013) and tried to predict this from initial clinical data (Kinon et al. 2010). Second, although both drugs possess antagonistic effects at various dopamine receptors, they differ strongly along the cholinergic dimension in that risperidone has no affinity to cholinergic receptors, while olanzapine is a strong muscarinic antagonist (PDSP K_i Database[1]). This means that any potential individual difference in treatment response might be attributable to individual differences in cholinergic function, which in turn may be detectable using model-based inference and used for predictions about treatment response.

Following EEG measurements and treatment switch, patients would require clinical follow-up examinations, with clinical assessment obtained at fixed intervals (e.g., two and eight weeks after treatment) according to the positive and negative syndrome scale (PANSS). These clinical symptom scores would represent the target of prediction by model parameters. From individual parameter estimates of the circuit described above, one would select parameters with empirically demonstrated sensitivity to pharmacological manipulations of NMDA receptors (e.g., parameters encoding the plasticity of glutamatergic connections from primary to secondary auditory cortex; Schmidt et al. 2012) and cholinergic receptors (e.g., parameters representing the postsynaptic gain of pyramidal cells in primary auditory cortex; Moran et al. 2013). Thereafter, one could test whether subject-specific parameter estimates predict clinical symptom scores following treatment switch.[2] Evaluating this putative predictive power could proceed in at least two ways. First, the individual parameter estimates of interest could serve as features for unsupervised learning (clustering), with the goal

---

[1]   http://kidbdev.med.unc.edu/databases/kidb.php (accessed July 10, 2016).

[2]   As a caveat, Brodersen et al. (2011) used galantamine (an acetylcholineesterase inhibitor) which also has allosteric action at nicotinic receptors. Thus it is not clear to what degree the empirically demonstrated sensitivity of DCM parameters to galantamine partitions into muscarinic and nicotinic effects. Generally, however, previous studies in humans (Pekkonen et al. 2001) and unpublished data from rats (based on selective muscarinic receptor manipulations) demonstrate sensitivity of the MMN to muscarinic receptor alterations.

of detecting patient subgroups delineated by differences in the parameters of interest (see Brodersen et al. 2014). Under this perspective, one could then try to validate the proposed subgroups by testing for significant differences in response to a treatment switch across patient subgroups. Alternatively, one could directly predict the change in symptom scores as a function of model parameter estimates, using conventional multiple linear regression. If successful, the former option would provide the clinician with a tool that would enable assignment of individual patients to a particular subgroup, and hence predict treatment response in a categorical fashion. The latter option, by contrast, would enable the clinician to predict the change in (continuous) symptom scores for an individual patient, following a treatment switch.

The above (so far hypothetical) application of a computational model to individual patient data represents an example of how a relevant clinical question could be addressed through existing modeling frameworks that are supported by pharmacological validation studies in humans and animals. We have spelled out this fictitious case study in some detail to showcase the motivation, potential, and limitations of a computational psychiatry approach.

An Allostatic Recovery Model

Recovery models in psychiatry and mental health refer to the personal restabilization of a patient's life through their participation in treatment, the development of coping strategies, and renewal of a sense of self. In other words, this process guides the patient to return to a new, nonharmful "normal" state of being (Ramon et al. 2007).

Selfhood has physiological representations through internal bodily states (Critchley and Seth 2012) that are sensed by the brain in the insular cortex (Simmons et al. 2004; Gu et al. 2013; Kirk et al. 2014). Dysregulation of this process is associated with anxiety disorders and depression (for a review, see (Paulus and Stein 2010). Mounting computational literature (Seth et al. 2011) casts these dysregulated interoceptive signals as errors of prediction, whereby "individuals who are prone to anxiety show an altered interoceptive prediction signal, i.e., manifest augmented detection of the difference between the observed and expected body state" (Paulus and Stein 2006:383). In this novel example, sketched purely for this chapter, we develop the idea of dysregulated stress responses in the hypothalamic-pituitary-adrenal (HPA) axis and its potential behavioral control through amygdala–hypothalamic projections. Specifically, given the ubiquity of serotonergic drugs in treating depression, we aim to model the dynamic balance of a positive feedback loop between excitatory serotonergic effects on HPA mediated by the amygdala (Weidenfeld et al. 2005), amygdala → HPA, and the amygdala's excitability dependency from corticosertone in the HPA (Stutzmann et al. 1998), HPA → amygdala. We aim to illustrate how a simple model could capture this network and, in doing so, quantify a biological marker of recovery where the patient's goal is to cycle through amygdala and HPA activity under stress. This example thus serves as a conceptual outline to a computational recovery model of anxiety and depression.

To build a generative model of this process, we utilize dynamical systems theory, employing a simple description which has state variables that represent the dynamic processes of homeostasis and allostasis (Figure 12.3). In this setting, we refer to the general definition of allostasis: "stability through change" (Sterling 2014), specifically stabilization to a new state that is not harmful, but is also not the "normal" state that patients were used to prior to their illness.

Homeostasis is a process by which system perturbations are reactively damped, ensuring return to a stable equilibrium. Allostasis is the process of reaching equilibrium through change and offers anticipatory control over homeostatic mechanisms (Schulkin 2010). Whereas homeostatic mechanisms have a fixed-point equilibrium, allostatic processes can yield more complex dynamics, including limit cycles and even chaos (Rodrigues et al. 2007) through, for example, positive feedback loops (Spiga et al. 2008). The dynamics of allostatic control over homeostatic dynamics begins in our model with a single state variable, x, a signal that returns to homeostasis through a mean reverting Ornstein–Uhlenbeck process (i.e., a nondelayed self-correcting random walk). In turn, this process is modulated by a second signal, y, which embodies a prediction of x. The introduction of a (negative) time lag between

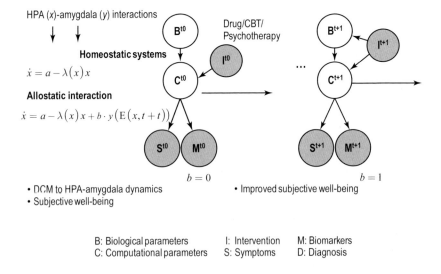

Figure 12.3 A dynamical systems model of allostatic equilibrium in the HPA-amygdala axis to test the effectiveness of mental health recovery. This app is dedicated to the memory of Xavier, the random walking spider.

prediction, measurement, and action corresponds to the extension from homeostasis to allostasis and allows for more complex behaviors. The goal of our hypothetical treatment intervention is to ensure that the homeostatic signal is receiving allostatic system inputs and that allostatic systems are exhibiting complex dynamic behaviors, such as periodicity and even multistability, hence achieving stability through change.

To make this more concrete, let us consider the case of depression. In this disorder, stress can act as a precipitating factor leading to elevated cortisol levels and a disruption of its rhythmicity (Johnson et al. 2006). This has been hypothesized to result in abnormal homeostasis in HPA and could arise from reduced excitatory amygdala inputs (Herman and Cullinan 1997). In our model, this downregulation of allostatic drive will break its rhythmicity without sufficient predictions and render the patient enslaved to homeostatic effects, as reflected in hyperactivity of the HPA axis (Pace et al. 2006). The model is designed to monitor the return of a patient's HPA axis from homeostatic reaction to allostatic control. The types of data which could be used to monitor these dynamics include metabolic and neurophysiological assays together with neuroimaging time series where connectivity assessments (e.g., DCM) could potentially be used to monitor rebalance. The advantage of this approach is its ability to quantify directly the return to allostasis. It would thus provide a metric for suitable change in a recovery model, which has received some criticism for its lack of an evidence base (Davidson et al. 2005).

Developing Community-Wide Standards for Model Development

Impacting standard clinical practice will require committed efforts from stakeholders, including psychiatrists and allied mental healthcare professionals as well as patients, insurance companies, and policy makers. The goal is to enable the type of computational prototypes developed here to be interactive, so that clinicians can access and probe precise simulations to better predict patient outcomes. For these systems to be realized, communities must adopt standards to ensure consistent reporting in terms of patient tests, specific models, and measurements. The important consideration is how, in practice, to best advance the agenda of applying computational approaches to psychiatry, so as to maximize the probability of the field having a tangible impact on psychiatric theory and practice. For this to happen, close collaborations must be forged between clinicians, theoreticians, and neurobiologists. Each area of expertise brings something essential to the table. For example, a theoretician may be able to build an elegant model of a computational process that may be altered in a particular disease or symptom. However, if such models are based on idealized characterizations of a disorder and fail to make contact with the real complexity and/or heterogeneity of a disease, then the model will most likely not have predictive or even face validity, and is therefore unlikely to be useful to the field as a whole. Conversely, to interpret findings about the meaning of specific effects observable at the molecular, cellular, circuit, or systems level in the brain, and to translate those findings effectively to psychiatry, neurobiologists need to interact closely with theoreticians and psychiatrists. One important overall strategic objective for advocates of this field will be to find ways to promote such close collaborations.

How can this be achieved? One obvious avenue is to bring psychiatrists, theoretical neuroscientists, and neurobiologists together in regular conferences, summer schools, and workshops focused on computational psychiatry. This would help facilitate a common language and provide a fertile environment for new collaborations to form. Because of the difficulty in understanding each other's terminology, such meetings should take the form of dialogue meetings, in the spirit of the Strüngmann Forum, and utilize a structured approach to address specific topics in computational psychiatry. Furthermore, the development of interdisciplinary training programs would serve a critical role in training the next generation of clinicians and basic scientists. Their goal should be (a) to train psychiatrists in computational techniques so that they are familiar with the computational language without necessarily becoming theoreticians in their own right and (b) to introduce neurobiologists and theoretical neuroscientists to the complexity inherent in the diagnosis and treatment in psychiatry. Efforts to facilitate the emergence of genuine hybrids who are both theoretical neuroscientists and practicing clinicians (psychiatrists or clinical psychologists) could have a major impact on the future of the field. Finally, we need to institutionalize computational psychiatry by establishing units where scientists with computational and biomedical backgrounds can work together, literally under the same roof and ideally in shared offices. Physical coexistence is perhaps one of the most powerful ways to facilitate the creation of a common language and mutual understanding.

Other practical considerations include working to convince funding agencies throughout the world as to the potential impact of the field, with the goal of encouraging these agencies to set up specific funding programs or mechanisms tailored to computational psychiatry. This could happen, for instance, by explicitly requiring collaborations between theoreticians, neurobiologists, and psychiatrists as an eligibility criterion in applications for a particular funding mechanism focusing on topics in computational psychiatry.

Finally, we need to consider how the field as a whole can work to facilitate progress in research in this domain. One possible avenue is to develop frameworks in support of data sharing, or the development of large-scale collaborations so that we can reach a critical mass in terms of number of participants, diversity of theoretical constructs, tasks and measures to test computational hypotheses in psychiatric populations, in a manner that ensures sufficient statistical power and robustness to variation across psychiatric populations.

One practical suggestion for how to go forward with all of these proposals is to convene a committee charged with developing strategies for making

progress in each of these domains over the next several years. Below, we outline a major component of community-wide models; namely, the stimulus
paradigms from which we elicit our signals and to which we apply our models. Thereafter we consider existing efforts, highlight the CNTRICS battery
developed for schizophrenia researchers, and conclude by suggesting a new
international consortium for resource and knowledge sharing.

Standard Tasks and Paradigms

Developing standardized experimental paradigms that provide reliable assessments of cognitive function, such as working memory or decision making,
will be crucial to get data for precise modeling results. The downside of these
general assessments may be twofold: First, they are potentially ignorant about
the current emotional state or symptom expression of the individual patient.
For example, specific aspects of decision making in an individual patient with
addiction can be fully intact in multiple domains and yet be highly impaired in
the context of drug-associated environments. Paradigms that fail to elicit relevant contextual states may suffer from small effect sizes and yield uninformative experimental outcomes. Second, using fixed stimulus sequences might not
account for the heterogeneity across individuals, particularly in patient populations. To address these issues, experimental setups could be tailored to individual patients by probing state- rather than trait-dependent aspects of cognitive
function and/or adjusting data acquisition online to optimize the feasibility of
model inversion (parameter estimation) and model comparison.

One concrete example of state-specific paradigms is symptom-provok-
ing stimuli in obsessive-compulsive disorder, where patients are exposed to
stimuli that relate to their individually expressed obsessions and compulsions
(Adler et al. 2000). Emotional stimuli (movies) or subject-specific biographi-
cal events are similarly effective tools in the context of mood disorders. In
addition to choosing stimuli with higher face validity, the design of the experiment itself can be optimized in further aspects. Game-inspired paradigms can
provide more naturalistic environments to provoke domain-specific behavioral
patterns: the use of slot machines to probe decision making in a gambling context (Clark 2010) or virtual environments to simulate real-world interactions
(Parsons and Rizzo 2008).

Optimized data acquisition can also be accomplished through adaptive experimental designs, whereby stimulus presentation is dynamically adjusted
throughout the experiment. This may involve online adjustment of stimulus
presentation, based on the past history of individual responses, to yield optimal data for parameter estimation and model comparison. A simple example
of this type of adaptive data acquisition is the staircase paradigm for deliver-
ing stimuli according to individual (and potentially time-varying), perceptual,
and performance-related properties (e.g., perceptual thresholds and task accuracy). More sophisticated approaches derive from probability theory and

would enable an optimal experimental design to be found for testing a specific hypothesis embodied by a particular model's structure. This could be done *a priori* (before the experiment) or online (Daunizeau et al. 2011b).

In summary, individually tailored paradigm designs have the potential of markedly improving model-based inference by targeting relevant state-dependent behavior and optimizing data acquisition in particularly heterogeneous patient groups. This would result in increased effect sizes and more sensitive statistical tests.

Example Task Battery: The Cognitive Neuroscience Treatment Research to Improve Cognition in Schizophrenia (CNTRICS) Initiative

During the 1990s, a growing awareness of the disabling nature and treatment refractoriness of cognitive impairment in schizophrenia highlighted the need to develop new treatments for this aspect of the illness. In developing a pathway to drug registration, a set of tools was developed with the support of the U.S. National Institute of Mental Health (NIMH). One initiative, the Measurement and Treatment Research to Improve Cognition in Schizophrenia or MATRICS, developed a battery of tests that consisted primarily of clinical neuropsychological tests already in use in drug development trials. During this process, it was proposed that experimental measures from cognitive neuroscience, as opposed to these clinical tests, would offer the advantage of targeting more specific cognitive systems that were linked more directly to discrete neural systems. Concerns included that there was no general consensus in the field as to what cognitive systems should be targeted, no standard, easy to administer tasks to be used for measurement, and no information about the psychometric properties (reliability, presence of ceiling, and floor effects). To address these concerns, and propel the field toward a neuroscience-based approach to measuring cognition and the impact of treatment on cognition in schizophrenia, the CNTRICS Initiative was launched in 2007, supported by an R13 conference grant from NIMH and led by Cameron Carter and Deanna Barch.

In all, seven conferences were held over a period of four years at a variety of locations across the United States. Each was informed by pre-meeting surveys of the larger field and brought together an international group of basic cognitive neuroscientists, clinical researchers, and those involved in treatment development, using a semi-structured consensus-based process. The initial three meetings developed a set of theoretical cognitive domains to be targeted and a set of experimental cognitive tasks with strong construct validity as measures of these domains. In all, twenty-three tasks across seven domains were recommended for development. The next four meetings focused on developing imaging and ERP biomarkers with strong construct validity for measuring the cognitive and neural systems associated with each domain, in addition to two meetings which focused on the development of more integrated animal model systems for use in the drug discovery process.

At about the midway through the CNTRICS process, funding was obtained to begin developing tasks (selected because of their construct validity, clinical importance, and other factors) into tools that could be used for standardized measurement of cognition in clinical and treatment research. This new project—Cognitive Neuroscience Test Reliability And Clinical Applications for Schizophrenia, or CNTRACS—is currently ongoing and involves five sites across the United States. Tasks have been adapted to ensure that specific deficits in cognitive mechanisms could be measured independently of generalized deficits (e.g., attention lapses, poor motivation), optimized (on factors such as numbers of trials, length of administration and maximizing effect sizes), and studied for their psychometric properties and relationship to symptoms and to measures of functioning. The first round of data collection has been completed, including a supplementary study using fMRI for three of the four tasks initially studied, and a number of publications have resulted. Importantly, brief, well-tolerated versions of measures of cognitive control (AX-CPT), episodic memory (Relational and Item Specific Encoding, or RISE task), perceptual integration (Jitter Orientation Visual Integration Task, or JOVI task), and early visual perception (Contrast-Contrast Effect, or CCE task) have been developed with acceptable test-retest reliability and predictable relationships with different sets of symptoms and functioning in the patients with schizophrenia. A number of the theoretical constructs and recommendations were subsequently incorporated into the Research Domain Criteria (RDoC) framework by leaders at NIMH. The presentations from each of the CNTRICS meetings, along with the papers documenting the results of each meeting are available on line (cntrics.ucdavis.edu). In addition, publications resulting from the cognitive neuroscience test reliability and clinical applications for schizophrenia (CNTRACS) consortium and scripts for running the four tasks studied in the initial round of this project are programmed in Eprime and available for free download from the site.

Despite this valuable activity and our optimism toward it, we end with a note of caution: The field of computational psychiatry is still very much in its infancy and the problem it aims to address is immense. Therefore, we need to take a long-term view and exercise patience, for progress may proceed inconsistently and irregularly. As with genetics, the original promise inherent in the field of computational psychiatry may take decades to be fully realized.

13

There Are No Killer Apps but Connecting Neural Activity to Behavior through Computation Is Still a Good Idea

P. Read Montague

Abstract

The quest to understand the relationship between neural activity and behavior has been ongoing for well over a hundred years. Although research based on the stimulus-and-response approach to behavior, advocated by behaviorists, flourished during the last century, this view does not, by design, account for unobservable variables (e.g., mental states). Putting aside this approach, modern cognitive science, cognitive neuroscience, neuroeconomics, and behavioral economics have sought to explain this connection computationally. One major hurdle lies in the fact that we lack even a simple model of cognitive function. This chapter sketches an application that connects neuromodulator function to decision making and the valuation that underlies it. The nature of this hypothesized connection offers a fruitful platform to understand some of the informational aspects of dopamine function in the brain and how it exposes many different ways of understanding motivated choice.

Introduction

Let's face it. Computational neuroscience and its fledging product, computational psychiatry, simply do not have a killer app—yet. Certainly nothing like Newton's laws, William Rowan Hamilton's transformative approach to dynamics, the Dirac equation, Darwin's evolution by natural selection and its rendering in the twentieth century modern synthesis (Mayr and Provine 1980/1998), or even Shannon's breakthrough efforts in what is now called information theory. Marshaling such a pantheon isn't quite fair, but it makes a point. We should require a lot from any account that calls itself a killer app, especially in an area that purports to connect mind and brain in a meaningful way. In the world of sustaining healthy human mental function and characterizing and treating unhealthy human mental function, the killer app will depend on a much more evolved body of constructs (models) surrounding cognition. The limiting factor is (at least) our woefully simple models of cognitive function. We simply do not yet have an evolved and integrated model of human cognition that can render a human-like model agent in a perceptual problem or learning problem to use such a set up to gain penetrating insight into a psychiatric disorder. Instead, the current situation bears the hallmarks of the early days of any discipline: some very provocative models exist, focused in particular areas and mapped with variable success to experimental data extracted from candidate neural systems.

In this chapter, I will sketch an application that connects neuromodulator function to decision making and the valuation that underlies it. The nature of this hypothesized connection has proved to be a fruitful platform for understanding some of the informational aspects of dopamine function in the brain and how it exposes many different ways of understanding motivated choice.

The Platform of Reinforcement Learning

For well over a 100 hundred years, models of learning have been dominated by psychological concepts about how animals adapt to the changing world around them. The foundational ideas emerged in the early nineteenth century from the work of physiologist Ivan Pavlov and his star student Jerzy Konorski on the conditioned reflex (Pavlov 1927; Konorski 1949). This work developed into an entire behaviorist movement that flourished through the twentieth century with its now familiar collection of names: Thorndike, Hull, Watson, Skinner and so on. One of the strictures of this movement was to remove all mention of variables that could not be observed, especially any mention of unobserved mental states. All behavior was to be rendered as stimulus and response, a view that modern cognitive science, cognitive neuroscience, neuroeconomics, behavioral economics, and their computational expressions toss aside. Apparently, there was something to be feared about positing unobserved states of mind as though such unobserved entities prevent the hard science from taking place. The behaviorists were likely just reacting to Freud's influence on psychology; however, it is noteworthy that unobserved or unobservable entities and states pervade physics and biophysics despite the fact that both areas are viewed as hard science. Biophysical models of ionic channel function have long and happily accepted unobserved states and state transitions, usually cast mathematically as hidden Markov models (Hille 2007). Latent states, latent variables, unseen fitness or hazard functions, hidden Markov models, and their more exotic congeners are now simply part of the inventory of modern computational approaches to mind and brain.

Capturing the Regularities of Learning: From Bush–Mosteller to Sutton–Barto

One key area where a rigid stimulus-response framing was very useful was learning, since it is here that experimental psychology first began to identify so-called learning rules—statistically lawful mappings between input, internal state, and the output of the entire creature. All mobile creatures need to learn because they move; movement ensures that the contingencies for survival change, and do so on multiple time and space scales. Sessile creatures (e.g., sea cucumber) have developed some very peculiar strategies for adapting to environmental threat (they partially eviscerate themselves as a defense), but a moving creature is where real learning action takes place. At the minimum, mobile creatures must deal with the environmental changes that result from their own movement. In this sense, movement and learning have always been partner processes, so behaviorist paradigms provide very nice and structured ways to probe simple learning and capture the results in equally simple laws.

The rules that characterize learning in mobile animals start with the work of Ivan Pavlov, who originated the modern interpretation of the conditioned reflex: ring the bell, feed the dog, rinse-and-repeat. Through this regular training, the originally neutral bell comes to elicit the features (orientating, salivating, secretion of digestive enzymes, and so on) of the unconditioned response to food. This is classical conditioning, and its "cousin," instrumental conditioning, contains the same regularities but requires an action on the part of the animal. Pavlov generated a tradition around this idea, and it was certainly the foundation for the behaviorism movement, as mentioned above. However, for modern computationalists, the important steps were taken just after World War II with the emergence of work by Robert Bush and Frederick Mosteller (e.g., Bush and Mosteller 1951a, b, 1953, 1955). At that time, these investigators were considered part of a new breed of mathematical psychologists—perhaps the first generation (were it not for Hermann Helmholtz's work in the late nineteenth century). They originated the idea of prediction learning and introduced the first rigorous account of the kinds of learning described by the behaviorists. Rendering learning as a problem of learning-to-predict was a departure from the correlational theories of Konorski (1949) and Hebb (1949). The problem with such correlational accounts are manifold, but the main impediment is that they do not provide a natural way for a "correlation-based" learner to learn chains of events. Both correlational accounts and prediction accounts for learning, however, viewed the animal as a statistical learner whose "learning job" was to extract regularities latent in the statistics of their experience.

In the Bush and Mosteller account, the conditioning described by Pavlov was rendered as a trial-based prediction of the unconditioned response and also provided a simple way to update that prediction from trial to trial:

$$p_{t+1} = p_t + \alpha \left(R_o - p_t \right). \tag{13.1}$$

The goal here is to associate stimuli with actions, and the Bush–Mosteller model updates the probability p that the action (salivation) occurs on trial $t+1$ as a function of its value in the previous trial t and the value of the observed reward R_o. This is the first good account of prediction learning to explain the learning associated with behavioral conditioning paradigms. But Bush and Mosteller went further and modeled the animal as a collection of probabilistic processes. In an obituary for the late Robert Bush, Mosteller (1974:170) aptly describes the modern flavor of their approach:

> In the models for learning that Bush and I developed, the fundamental representation was that prior to a trial an organism was a vector of response probabilities. A stimulus corresponded to a mathematical operator that replaced the organism's current vector by a new probability vector. In the models of Bush and Mosteller (1955), the effects of previous responses were summed up in the current vector, independent of the path to the present state. The operators had a linear form, so that if p is a vector of probabilities $(p_1, p_2,..., p_k)$ and Q is applied to p the new vector is:
>
> $$Qp = \alpha p + (1 - \alpha)\, \lambda,$$
>
> where λ is also a probability vector $(\lambda_1, \lambda_2,..., \lambda_k)$ and α is a scalar, $0 \le \alpha \le 1$. If Q is repeatedly applied, the limiting vector is λ, when $\alpha \ne 1$.

This is an extremely rich model of the processes putatively at work inside the learner. Notice that even the history-independent assumption (the Markovian assumption) is present in their papers as of the early 1950s, "…independent of the path to the present state" (see also Rescorla and Wagner 1972). As forward looking as the Bush–Mosteller approach was, it still missed some important aspects of learning, such as the detailed dependence on timing of stimuli and other well-known conditioning effects such as secondary conditioning: If A predicts reward, and B is trained to predict A, then B will also predict reward. From the psychological and computer science literature there emerged another approach to this problem offered up by Richard Sutton and Andrew Barto (1981, 1987, 1998; for complete references, see Sutton 1988). Their work focused on an incremental learning algorithm, called the method of temporal differences, which exploited differences between successive predictions rather than simply the difference between a prediction and an outcome. This difference is crucial, as it framed the "goal of learning" as the problem of *learning to value the future of the states* of the agent. This agent is portrayed as moving about in some kind of high-dimensional state-space, making transitions from one state S_t at time t to another state S_{t+1} at time $t + 1$. There were two basic assumptions to this approach. First, the *goal of learning* is to learn the value of states taken as the discounted amount of future reward expected from that state forward into the distant future. The other assumption was that it did not matter how the state was reached:

$$V\left(S_t\right) = E\left(r_t + \gamma r_{t+1} + \gamma^2 r_{t+2} + \cdots\right) \text{ for } 0 < \gamma \leq 1. \qquad (13.2)$$

Here the expected value operation "*E*" is slightly bad notation. The expectation *E* is taken for each "tic" forward and so we should read it as expressing that each *r* is a separate expected value of reward at each step to the future of . So the value of the current state of the agent depends on its future. The second bit of ambiguity in Equation 13.2 is that the expectation implicitly includes the rule that the agent uses to transition from state S_t to state S_{t+1} The single *E* symbol does not specify this clearly in Equation 12.2, but these details do not matter here.

Overall, the Sutton–Barto account appears to be a small change from the Bush–Mosteller approach; moreover, it mimicked approaches from the late 1950s by Samuel, who made automatic checker-playing programs (Samuel 1959). The Sutton–Barto effort did account, however, for secondary conditioning, as well as the way that an agent learns to chain events together. It also connected to animal conditioning (Sutton and Barto 1987) and to an independently developing area of optimal control called dynamic programming (Bellman 1957). As I review below, it also reached down to important biological observations. This multidimensional reach, which crossed levels of description, is and was what makes this work so important. Sutton and Barto understood the connection of their work to previous approaches but they also understood that they had added crucial insights. Quoting from Sutton (1988:9):

> This article introduces a class of incremental learning procedures specialized for prediction – that is, for using past experience with an incompletely known system to predict its future behavior. Whereas conventional prediction-learning methods assign credit by means of the difference between predicted and actual outcomes, the new methods assign credit by means of the difference between *temporally successive predictions*. Although such *temporal-difference methods* have been used in Samuel's checker player, Holland's bucket brigade, and the author's Adaptive Heuristic Critic, they have remained poorly understood. Here we prove their convergence and optimality for special cases and relate them to supervised-learning methods.

The Valuation of the Future

Let us turn back to the central idea of Sutton–Barto: the value of a state scales according to the value of the discounted future it portends. Why should a mobile creature need to value the future? One word: uncertainty. It appears that in our world a vast amount of uncertainty lies in the future with the more important bits of uncertainty rolling out in the near future. Therefore, the valuation of a state based on its expected future, when combined with the assumption that the system is Markovian (history independent), animates the power of this simple approach. Let us take Equation 13.2 and step forward one tic to a new state S_{t+1} and value this new state in exactly the same way: as the expected value of reward from that state forward:

$$V(S_{t+1}) = E(r_{t+1} + \gamma r_{t+2} + \gamma^2 r_{t+3} + \cdots). \tag{13.3}$$

In principle, Equations 13.2 and 13.3 would require a system to run through a state infinitely often and sum a long (infinite) series of numbers to estimate the value of the state. Herein lies the very nice way that this valuation function is formulated. If we scale Equation 13.3 by the discount factor γ (the rate that the future becomes less valuable at each "tic") then:

$$\gamma V(S_{t+1}) = E(\gamma r_{t+1} + \gamma^2 r_{t+2} + \gamma^3 r_{t+3} + \cdots). \tag{13.4}$$

Notice now that there is a way to relate the value function of time t to the value at time $t+1$:

$$V(S_t) = E\{r_t\} + \gamma V(S_{t+1}). \tag{13.5}$$

If the agent (the learner) had perfect estimates of the value of all its states, then this expression would be exact. The real world is not exact so this condition never holds perfectly; however, it does give a natural way to define an error signal. Simply subtract the left-hand side from the right-hand side:

$$\text{``0''} = E\{r_t\} + \gamma V(S_{t+1}) - V(S_t). \tag{13.6}$$

The quotes here indicate that this difference is never really 0 in the real world. However, Sutton and Barto exploited this temporal difference (TD) to build an approach to learning that has wide-reaching applications and implications (Sutton and Barto 1998). An error signal of just this type was first hypothesized to be a general mechanism for biological systems to learn how to value states, store predictions, and link value to action based on predictions (Montague et al. 1993, 1995, 1996, 2004; Schultz et al. 1997; Dayan et al. 2000; Dayan and Abbot 2001; Dayan and Daw 2008; Niv and Montague 2008; Dayan 2012).

Connecting Levels: Computation, Behavior, and Neuronal Activity

An early sign of a strong connection of a TD error signal to a biological system was the connection of octopaminergic neuron activity to odorant conditioning in honeybees (Real 1992; for physiology, see Hammer 1993). Behaviorally, honeybees show many sophisticated computations in the way they sample flowers yielding variable returns. For two flowers yielding the same mean return (let's say in nectar units), bees will sample more frequently from the flower with lower variance in the nectar return despite the matched average return (Real 1992; Hammer and Menzel 1995). All things being equal, bees avoid the flower color that predicts the more variable return. This is just one of many sophisticated computations available to the honeybee for adjusting its behavior in pursuit of maximizing its returns on nectar. What is remarkable is the presence and functional role of aminergic neurons in the bee's subesophageal ganglion that project to widespread targets throughout the bee brain and deliver the neuromodulator octopamine (a close chemical cousin to dopamine). Without diving into too much detail, let me summarize by saying: (a) it is known now that octopamine action at target neural sites is crucial for conditioning, (b) the physiological behavior of these neurons is consistent with a TD error signal, and (c) this error signal can be mapped simply onto action choice by the bee in a manner (somewhat artificial though it may be) that can account straightforwardly for how the bee trades off mean and variance of nectar returns (see also Douglas 1995; Montague et al. 1995). The TD account was provocative because of the way it linked levels of description of the bee: the physiology of the octopamine neurons under a rigorous behavioral challenge, the behavioral consequences of the TD error computation putatively performed by these aminergic neurons, and the computations that reached from the level of the neurons to the choice behavior of the bee.

A richer connection between the TD algorithm (the computation), behavioral choice, and detailed recordings of neuronal behavior arose in the early 1990s from the work of Wolfram Schultz and his colleagues. In their work, nonhuman primates were trained on simple conditioning tasks where a light would indicate which of two levers were to be pushed to receive a juice reward. Early in the training, dopamine neurons give phasic responses only to the delivery of reward; these responses disappear, however, with training after which the neurons give transient responses only to the earliest consistent predictor of future reward (here the initial cue light). These data are now over twenty years old, but the guidance of the Sutton–Barto TD algorithm in understanding the features of these kinds of data is made clear by the then novel interpretation that the algorithm provided. These features include:

1. Temporal consistency was key and clearly being encoded into the response of the dopamine neurons. For the nearly identical behavioral paradigms, the neurons lose their initial response to reward delivery, develop a transient response to the earliest predictive cue, but differ in the way they respond at the trigger cue. When the timing of the trigger cue is completely predictable, neurons give no response; when the trigger cue has temporal uncertainty, neurons continue to modulate at the trigger. This is to be expected quite naturally from a TD error signal-based account; however, a trial-based account such as the Bush–Mosteller rule (and also of course the Rescorla–Wagner rule) would have to add extra detail to explain these data.

2. Sensory-reward prediction and sensory-sensory prediction are unified in a TD error-based account. The sensory-sensory prediction piece is exemplified by the response (or lack thereof) of the dopamine neurons to the trigger cue.

3. Dopamine neurons emit information even when they do not modulate their activity. This means, for example, that the neurons are emitting information throughout the duration of the trial. This is a new idea, certainly for the conditioning literature (though the weakly electric fish literature may be a good counterexample), and the lack of modulation during the interval between cues and reward outcome were vexing for the experimentalists who uncovered these data. In fact, this problem blocked their understanding of what the dopaminergic modulation could mean. As Schultz et al. (1993:900) stated: "None of the dopamine neurons showed sustained activity in the delay between the instruction and trigger stimuli that would resemble the activity of neurons in dopamine terminal areas, such as the striatum and frontal cortex.…The lack of sustained activity suggests that dopamine neurons do not encode representational processes, such as working memory, expectation of external stimuli or reward, or preparation of movement. Rather, dopamine neurons are involved with transient changes of impulse activity in basic attentional and motivational processes underlying learning and cognitive behavior."

Such delay period nonmodulation is to be expected from a TD error-based account of the results. However the conclusion is understandable. At the time (1992–1993), results using working memory tasks in nonhuman primates showed delay period activity in dopaminergic terminal regions of cortex (e.g., Brodman area 46; Goldman-Rakic et al. 2004) that depended on intact dopamine transmission through identified receptor types. It seems likely that this inspired Schultz and colleagues to look for delay period activity at the level of parent dopamine neurons in the midbrain that gave rise to this cortical input. The computational model (even the simplest Sutton–Barto TD error account) was essential to see these data in a different light. Furthermore, while trial-based accounts, such as the Rescorla–Wagner rule, have been offered to account for these physiological data, they cannot account for the signature temporal features. This point has been made clearly in a review of the dopamine prediction error hypothesis by Glimcher (2011). Glimcher also makes a very nice social case for the precedence and farsightedness of the Bush–Mosteller approach to modeling animal conditioning, when set beside the very popularly quoted Rescorla–Wagner rule.

Schultz and colleagues have now provided strong evidence for the TD error model and have shown (among a variety of new findings) that midbrain dopaminergic responses encode the expected value of the future predicted reward in similar experiments (see Schultz et al. 1997; Montague et al. 2004; Tobler et al. 2005; and the sober warnings of Dayan and Niv 2008). This summary suggests that mammals possess an efficient prediction system that can be deployed in a variety of behavioral settings to learn to value the near-term future and act reasonably based on those valuations. Here, I have avoided all discussion of the interesting complexities that arise when mapping such valuations to action choice and the further connections that can be made to optimizing control models (e.g., Kalman-filter models and models that require more complex representations). Let us finish here with some forward-looking pointers to the way that reinforcement-learning models can be applied to disease states like addiction (McClure et al. 2003a; Redish 2004), human neuroimaging data through model-based approaches to reinforcement learning (McClure et al. 2003b; O'Doherty et al. 2003, 2004), and to "exceptions" where evidence in rodents and nonhuman primates suggests that such models are incomplete (of course they are) or misleadingly wrong (Dayan and Niv 2008; Niv and Schoenbaum 2008). In my opinion, the big questions for reinforcement-learning models involve the nature of the representations used to control midbrain dopamine neurons and the use of these representations in cognitive control (Carter et al. 1998; Botvinick et al. 1999, 2001, 2009).

Reaching toward Humans

Thus far, this account has focused on the valuation aspect of the TD model and ignored the action-selection piece, except for simple conditioning paradigms which help highlight the main points. With the advent of functional MRI, many functional questions can now be asked that would probe directly the claims or extensions of the TD model. In the first direct test of the dopamine prediction error hypothesis in humans, McClure et al. (2003b) and O'Doherty et al. (2003) found that in terms of BOLD signals, the striatum shows activation and deactivation in accord with the TD model. These results are comforting but not definitive. There are many signals that may combine at the level of the striatum to elicit BOLD responses consistent with prediction error signals. Direct measures of dopamine and other neuromodulator delivery during similar behavioral probes will be required to expose the exact contributions of dopamine delivery to such BOLD measurements. One of the novelties opened up by the O'Doherty et al. and McClure et al. work is the possibility of using the computational models to define a computational process encoded throughout some behavioral challenge, and then use estimates of this process to seek its physical correlates. This approach is now called model-based fMRI.

Hypervaluation Disease

Another reach toward humans can be seen in the work of Redish (2004), who used TD-type models to address addiction as (in part) a valuation disease. The temporal-difference reinforcement-learning (TDRL) model uses an error, the TD error, putatively encoded by transient changes in dopaminergic activity (and presumably dopamine delivery) to adjust available parameters to estimate the expected value of discounted future rewards predicated on the current state of the animal. As expressed by Redish (2004):

$$V(t) = \int_{t}^{\infty} ds\, \gamma^{s-t} E[R(s)].$$

(13.7)

This is just a continuous version of Equation 13.2, where the dependence on state is replaced with a time variable (in simple settings such an equivalence is fine but potentially confusing). As the animal learns to associate sensory cues with receipt of rewards, the dopamine systems adjusts its value function using the TD error and does so until the error is driven to 0. What if one could induce an error without going through the entire cue-predicts-future-reward machinery? Wouldn't such an "outside" error induce the system to learn the wrong valuation function? The answer is yes. The idea proposed by Redish is that addictive drugs, such as cocaine and methamphetamine, produce "bumps" in dopamine release (bumps in the TD error term) through mechanisms that escape the cue-predicts-reward setting captured by the TDRL model. In doing so, the system cannot learn a value function; temporal differences in this value function cancel the impact of cues associated with drug taking. In short, the proposal is a value function "run away" where uncompensable changes in the value function, induced by pharmacologically mimicked error signals, create a kind of valuation disease condition. In his analysis, Redish is careful to point out that this feature is only one of many aspects of addiction, but the entire pro-posal centers around the model and its mapping on physical substrates in the brain and aberrant behaviors that can ensue. This way of thinking has opened up many new questions in the area of addiction, and the models here have ex-panded immensely in recent years. Computationalizing addiction will certainly lead to new insight and likely better models of it.

Flat Valuation Diseases and Rational Freezing Responses

TDRL models also provide a new way to understand some aspects of Parkinson disease, a neurodegenerative disorder associated with a profound loss of mid-brain dopamine neurons. By the time symptoms appear to warrant diagnosis, dopamine neuron loss ranges from 70–90%. There is virtually nothing known about how either dopamine systems or downstream targets adapt their dynam-ics as this loss occurs. However, one possible consequence of dropping the number of neurons is to increase the "dopamine noise" at the target structures. If dopamine delivery fluctuations are to communicate reward prediction er-rors in their rapid (subsecond) transients and possibly other important com-putations in their tonic (mesoscale) averages, then decreasing the number of neurons increases dopamine noise. A downstream target may not be able to distinguish the value of one state from another because the dopamine noise level is high. Ultimately this could look like a relative "flat" value function to these downstream targets.

Let us use this caricature and hypothesize that the best response to a flat value function is to commit no new resources—do nothing. In fact, why not just freeze? Perhaps one part of the syndrome of Parkinson disease is a kind of rational freezing response to a flat value function. In this sense, the small dopamine fluctuations are buried in noise. Therapies that would raise baseline dopamine (e.g., taking L-DOPA or perhaps putting engineered dopamine-secreting cells in the striatum) would make those fluctuations significant and perhaps "readable" by the otherwise confused downstream processes. This is, of course, wild speculation, but its possibilities are suggested by the model and thus support modern efforts in computational neuroscience.

Breaking the Reinforcement-Learning Hegemony?

This brief summary has been, in large part, rearward-looking and very focused on the valuation part of the simplest reinforcement-learning models. I close with apologies to those investigators whose work is not mentioned here: reinforcement-learning approaches are now so vast that it was not possible to include all relevant species of animal and explanation in this succinct summary. However, the best outcome for any class of model is to be very, very wrong in some productive way. There are already indications that reinforcement-learning models have guided work to some of these creaky zones (Dayan and Niv 2008; Gershman et al. 2009; Dayan 2012).

14

Call for Pragmatic Computational Psychiatry

Integrating Computational Approaches and Risk-Prediction Models and Disposing of Causality

Martin P. Paulus, Crane Huang, and Katia M. Harlé

Abstract

Biological psychiatry is at an impasse. Despite several decades of intense research, few if any, biological parameters have contributed to a significant improvement in the life of a psychiatric patient. It is argued that this impasse may be a consequence of an obsessive focus on mechanisms. Alternatively, a risk-prediction framework provides a more pragmatic approach, because it aims to develop tests and measures which generate clinically useful information. Computational approaches may have an important role to play here. This chapter presents an example of a risk-prediction framework, which shows that computational approaches provide a significant predictive advantage. Future directions and challenges are highlighted.

Biological Psychiatry: What Have You Done for Us Lately?

Biological psychiatry is in a crisis (Insel and Cuthbert 2015) for a number of different reasons. First, despite profound advances from molecular to systems neuroscience, these insights have had relatively little influence on practical psychiatry. Second, the development of new therapeutics, based on neuroscience approaches to understand the pathophysiology of these illnesses, has stalled (Insel 2012). Third, in the development of a new diagnostic classification for mental disorders (APA 2013), neuroscience had virtually no impact on contributing to the delineation and definition of the disorder categories. Fourth, there are no clinical tools for prognosis, diagnosis, and treatment monitoring that are based on neuroscience approaches (Prata et al. 2014). Taken together, the fundamental insights into basic neuroscience have not translated into practical and clinical tools or treatment in psychiatry. Here we argue that computational approaches may play an important role in linking behavior (including emotion and cognitive processing) to neural implementations of these processes in the brain.

The lack of impact that neuroscience has had on practical psychiatry may be due to several reasons. First, one might postulate that mental health conditions, which are complex constellations of symptoms and social conditions, are fundamentally not reducible to simple biological processes. This topic deserves a thoughtful discussion, which might focus on the level of reductionism possible when observing complex clinical phenomena. Such discussion, however, is beyond the scope of this chapter.

Second, we may not have sufficiently developed technologies and approaches to map psychiatric diseases onto biological processes. This perspective is useful in generating incentives to develop new techniques in the future to advance biologically based research in psychiatry. However, the lack of progress, despite decades of increasingly sophisticated technologies, might cast doubt over the argument that it is simply a "technology problem."

Third, making biology useful for clinical psychiatry is an extremely difficult problem to solve. Given the complexity of the human brain—in terms of its amazing array of topographically organized units, which are highly interconnected, the complex orchestration of molecular events that accompany even "simple" psychological processes, and the multilevel organization that occurs from a molecular to a circuit level—this argument is hard to dispute. One would expect, however, that predictable relationships would have emerged by now between different levels of brain functioning and clinical problems.

Fourth, operational, institutional, and procedural aspects of biological research in psychiatry have not provided the appropriate environment and incentives within which biological approaches could be developed to solve clinical problems. This argument focuses on the "doing of biological psychiatry research" and might need to be addressed by leaders of funding agencies, interest groups, and research organizations.

Lastly, by focusing the search on "mechanisms" that underlie psychiatric illnesses, progress has been directed toward understanding dysfunctional processes and symptoms, rather than on clinical course and risk/protective factors. Implicit in this approach, however, is the assumption that mechanistic understanding will provide better diagnosis or treatment. We argue here that the current mechanistic viewpoint may be insufficient at this stage, and that a predictive framework may be equally fruitful to bring neuroscience to contribute to clinical psychiatry. In this context, a computational approach can provide an important framework to link behavior to neural systems processes.

Mechanisms

The notion of a mechanism is tightly linked to causation, which can be defined as an antecedent event, condition, or characteristic that was necessary for the occurrence of the disease at the moment it occurred, given that other conditions are fixed (Rothman and Greenland 2005). Alternatively, a mechanism is, roughly speaking, a set of entities and activities that are spatially, temporally, and causally organized in such a way that they exhibit the phenomenon to be explained (Menzies 2012). Moreover, it has been highlighted that causal analyses aim to extract beliefs or probabilities that underlie observed data in both static and dynamic environments (Pearl 2010). However, causal relationships in complex systems are difficult to establish. In the context of disease and environment, Hill (1965) suggested a number of criteria in an attempt to distinguish causal from noncausal associations: strength, consistency, specificity, temporality, biological gradient (i.e., a dose-response curve), plausibility, coherence (i.e., consistency with the natural history and biology of the disease), experimental evidence, and analogy (i.e., similarities across diseases). A closer examination of examples of these criteria clearly shows that none of them are both necessary or sufficient to establish a clear causal relationship (Rothman and Greenland 2005). Moreover, there is clear evidence from carefully conducted clinical studies that causal relationships in psychiatry are difficult, if not impossible, to establish, even if many factors are considered. For example, Kendler used a propensity analysis approach to delineate covariates from causal risk factors for depression (Kendler and Gardner 2010). He concluded that dependent stressful life events, which were found to be most strongly associated with depression onset, had only a weak, if any, causal effect on the emergence of a depressive episode in the subsequent year. Further, a comprehensive analysis of the factors that influence the onset of a depressive episode shows that these factors cut across many different levels (genetic, psychological, social, economic), are highly interconnected, and differ between males and females (Kendler and Gardner 2014). These, and other results, led Kendler (2012:385) to conclude that "to develop an etiologically based nosology for psychiatric disorders is deeply problematic." Finally, in the development of increasingly sophisticated molecular approaches to understand psychiatric disorders, a silent assumption has been that one needs a more refined scale to clearly differentiate the pathophysiological processes that underlie these disorders. However, in a recent theoretical analysis of causal relationships between variables, Hoel et al. (2013) emphasized that the continued search for the "molecular cause" of a psychiatric illness may be fundamentally flawed. Specifically, they showed that one can construct interacting systems such that causal relationships emerge on a macro level but may not hold on a micro level and vice versa. Taken together, these findings suggest that at this stage a mechanistic emphasis to understanding mental disorders may delay neuroscience from making an impact for psychiatry.

We do not propose to do away completely with causal analyses, which are at the basis of mechanistic understanding of a process. Our standard statistical approaches are insufficient to clearly differentiate causal from noncausal associations. Recent attempts have been made to generate a more reliable mechanistically based quantitative theoretical framework (Pearl 2009b). Specifically, Pearl (2009a) contrasts standard statistical analyses (which aim to infer associations among variables and estimate beliefs, or probabilities of past and future events) and updates those probabilities in light of new evidence or new measurements with causal analysis. Causal analysis aims to infer not only beliefs or probabilities under static conditions, but also the dynamics of beliefs under changing conditions (e.g., induced by treatments or external interventions). Critical for this distinction, however, is to differentiate associational concepts; that is, any relationship that can be defined in terms of a joint distribution of observed variables against a causal concept, which is any relationship that cannot be defined from the distribution alone (randomization, influence, effect, confounding, "holding constant," disturbance, spurious correlation, faithfulness/stability, instrumental variables, intervention, explanation, attribution). It is important to emphasize that in psychiatry, it is very difficult to isolate causal relationships. Nevertheless, by implementing these advanced mathematical tools, we may be better able to delineate causation and, as a consequence, mechanistic frameworks for psychiatry. In the interim, however, a complementary framework may yield productive results.

Risk-Prediction Framework

One complementary approach to the sometimes elusive search for mechanisms is to embed a program of research into a risk-prediction framework. Risk-prediction models use predictors (covariates) to estimate the absolute probability or risk that a certain outcome is present (diagnostic prediction model) or will occur within a specific time period (prognostic prediction model) in an individual with a particular predictor profile (Moons et al. 2012b). The components of risk prediction (Gerds et al. 2008) consist of (a) a sample of *n subjects*, (b) a set of *k markers* obtained for each subject, (c) an individual subject *status* at some later time *t*, which can be a scalar or vector variable, and (d) a *model* which takes the sample and markers and assigns a probability *p* of the status at time *t* for each individual.

To be useful, a prediction model must provide validated and accurate estimates of the risks; the uptake of those estimates should improve subject (self-) management and therapeutic decision making, and consequently, (relevant) individuals' outcomes and cost-effectiveness of care (Moons et al. 2012a). Risk-prediction models can be derived with many different statistical approaches. To compare them, measures of predictive performance are derived from receiver operating characteristic (ROC) methodology and probability forecasting theory. These tools can be applied to assess single markers, multivariable regression models, and complex model selection algorithms (Gerds et al. 2008). The outcome probabilities or level of risk and other characteristics of prognostic groups are the most salient statistics for review and perhaps meta-analysis. Reclassification tables can help determine how a prognostic test affects the classification of patients into different prognostic groups, hence their treatment (Rector et al. 2012).

According to Cook (2007), one can compare the global model fit using a measure such as the Bayes's information criterion, in which lower values indicate better fit and a penalty is paid if the number of variables is increased. Moreover, one can compare general indices of calibration (e.g., the Hosmer–Lemeshow statistic, which compares the observed and predicted risk within categories) and discrimination (e.g., the c-statistic). In addition, if the overall fit for one model is better than another, but general calibration and discrimination are similar, one can assess whether the fit would be better among individuals of special interest. This would help to determine how many individuals would be reclassified in clinical risk categories and whether the new risk category is more accurate for those reclassified. Finally, one can assess utility of the risk-prediction model if it is based on an invasive or expensive biomarker, by determining whether a higher or lower estimated risk would change treatment decisions for the individual subject.

This general approach is similar to one proposed by Pencina and D'Agostino (2012), who argued that the incremental predictive value of a new marker should be based on its potential in reclassification and discrimination. In that sense, new potentially predictive (bio)markers should be assessed on their added value to existing prediction models or predictors, rather than simply being tested on their predictive ability alone (Moons et al. 2012b). The ultimate test of the effectiveness of a risk-prediction tool, like any other intervention, is a randomized clinical trial in which groups of doctors are randomized to use the tool in addition to usual care versus usual care alone (Scott and Greenberg 2010). In summary, the risk-prediction model framework has a number of advantages over a mechanistic framework: (a) a clear utilitarian approach, (b) sound statistical background, (c) a framework of iterative improvement, and (d) the ability ultimately to connect to and coexist with a mechanistic understanding of psychiatric disease. In the context of computational approaches, these aspects of the risk-prediction model framework provide clear guidance for the modeling approach: Can the underlying computational approach contribute substantially to the predictive value of the model?

Machine learning (Hastie et al. 2001) consists of a set of tools (e.g., support vector machines, random forest, recommender systems) that uses large data to understand the underlying structure (James et al. 2013). One can differentiate machine-learning tools into those that are supervised (i.e., models derived from inputs and outputs that are built for prediction or estimation) and unsupervised (i.e., models used to extract relationships and structure from multidimensional data). Machine-learning tools have found their way into the medical field for a large number of different applications: from the prediction of healthcare services (Padman et al. 2007) to clinical predictions of the progression of Alzheimer disease (Kohannim et al. 2010; Maroco et al. 2011). Random forest is one machine-learning tool that uses predictor variables to classify members of a sample into categories (e.g., relapse or abstinent). The forest is constructed from a multitude of decision trees (Breiman 2001). Whereas a single decision tree is susceptible to noise, the average of many trees, obtained by a forest, is not, so long as the trees are uncorrelated. A random forest performs as well or better than alternative classification techniques in terms of accuracy and robustness; even in the presence of noise, the model does not overfit to a given sample (Breiman 2001). One potential downside of random forest modeling is the black box nature of the model (Strobl et al. 2009). In fact, as Breiman (2001:23) states: "a forest of trees is impenetrable as far as simple interpretations of its mechanism go." Thus, similar to other machine-learning approaches, predictive utility may, in some circumstances, come at a cost of simple mechanistic interpretations.

Computational Approaches

While the above examples of predictive methods can be applied to all types of predictors (e.g., self-report and behavioral measures, including indicators of clinical severity and symptom types), we propose that they may be most powerful when used in combination with sophisticated inference models of beliefs and behavior. For instance, such generative models may be used at a first stage analysis to infer latent mental processes and states (associated with individual-level parameters). Such inferred states may include individuals' beliefs (e.g., about hidden reward rates of various choice options in the environment) or their decision policies (i.e., functions describing how they translate their expectations/beliefs about hidden variables into action). Recent contributions from machine learning and neuroeconomic research have highlighted several different computational approaches that can be used to infer underlying processing states from observed behavior. In our group, we have focused on two techniques, described below: optimal control and Bayesian ideal observer models.

Optimal Control

Inverse optimal control is a computational approach used to infer an individual's reward function of a goal-directed motor task, given observed behavior. Optimal control theory has been show to be an effective computational framework to explain human movements in continuous time (Todorov and Jordan 2002). This framework is particularly valuable when examining motor behavior; however, it can be extended readily to help understand the reward functions that drive motor behavior as part of a cognitive or affective paradigm. In this context, motor control in a goal-directed task is a dynamic process of sensorimotor integration, in which the brain takes sensory information, which includes paradigmatically specific instructions, and uses it to make continuous motor actions. Optimal control theory frames this dynamic process in a feedback control loop: the optimal controller estimates the current state at time *t*, produces a motor command based on the goal and keeps an efference copy (i.e., the expected outcome of the motor command) as the state estimator, then sends the motor command to muscles to generate the movement. The agent is thought to select actions which optimize performance on a task—the critical component of optimal control theory. In particular, the performance criterion is defined as a reward function that includes task-related performance measure and action cost. For example, in a task that instructs subjects to drive to a location A as quickly as possible, the performance measure can be the stopping distance to A; the action cost can be the accumulated effort of accelerating and decelerating controls. Individual differences are thought to emerge because subjects may have different target stopping distances and different weights to assess the ratio of the closeness to the target location over the action cost. The latter ratio defines the amount of effort one is willing to spend to achieve the intended stopping distance. Taken together, there are three components in the optimal control framework: (a) a *dynamic system* that describes how the states of the system evolve based on the action input, (b) an *action policy* that determines which action to take given the current state, and (c) a *reward function* that specifies the goal of this task (balance between goal state and action cost). A forward optimal control model generates a sequence of (optimal) actions, which maximizes the reward in the task. The goal of inverse optimal control model is to uncover the reward function assuming the optimality in observed action sequences. Proposed by Kalman (1964), inverse optimal control has been applied to study apprenticeship learning (Abbeel and Ng 2004), drivers' intention in simulated highway driving, and parking lot navigation Abbeel et al. (2008).

Using an optimal control framework (Figure 14.1) to study the behavioral processes and their dysfunction in subjects with psychiatric disorders has three main advantages. First, individuals with psychiatric disorders have been shown to have sensorimotor deficits, such as psychomotor disturbance in depressed individuals (Sobin and Sackeim 1997), which may have significant effects on the performance of effortful cognitive or affective tasks. In optimal control theory, sensory delay (i.e., the latency of an individual to respond to a visual or auditory stimulus) and motor delay (i.e., the speed with which an action plan can be carried out once the individual has selected a motor plan) can be incorporated in the dynamic system.

Second, individuals with certain psychiatric disorders may have altered reward processing; for example, individuals with depression are more sensitive

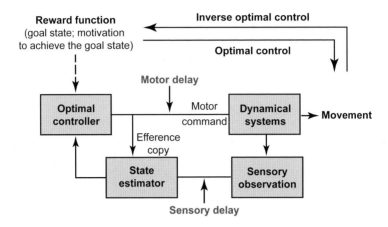

Figure 14.1   This schema shows the basic components of the inverse optimal control framework. Each component can be estimated from a subject's motor data.

to punishment than to reward (Must et al. 2006; Eshel and Roiser 2010). The imbalance of reward and punishment sensitivity in these subjects may affect the goal state, which may differ substantially from the experimenter-instructed target state. In optimal control theory, the goal state is a parameter in the reward function that corresponds to the individual's intended state of the object in control (e.g., the position and velocity of a car in a driving task). The closer the current state is to the goal state, the higher the reward.

Third, individuals with psychiatric disorders may lack the motivation (i.e., the amount of effort to spend to achieve the subjective goal state) to perform the task (Treadway and Zald 2011; Der-Avakian and Markou 2012). In optimal control theory, motivation is also a parameter in the reward function, which measures the ratio of the weights between the accuracy to achieve goal state and the action cost in the task. The higher the motivation, the more effort one is willing to spend to achieve high accuracy toward the intended goal state. Taken together, with the appropriate experimental manipulation, we can investigate if individuals can learn and adapt their action policies to different environments by changing the dynamical system, and if their reward function will change and thus improve their performance by changing the feedback provided (e.g., reward vs. punishment).

Bayesian Ideal Observer Models

A second computational framework which may help to extract predictive and potentially causative relationships in patients with psychiatric disorders is the

Bayesian ideal observer model or dynamic Bayesian model (DBM). DBM provides a computational framework which enables one to generate fine-grained quantification of emotion and cognitive processing as well as their interactions. The approach is to divide the observed behavior into several subprocesses, which can be submitted to test subtle hypotheses about changes in optimizing behavior. Similar to the inverse optimal control theory, DBM shares the basic assumption that changes in behavior observed in individuals with psychiatric disorders are a consequence of an altered optimization of available actions within the constraints of specific affective and cognitive processing dysfunctions. In particular, DBM is based on the notion that an individual has underlying beliefs and expectations about the situation at hand. This approach aims to quantify an individual's belief and expectation about their environment as a function of behavioral context and experienced choices and outcomes. DBM models provide a quantitative and explicit way to delineate how the brain processes complex environments, and how the breakdown of this process can contribute to the development of psychiatric disorders. Using this approach, we *infer* otherwise unknown beliefs in individuals regarding upcoming events and how such beliefs are updated based on past events experienced by the observer. This may be particularly important in a context of a risk-prediction model, when target populations exhibit very subtle or nondetectable behavioral differences on standard behavioral paradigms.

One example that shows how a simple experimental paradigm can be used to extract subtle but important cognitive control differences in healthy individuals and psychiatric subjects is the application of DBM to inhibitory control using the stop signal task. Specifically, Yu and colleagues used DBM to capture behavioral adjustments on a trial-by-trial basis of stopping behavior (Shenoy et al. 2010; Shenoy and Yu 2011; Ide et al. 2013). This model is based on the assumption that an individual updates the prior probability of encountering Stop trials, $P(\text{stop})$, on a trial-by-trial basis, based on trial history; it adjusts decision policy as a function of $P(\text{stop})$, with systematic consequences for Go response times and Stop accuracy in the upcoming trial. An optimal response when an individual assumes that it is more likely to encounter a Stop trial—that has a higher $P(\text{stop})$—is to slow one's response latency (i.e., exhibit a slower Go response time), which would result in a higher likelihood of correctly stopping on a Stop trial. This adjustment has been shown in two different experiments in healthy subjects (Ide et al. 2013; Harlé et al. 2014). To model the trial-by-trial adjustment of prior expectations, we used a Bayesian hidden Markov model adapted from the DBM (Yu and Cohen 2009; Ide et al. 2013). The model makes the following assumptions about subjects' internal beliefs regarding task structure: on each trial k, there is a hidden probability r_k of observing a Stop signal ($s_k = 1$ for Stop trial) and a probability $1 - r_k$ of observing a Go trial ($s_k = 0$); r_k is the same as r_{k-1} with probability α and is resampled from a prior beta distribution $p_0(r)$ with probability $1 - \alpha$. The predictive probability of trial

k being a Stop trial, $P_k(\text{stop}) = P(s_k=1 \mid \mathbf{s}_{k-1})$, where $\mathbf{s}_k = (s_1, ..., s_k)$ is a vector of all past trial outcomes, 1 for Stop trials and 0 Go trials, can be computed as:

$$P(s_k = 1|s_{k-1}) = \int P(s_k = 1|r_k)p(r_k|s_{k-1})dr_k$$
$$= \int r_k p(r_k|s_{k-1})dr_k = \langle r_k|s_{k-1}\rangle. \tag{14.1}$$

Predictive probability of seeing a Stop trial, $P_k(\text{stop})$, is the mean of the predictive distribution $p(r_k|s_{k-1})$, which, by marginalizing over the uncertainty of whether r_k has changed from the last trial, becomes a mixture of the previous posterior distribution and a fixed prior distribution, with α and $1 - \alpha$ acting as the mixing coefficients, respectively:

$$p(r_k|s_{k-1}) = \alpha p(r_{k-1}|s_{k-1}) + (1-\alpha)p_0(r_k). \tag{14.2}$$

Posterior distribution over Stop trial frequency is updated according to Bayes's rule:

$$p(r_k|s_k) \alpha P(s_k|r_k)p(r_k|s_{k-1}). \tag{14.3}$$

The DBM model further assumes a positive linear relationship between trial-wise $P(\text{stop})$ and reaction times at the individual level. That is, on a given trial, the higher the expected likelihood of encountering a Stop signal, the more a person should slow down to avoid making a Go error. For each parameter setting (i.e., each pair of alpha and the prior distribution mean), the corresponding $P(\text{stop})$ sequence can be inferred, and linear regression can be used to determine the optimum parameter values providing the strongest correlation coefficient or R square coefficient between $P(\text{stop})$ and reaction times. In our previous work, we have found that subjects' prior resampling rates to be best captured with alpha values between .6 and .8 (Shenoy and Yu 2011; Ide et al. 2013; Harlé et al. 2014).

An Example Study: Predicting the Emergence
of Problem Stimulant Use

While significant executive deficits have been demonstrated in chronic stimulant dependence (Salo et al. 2002; Monterosso et al. 2005; Hester et al. 2007; Tabibnia et al. 2011), only subtle behavioral impairments in error monitoring and inhibitory control have been observed in individuals at risk for stimulant dependence (Colzato et al. 2007; Reske et al. 2011). Thus, in the following example, we applied DBM to the analysis of event-related functional magnetic resonance imaging (fMRI) data associated with baseline inhibitory function during a stop signal task to predict clinical status three years later. Previously, healthy volunteers (Ide et al. 2013) and individuals at risk for stimulant use disorder (Harlé et al. 2014) were shown to adapt their response strategy in this inhibitory control paradigm. Thus, we hypothesized that computational models might help identify neural substrates that contribute to subtle inhibitory deficits. Such computational neural variables were hypothesized to perform significantly better than other variables, such as noncomputational task-based brain activity and clinical measures (e.g., cumulative drug use), in predicting long-term clinical status (Harlé et al. 2014).

We recruited occasional stimulant users (OSUs) from the student population of different local universities. OSUs were defined primarily as having (a) at least two off-prescription uses of cocaine or prescription stimulants (amphetamines and/or methylphenidate) over the past six months and (b) no evidence of lifetime stimulant dependence. Participants completed a baseline interview session to evaluate clinical diagnoses and determine current patterns of drug use as well as a neuroimaging session that examined brain and behavior responses during decision making; they completed a stop signal task while being scanned. We were able to follow these OSUs for three years, after which they completed another standardized interview (phone or in-person) which examined the extent of drug use over the three-year interim period (using the SSAGA II). Two groups of interest were identified: problem stimulant users (PSUs) and desisted stimulant users (DSUs). PSUs were *a priori* defined by (a) continued stimulant use since baseline interview and (b) endorsement of 2+ symptoms of DSM-IV amphetamine and/or cocaine abuse and/or dependence criteria occurring together 6+ contiguous months since the initial visit. DSUs endorsed (a) no 6-month periods with 1+ stimulant uses and (b) no symptoms of interim stimulant abuse or dependence.

Using a split-sample approach, we first identified potential predictive neural regions with voxel-wise robust logistic regressions to predict three-year follow-up status (coded 1 = PSU vs. 0 = DSU) in a randomly selected "training" subset of our sample. The remaining "test" subset was used to assess the relative predictive power of the activation clusters identified with the training sample by using random forest analysis (Breiman 2001). In this study, we ran three random forest analyses, each with a distinct set of baseline variables to compare the overall performance of (a) drug-use measures (total uses of stimulants, cocaine, and marijuana, based on self-report), (b) categorical fMRI regressors (task-based contrasts such as Stop vs. Go, Stop Success vs. Stop Error), and (b) Bayesian/computational fMRI regressors, respectively. To construct those regressors, we first convolved three types of trials (Go, Stop Success/SS, and Stop Error/SE) with a canonical hemodynamic response function in a general linear model (GLM). Each of these predictors were entered both as linear regressors and parametrically modulated by trial-level P(stop) estimates. This model allowed us to isolate neural activations associated with both trial type alone (i.e., categorical regressor) and P(stop). Thus, after deconvolution, this model included six task regressors. Three were categorical: Go, SS, SE. Three were model-based parametric: Go $\times P_k$(stop), SS $\times P_k$(stop), SE $\times P_k$(stop). A second GLM was created with trial-wise Bayesian signed prediction error

(SPE), defined as Outcome – *P*(stop), and unsigned prediction error (UPE), defined as |Outcome – *P*(stop)|, included as parametric regressors of interest. Individual subjects' percent signal change (%SC) scaled beta weight values for five regressors; contrasts of interest from these two GLM models were extracted and used as independent variables in the prediction analyses. The categorical regressors included two contrasts: (a) (Stop – Go), that is, (SS + SE)/2 – Go and (b) (SE – SS). The Bayesian regressors included three computational predictors: (a) *P*(stop), that is, ½*Go × P_k(stop) + ¼ *SS × P_k(stop), + ¼*SE × P_k(stop); (b) UPE; and (c) SPE. For full description of these first-level fMRI analyses, see Harlé et al. (2014).

Based on logistic regressions in the training sample, predictors in the full model included activations extracted from 21 ROIs identified with robust logistic regressions, including three ROIs for trial type-independent *P*(stop) activation, six ROIs associated with Bayesian UPE activation—UPE: Outcome – *P*(stop)—and twelve ROIs associated with SPE activation—SPE: |Outcome – *P*(stop)|. Based on random forest analyses in the test sample, we found that:

1. no variable met criteria for inclusion in the drug-use model, which had an overall accuracy of 52%;
2. only one variable met criteria for inclusion in the fMRI categorical predictor model (i.e., SE – SS contrast activation in the rostral anterior cingulate cortex), but with an overall accuracy of 64%, which was not significantly different statistically from the no-predictor model based on response rate alone; and
3. four variables met criteria for inclusion in the fMRI computational predictor model, including UPE activation in right thalamus as well as SPE activation in right anterior insula/inferior frontal gyrus, in the right superior medial prefrontal cortex/dorsal anterior cingulate cortex (BA32), and in right caudate (BA25).

Notably, this final model yielded an overall accuracy of 74%, which represents a statistically significant improvement in accuracy from the model based on response rate alone.

The utility of the computational approach can be most easily displayed using a Bayesian nomogram (Figure 14.2). The vertical axis of the left-hand side of the nomogram shows the prior probability of developing problem use, or the proportion of the total sample that showed problem use. The vertical axis of right-hand side shows the posterior probability of problem use given a positive or a negative test result, respectively. The vertical axis in the center displays the positive or negative likelihood ratio, which is the most important characteristic of a test in terms of linking the knowledge before applying the test to the knowledge once the test has been conducted and found to be either positive or negative. In this instance, we used the random forest model as the basis for the testing procedure. The upper and lower brackets around the central estimate represent the 95% confidence interval of the post-test probability, providing

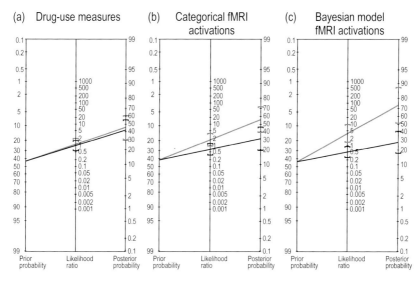

Figure 14.2 Bayesian nomogram for (a) drug-use measures, (b) categorical fMRI activation measures, and (c) Bayesian model-based activation measures. The positive likelihood ratio (gray) and negative likelihood ratio (black) show that the test based on the Bayesian ideal observer model clearly provide the best separation. For example, based on a base rate of approximately 57%, a positive test indicated a 74% chance of becoming a problem user, whereas a negative test reduces the chance to approximately 28%.

a graphical means of indicating whether the test measurably improved our knowledge; that is, yielded a higher or lower post-test probability without the confidence interval including the pretest probability. Thus, when the 95% confidence intervals do not intersect, positive and negative tests are statistically significantly different.

For all four computational predictors, larger neural responses negatively correlated with Bayesian prediction errors were associated with a higher likelihood to be categorized in the PSU group three years later. Specifically, for every standardized unit increase in UPE deactivation in right thalamus, one was about three times as likely to develop a future stimulant-use disorder (odds ratio = 3.45, $p < .05$). In addition, an individual was two to three times as likely to be categorized in the PSU group for every standardized unit increase in SPE deactivation in medial prefrontal cortex/anterior cingulate cortex (odds ratio = 2.44, $p < .05$), anterior insula/ inferior frontal gyrus (odds ratio = 3.19, $p < .05$), and caudate (odds ratio = 3.02, $p < .05$). To summarize visually the relative predictive power of the three predictive models considered, we used bootstrapped robust logistic regressions to produce cumulative ROC curves associated with each added layer of predictor types. As seen in Figure 14.3, the computational predictors added in the last layer (black line) significantly increased accuracy, including both sensitivity and specificity.

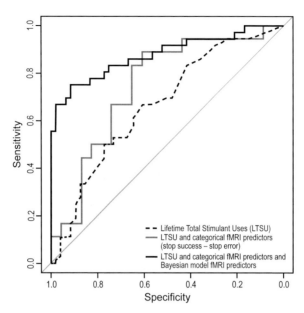

Figure 14.3   Receiver operator curve for three different predictor models. Lifetime stimulant use is indicated by the dotted black line; lifetime stimulant use plus fMRI and behavioral measures in gray; and lifetime stimulant use, fMRI, and Bayesian model parameters are shown in black. The Bayesian ideal observer model parameters add significantly to increased sensitivity and specificity of the model.

In this study we sought to determine whether the combination of functional neuroimaging and computational approaches to behavior are able to generate predictions that can help to determine whether an individual will progress to problem use. Using the combination of Bayesian ideal observer model, stop signal task as a measure of inhibitory control, and fMRI imaging, we show that those individuals who demonstrated greater neural processing, related to a Bayesian prediction error, are more likely to develop a future stimulant-use disorder over the subsequent three years. For this, we used a Bayesian ideal observer model to quantify an individual's belief about the likelihood of an upcoming Stop trial (i.e., the person's probabilistic expectation of having to mount an inhibitory response during a stop signal task). Importantly, these data were collected in individuals at risk for stimulant-use disorder three years prior to the assessment of the outcome (i.e., whether the subject would progress to problem use or desist using). Cross-validated robust regression and random forest analyses showed that neural responses associated with Bayesian model-inferred prediction errors (representing the trial-wise discrepancy between expectation of a Stop trial and actual trial outcome) in right anterior cingulate cortex, anterior insula, caudate, and thalamus most robustly predicted three-year clinical status (i.e., meeting criteria for stimulant-use disorder vs. desisted-use status). These computational neural variables significantly contributed predictive validity above the base rate, which was not the case for other baseline predictors *a priori* thought to be promising, such as reported lifetime drug use or non-model-based neural predictors. Therefore, this study shows that, in principle, a Bayesian cognitive model applied to an event-related neural activity can be used to predict long-term clinical outcome. Taken together, these results are consistent with the notion that the combination of functional neuroimaging and computational modeling can provide better predictions of future clinical states.

Future Directions

This example is one among a series of emerging studies in neurology and psychiatry that use machine-learning approaches in the context of risk-prediction models to generate individual-level predictions (Perlis 2013; Moradi et al. 2015). The emphasis of computational approach, up to now, has been on improving a mechanistic understanding of behavior in complex situations. For example, temporal difference models (Schultz et al. 1997) provide a clear and convincing framework for the acquisition of reward (Daw and Touretzky 2002) and aversive learning (Iordanova 2009) in both animals (Schultz and Dickinson 2000) and humans (Klein-Flugge et al. 2011). The extension to Bayesian models has been based on the notion that humans utilize not only information about experienced averages but also about the underlying distribution; that is, the degree of uncertainty associated with the experience (Behrens et al. 2007). The use of computational models provides a powerful technique to disambiguate processes that result in an observable behavior and can thus be used to make inferences about processes which constrain behavior in a way that is observed in psychiatric populations (Huys et al. 2011).

There are, however, several caveats that one must keep in mind when applying these models in psychiatry. First, our diagnostic descriptions of patients are at best initial phenomenological approximations of heterogeneous subgroups of individuals (Insel and Cuthbert 2015). Consequently, behavioral dysfunctions are likely to result from different underlying pathologies associated with different computational processes. In other words, it is unlikely that individuals with anxiety or depression will show a uniform computational dysfunction.

Second, psychiatric illnesses are a mixture of long-range dysfunctions, best captured by trait variables and momentary dysregulation, which are assessed using state measures. Moreover, there may be different stages of psychiatric illnesses based on the recovery from the illness process itself. For example, individuals who show substance-use disorder might undergo prolonged recovery of function, which may result in changes of the computational process that guides the decision making. Thus, it will be important to examine psychiatric populations at different stages of illness to better understand the dynamics of the underlying process dysfunction.

Third, Paulus (2007) has previously proposed that decision making is a homeostatic process closely related to the physiological state of the body. As a consequence, computational processes that underlie the selection of an option might be sensitively affected by the individual's body state. For example, decision making focused on the selection of food items are highly dependent on the satiety state of the individual (Haase et al. 2009). Therefore, it should be clear that computational processes may need to be examined in the context of different motivational states. Taken together, the use of computational models in psychiatry to explain the underlying mechanism of behavior is promising, but at an early stage and will require many future studies to delineate some of the issues raised above.

Alternatively, computational approaches may have a more immediate impact on psychiatry by providing better prediction models. In this context, the value of the computational approach is to modify the receiver operator statistic that determines where to set a cut off for a positive or negative test, or to improve the likelihood ratio of the test being developed. The goal here would be to use underlying belief updating models to improve the prediction of future behavior or future clinical outcomes. This approach does not rely on a particular disease category (i.e., whether an individual has major depressive disorder, dysthymia, or bipolar depression). Instead, the risk-prediction model framework aims to exploit individual differences to make better predictions. However, several issues need to be taken into account:

Clinically relevant and robust predictions require large data sets. Currently, most studies, with a few exceptions (Whelan et al. 2014), are based on relatively small samples. Thus, it will be important to collect data sets that are based on "real" patient populations of sufficient size to be able to make robust predictions.

In addition, it is unclear at which level one is best able to delineate a causal pathway to the pathology in psychiatric illnesses. The Research Domain Criteria approach (Insel et al. 2010) relies on the assumption that more basic molecular levels will eventually result in a better causal prediction of the emergence and maintenance of psychiatric illnesses. This assumption may, however, be deeply flawed, as greater clarity at the molecular level might not necessarily yield stronger causal relationships (Hoel et al. 2013). In fact, the presence of a romantic relationship in an adolescent's life resulted in the single strongest predictor of the emergence of binge drinking (Whelan et al. 2014), whereas genetic and neuroimaging markers were only weakly predictive. The use of a risk-prediction framework will act as an arbiter of what is the best information to be clinically useful. Moreover, it may also act as a pointer to show us where to carve nature at its joints. Computational approaches might well play an important role here, but we are still early in this endeavor.

15

A Valuation Framework for Emotions Applied to Depression and Recurrence

Quentin J. M. Huys

Abstract

The burden of depression is substantially aggravated by relapses and recurrences, and these become more inevitable with every episode of depression. This chapter describes how computational psychiatry can provide a normative framework for emotions and an integrative approach to core cognitive components of depression and relapse. Central to this is the notion that emotions effectively imply a valuation; thus they are amenable to description and dissection by reinforcement-learning methods. It is argued that cognitive accounts of emotion can be viewed in terms of model-based valuation, and that automatic emotional responses relate to model-free valuation and the innate recruitment of fixed behavioral patterns. This model-based view captures phenomena such as helplessness, hopelessness, attributions, and stress sensitization. Considering it in more atomic algorithmic detail opens up the possibility of viewing rumination and emotion regulation in this same normative framework. The problem of treatment selection for relapse and recurrence prevention is outlined and suggestions made on how the computational framework of emotions might help improve this. The chapter closes with a brief overview of what we can hope to gain from computational psychiatry.

The Candidate Clinical Issue: Recurrence

One episode of an acute mental health illness is bad enough. Unfortunately, many psychiatric disorders are chronic, with episodes of wellness repeatedly punctuated by relapses in which symptoms reemerge. The relentless repetition is an important factor in the burden these disorders impose globally and individually, and it means that treatment of the early illness stages is crucial in avoiding an unremitting decline of well-being.

In this chapter, I focus on depression: a disease that is highly recurrent and imposes a heavy burden on the world's citizens (Whiteford et al. 2013). Of those who experience a first episode of depression, around 40% will remain well and never go on to have a recurrence (defined as a reemergence of symptoms after a substantial period of absence) (Frank et al. 1991). The remaining 60% with a single episode will go on to develop further episodes. Thereafter, the proportion with unremittingly recurrent disease rises: 70% of those with two episodes will experience a third, and 90% of those with three episodes will experience a fourth (APA 2000). This process has been recognized since the time of Kraepelin. Hence, around 40% will have a single lifetime episode, at most 20% two episodes, 5–10% three episodes, and the remainder—a staggering 40% or more—will have more than three episodes (Angst 1992). At an average duration of 6–9 months per episode, this entails many ill years, and these numbers might even be optimistic (Hollon et al. 2006; Posternak et al. 2006). In addition, any one episode has about a 5–10% risk of becoming chronic, i.e., lasting for two years or more, with the risk for chronicity and recurrence being partially independent (Hollon et al. 2006). Hence, the early episodes are of fundamental importance as they mark a transition between those whose depression will have a relatively benign outcome and those who will experience a malignant course (Keller et al. 1983; Monroe and Harkness 2011).

Both antidepressant medication (ADM) and psychotherapy have proven utility in preventing relapses and recurrences. However, their differential indication is poorly understood, and it is likely that currently much can be gained from improving the targeting of treatments. For instance, DeRubeis et al. (2014) examined the ability to differentially predict treatment response to antidepressants and cognitive therapy based on standard clinical characteristics. Some variables were prognostic, predicting treatment response independent of modality (e.g., severity, age, IQ, chronicity), whereas others were prescriptive and differentiated between the modalities (e.g., marriage, employment, comorbid personality disorder, numbers of life stressors, and previous ADM trials). Using a simple general linear model combined with a standard machine-learning approach, DeRubeis et al. suggested that they could improve allocation so as to produce an average improvement of over 3 points on the Hamilton Rating Scale for Depression, which is clinically significant. McGrath et al. (2013) reported that insular metabolism measured by FDG-PET differentiated those who would respond to an ADM from those who would respond to psychotherapy. Others, however, have suggested that the subgenual anterior cingulate activity may differentiate those likely to respond to cognitive therapy from those likely to respond to ADMs (DeRubeis et al. 2008; Roiser et al. 2012). Using quantitative EEG to allocate patients to antidepressant treatment, DeBattista et al. (2011) reported an improvement of about 3 points on the Quick Inventory of Depressive Symptomatology—again, clinically a very significant finding. To date only a few attempts have been made to use such methods to address the problem of predicting relapse. Among these, the induction of dysfunctional attitudes with sad mood induction (Teasdale 1988; Segal et al. 1999, 2006) and variations in neuroimaging responses to sad movies (Farb et al. 2011) stand out. However, there appears to be no work on predicting specifically who will relapse after discontinuing ADMs. In addition to their clinical value, robust predictors of differential response might lead the way toward a novel understanding of the neurobiology of psychiatric disorders in terms of neurobiological features that are directly relevant for treatment.

Here, I propose that computational techniques are well placed to help us attain a better understanding of affective processes that are crucial in transforming depression from a mild and temporary into a chronic and disabling condition. I begin by introducing a computational framework for emotions that focuses particularly on model-based evaluations. Then I discuss aspects of model-based valuations relevant to depression relapse and attempt to outline a normative framework for phenomena such as stress sensitization and emotion regulation. Thereafter, a brief overview of the clinical management of relapses is provided, followed by specific candidate applications. I conclude with a critique of this approach and discuss its key limitations.

Before beginning, it is worth noting the distinction between recurrence and relapse: recurrences occur further apart in time and with more profound resolution of symptoms between the two exacerbations (Frank et al. 1991). The statistics of depressive episodes, however, do not reveal obvious points for such a categorization (Burcusa and Iacono 2007; Monroe and Harkness 2011). Instead, they point to degrees of risk for future exacerbations (Judd and Akiskal 2000; Judd et al. 2000; Dunlop et al. 2012). Variations in the definitions of relapse and recurrence are thus an important caveat, in terms of comparing and integrating across studies.

Valuation as a Framework for Emotions

Computational psychiatry provides broadly two approaches (Huys et al. 2016). The descriptive approach employs general, neuropsychologically agnostic methods to describe the relationship between neuropsychological measurements and some outcome of interest (e.g., applying regression to predict treatment response) (DeRubeis et al. 2014). The mechanistic approach is very different as it aims to describe how phenomena of interest come about using far more complex accounts of the data, ranging from biophysically realistic models of network dynamics to processes describing learning and decision making. Both of these complex accounts can be used to identify key variables, and by fitting models to data, these variables can be efficiently measured and improve the performance of descriptive approaches (Wiecki et al. 2015; Huys et al. 2016).

In approaching depression mechanistically, the first issue that needs addressing is how computational techniques could help explain such subjective and ethereal phenomena as emotions (Huys et al. 2011; Maia and Frank 2011; Montague et al. 2012). At the core of this suggestion is the link between

valuation and emotion (Huys et al. 2015a). Emotions are strongly characterized by subjective qualia which are easy to catalog but hard to comprehend.
Following the animal literature, however, one can focus on their behavioral
correlates and refer subjective aspects to future investigations, pending a better
understanding of consciousness itself. The key argument is as follows: To the
extent that emotions influence behavior, they implicitly implement a process
that assigns some behaviors a higher value than others. This opens up the possibility to describe emotions by considering the processes that give rise to valuations. Accordingly, the subjective phenomena entering emotional awareness
are introspective correlates of the primary valuation processes.

Accounts of valuation have benefited immensely from reinforcement-learning approaches (Sutton and Barto 1998), and nowhere most clearly than in the
elucidation of the phasic dopamine signals in learning (Montague et al. 1996),
the puzzling relationship of which was found to relate closely to a computationally precise learning signal. Reinforcement-learning techniques address
one of two fundamental inferential problems that the brain faces: actions taken
now influence which options will be available in the future. For instance, the
house you buy now influences how easily you will be able to accept an exciting
job offer at a faraway university next year. Hence, optimal behavioral choice
requires the appropriate assessment of vast future consequences. Approaches
to this problem come in at least two fundamentally different shapes (Sutton
and Barto 1998) that are behaviorally, neurobiologically, and computationally
discernible (Killcross and Coutureau 2003; Daw et al. 2005, 2011; Deserno et
al. 2015). *Model-based valuation* depends on an understanding of the structure
of the world. Choices are valued by inferring their future consequences according to one's understanding encapsulated in a model *M* that describes the
consequences of taking specific actions in different situations. This requires
processing power, but it is powerful and flexible. *Model-free valuation*, by contrast, assigns values to states or stimuli by virtue of their past association with
rewards or losses. At the time of choice, model-free values are computationally
cheap, but they demand substantial experience to be accurate. Hence, these
two systems trade experiential for computational costs. One changes slowly
with experience; the other does so rapidly but requires substantial cognitive
resources.

Generally speaking, these two theoretical components can be mapped onto
two broad currents of current thinking about emotions: automatic and cognitive accounts, respectively. The automatic account emphasizes that stimuli
can activate emotional centers and dictate responses, largely foregoing any
contact with cognition, such as when innately relevant stimuli result in reflex-
ive approach, fight, flight, or fright responses. Hirsch and Bolles (1980), for
instance, showed that innate defensive responses to predators can breed true
over multiple generations without contact to the predator. In contrast, cognitive theories of emotions (e.g., Beck 1967; Lazarus 2006) argue that many
emotions experienced by humans *follow* cognitive appraisals. The emotions

evoked by complex, ambiguous stimuli that are devoid of evolutionary importance depend on attributions and interpretations, linking them to causes that have meaningful consequences or relate to innately relevant stimuli. This insight arguably forms the cornerstone of cognitive therapies for a variety of psychiatric disorders (Beck 1967) and provides a way by which emotions (and their consequences) can be regulated.

Distinguishing between different types of values renders this argument more concrete. In Pavlovian scenarios, values $V(s)$ are attached to stimuli, states, or situations s independently of behavior; in instrumental scenarios, $Q(s, a)$ values are attached to a combination of stimuli (or states or situations) in combination with particular behaviors a; that is, values are attached to stimulus-action pairs s, a (Mowrer 1947; Dayan and Balleine 2002; Daw et al. 2005, 2006). In this theory, both of Q and V values can be derived through model-free or model-based mechanisms, leading to a quartet of values (Huys et al. 2014): model-based and model-free Pavlovian values $V^{MB}(s)$ and $V^{MF}(s)$, and model-based (MB) and model-free (MF) instrumental values $Q^{MB}(s, a)$ and $Q^{MF}(s, a)$. Importantly, however, only the latter Q values directly inform action choice. The former, Pavlovian V values, do not as they make no reference to actions a, only to stimuli s. The mapping from stimulus values s to actions depends on mappings $m^{Fix}(s, a)$ between the stimuli and responses, which must be determined separately. Hence, while instrumental Q values can be modified to theoretically implement any kind of behavior, Pavlovian values V are effectively restricted to modulating otherwise specified fixed-response mappings $m^{Fix}(s, a)$. Consider the innate link between positive valuation and approach. In a striking example of the possibly maladaptive nature of auto-shaping, Hershberger (1986) had a hungry chick on a linear track facing a food tray. The food tray was mobile, and moved in the same direction as the chick, but at twice its speed. To get close to the food, the chick had to run away from the food as quickly as possible—an impossible task. This suggests that the stimulus food was innately linked to approach behavior; that is, $m^{Fix}(food, approach)$ had a high effective "value." Akin to Hershberger's chicks, humans are easily able to learn to emit a Go action for rewards and to withhold active responses to avoid losses, but they perform poorly when they have to choose No-Go to earn a reward or Go to avoid losses. The extent to which rewards interfere with No-Go, and losses with Go, is well captured by allowing the expected value of the stimulus, throughout learning to evoke Go and No-Go behaviors (Guitart-Masip et al. 2012).

Here I propose that model-based Pavlovian stimulus values might capture important aspects of the cognitive view of stimulus-bound emotional responses. Situations or stimuli are examined within a greater interpretative framework (the "model" $\mathcal{M}$). The resulting model-dependent valuation leads to the recruitment of particular species-specific "emotional" response patterns (and, as a correlate, also subjective qualities). Through the process of model-based valuation, a stimulus s might also become predictive of other future stimuli s'

and thereby come to recruit the fixed-response mappings associated with these. Second, conversely, model-free stimulus values $\mathcal{V}^{MF}(s)$ might map better onto more automatic routes to acquiring emotional responses, whereby repeated exposure to a contingency between a conditioned stimulus and an unconditioned stimulus leads the conditioned stimulus to acquire the same value as the unconditioned stimulus, thereby triggering the same or similar innate response patterns by reference to past experience rather than by reference to an interpretative model. Third, it is important to consider the interaction between these processes. Finally, theories of valuation have been very useful in understanding the function of neuromodulators (i.e., the putative substrates of the majority of psychotropic medications). Due to space constraints, this will not be discussed here, and the reader is referred to existing work (Montague et al. 1996; Frank et al. 2004; Frank 2005; Yu and Dayan 2005; Niv et al. 2007; Dayan and Huys 2009; Cools et al. 2011).

Model-Based Valuation in Depression

This computational framework for emotions can capture facets of the cognitive features of depression, particularly those with relevance to relapse: hopelessness, stress sensitization, rumination, and emotion regulation.

Stress and Hopelessness

Stressors have an important and causal role in the onset of depressive episodes (Kendler et al. 1999, 2000). Just like initial episodes, recurrences can also be provoked by stress: subjects who score higher on measures of severe life events after recovery have a higher risk of enduring a recurrence (Monroe et al. 1986). However, the experience of depression also leaves a "scar" (Burcusa and Iacono 2007) and stress maligns more with every further episode. Specifically, while the first episode is strongly associated with prior severe life events, subsequent episodes are less dependent on a severe life event (Kendler et al. 2001) and occur more autonomously or with less severe events (Kendler et al. 1995; Wichers et al. 2007). In addition to this "kindling effect," depression also influences people's behavior such that they self-select into high-risk situations or manage difficult situations poorly. In an effect known as *stress generation* (Hammen 1991), with every episode the rate of further severe life events increases (Harkness et al. 1999; Kendler et al. 2001; Liu and Alloy 2010); this increases the risk of further episodes, even in the absence of sensitization.

Stress arises from an interaction between stimuli that are potentially threatening and individual situational appraisals (Lazarus 2006). Whether any one particular stressor (e.g., a relationship breakup) results in a depressive episodes depends on the individual's perception of it (it might come as a relief!). Thus models which formalize the complex impact of stressors play an important role in our understanding of depression, and several theories of depression have emphasized that depression is characterized by cognitive features (Abramson et al. 1989; Alloy et al. 1999) that have the potential to transform stressors malignantly. A particularly useful model is that of learned helplessness (Maier and Watkins 2005), where healthy animals are exposed to controllable or uncontrollable stress. Depression-like signs and symptoms only arise if the shock is perceived as uncontrollable, and this is learned from the experience of the shocks. In terms of the current argument, these inferential processes can be readily captured through the use of model-based valuation $Q^{MB}(s, a)$ (Huys and Dayan 2009; Lieder et al. 2013). As animals experience the shocks, they also learn about the relationship between shocks and behavior. To the extent to which this learning involves modifying a general prior belief about how controllable desirable outcomes are, this experience will generalize to model-based valuations in other situations. Normatively speaking, generalization between different situations should depend on similarities between these situations. One similarity is the presence of the agent, and hence inferences about the agent's own abilities support generalization and are naturally global, stable, and internal, which are the core features of hopelessness (Abramson et al. 1989).

The influence of a prior on controllability in model-based valuation also provides reasonable accounts for depressive symptoms more generally. A belief about a lack of control impacts the expected exploitability of options in the world, and thereby the opportunity cost for behavior. This variable, which has been linked to tonic dopamine (Niv et al. 2007; Hamid et al. 2016) and affects the level of energy expenditure, is a reasonable substrate for the core components of perceived loss of energy, psychomotor retardation, and diminished drive, all of which have similar discriminatory power as anhedonia (McGlinchey et al. 2006; Mitchell et al. 2009). A wealth of other features of reward and loss sensitivity also attests to the involvement of model-based valuation and has been reviewed elsewhere (Huys et al. 2015a).

The modification of beliefs and their influence on valuation in model-based systems provides a formalization of cognitive theories of depression. In terms of relapse, this formulation could also account for the "scar" left by previous episodes. As the prior is updated and learned, it becomes stronger and can now be combined with a weaker likelihood term (fewer data) to provide strong beliefs that a particular situation is uncontrollable. Such a process may capture the finding that the dependence of depressive episodes on severe life events weakens because depressive episodes would more easily be triggered by weaker events (Kendler et al. 2001). A helpless prior belief would also impair the ability to exploit opportunities in the environment, both in terms of reward seeking and loss avoidance. If the immediate future is poorly controllable, the more distant future is entirely uncertain and the immediate cost prevails over long-term gains (Huys et al. 2009). This impairs the ability to make appropriately farsighted decisions and would particularly hurt in difficult situations that demand careful, farsighted behavior, hence increasing the risk of further severe life events (Hammen 1991; Liu and Alloy 2010).

This account hinges on the impact of stress on recurrence being mediated by the worsening of cognitive biases, which is a controversial topic. On one hand, cognitive distortions measured by the Dysfunctional Attitudes Questionnaire appear to be sensitive to the depressive state, to recover after an episode (Teasdale 1988), to have predictive value for recurrence after induction of sad mood (Segal et al. 1999, 2006), and to interact with stress in predicting depressive onset (Lewinsohn et al. 2001). On the other, negative cognitive styles—a more direct measure of helpless thoughts and beliefs (Alloy et al. 2000; Haeffel et al. 2008)—mediate medium-term effects of stressful experiences on depressive symptoms in (Haeffel et al. 2007), differ between remitted and never depressed subjects (Haeffel et al. 2005), and predict longitudinally depressive symptoms (Iacoviello et al. 2010; Pearson et al. 2015). In the study by Lewinsohn et al. (2001), however, attributional style predicted depression only for mild levels of stress, and neither this nor dysfunctional attitudes interacted with a history of depression in predicting onset. Hence, cognitive styles motivated by helplessness theory appear, overall, to satisfy these requirements, although whether and to what extent they mediate the kindling effects of stress or the self-generation of stress is as yet unknown.

Metareasoning, Rumination, and Emotion Regulation

The computational models discussed thus far have made little reference to the specifics of the algorithmic implementation of the biases, but have just described their overall existence and consequences. One overarching point is that model-based inference is computationally extremely costly. Because the future ramifications of current choices are so exponentially vast (looking d steps ahead when there are n options at every steps involves evaluating n^d combinations), their extensive consideration either requires (a) unreasonably large computational resources, thus relying on a plethora of approximations and shortcuts (Huys et al. 2012, 2015c) that are potentially relevant for psychiatry (Huys et al. 2015b), or (b) sacrifices to be made in terms of some of the key features of model-based cognition (Daw and Dayan 2014).

In addition, this resource constraint leads to the metareasoning problem of how to allocate internal computational resources optimally (Anderson and Oates 2007; Hay and Russell 2011). The subject faces both an external problem of to evaluate actions, and an internal problem of how to allocate cognitive resources to the evaluation of particular behavioral options (e.g., to find the best action). The latter problem implies a decision problem about the decision problem: having considered a particular option, what option should next be considered?

This view of metareasoning allows us to consider further aspects of depressive cognition, including rumination, emotion regulation, and cognitive control, in the formal framework of valuation. Rumination is an internal predilection to focus on the causes, meanings, and consequences of depressive symptoms (Nolen-Hoeksema 1991) and is driven in part by the kinds of features which drive helplessness, being particularly prominent in those with high level of chronic burden (e.g., housework, caring for children or elderly) and a low sense of mastery (Nolen-Hoeksema et al. 1999). Hence, this raises the tantalizing possibility that a prior belief about lack of control could render all possible options equally unappealing (Huys and Dayan 2009), and thereby impair the ability to identify useful options to consider, with attempts to identify a valuable option resulting in a persistent preoccupation with the status quo.

Aspects of emotion regulation can be viewed similarly. Emotion regulation refers to cognitive, behavioral, and other strategies used to control which emotions are experienced (Gross 1998). Reinterpretation, for instance, turns a half empty glass into a half full glass. If emotions are viewed as the (automatic) recruitment of innate fixed-response patterns $m^{\text{Fix}}(s, a)$ by model-based valuations $\mathcal{V}^{\text{MB}}(s)$, then emotion regulation can either refer to the modulation of the link between valuation and the response pattern, or to the modulation of the valuation itself. The former would involve altering the metareasoning selection process, for instance by focusing internal evaluations more on those values likely to be positive. This distinction would parallel that drawn between response-focused and antecedent-focused emotion regulation: Gross (1998) has argued that the former is maladaptive in terms of a wide variety of emotional correlates, whereas the latter is adaptive. Viewing them as an internal focus on different valuations provides one rationale for why this might be. The focus on positive values increases the chances of selecting behavioral patterns with positive consequences, while the latter does not and, if anything, possibly impairs evolutionary set automatic adaptive responses (see also Coan and Allen 2007).

Interestingly, some of these metareasoning strategies seem to be both directly accessible through conscious report (Papageorgiou and Wells 2002, 2003) and amenable to treatment (Wells et al. 2012).

Algorithmic Accounts of Model-Based Evaluations: Pruning and Memoization

The precise processes which underlie model-based evaluation are currently of great interest in neuroscience (Johnson and Redish 2007; Pfeiffer and Foster 2013; Daw and Dayan 2014; Doll et al. 2015; Kurth-Nelson et al. 2015). By formalizing processes such as rumination and emotion regulation in the valuation framework, one can hope to benefit from these advances. I note, however, that these investigations are in their infancy. Nevertheless, because the internal evaluation processes are so rich and complex, computational modeling of decision making might come to be a helpful tool.

Detailed computational models of planning have identified at least two aspects in the regulation of the internal search that are potentially relevant for depression: pruning and memoization. When subjects are asked to plan ahead, they partially solve the metareasoning problem by reflexively pruning; that is, they stop thinking about an option when that option raises the possibility of a salient loss, even when forfeiting larger rewards behind the large losses (Huys et al. 2012). This internal thought inhibition has been suggested to be analogous to behavioral inhibition and related to serotonin (Dayan and Huys 2008; Crockett et al. 2012; Geurts et al. 2013). Preliminary results suggest that it involves the subgenual anterior cingulate cortex, a substrate known to be important in rumination (Sheline et al. 2009) as well as treatment response (Mayberg 2009; Fu et al. 2013), and to mediate the impact of aversive events on choices (Amemori and Graybiel 2012). Pruning might also be related to the notion of inhibitory control in depression, which refers to the ability to inhibit the processing of aversive information (Gotlib and Joormann 2010). Healthy controls show an impairment or delay in processing an affectively negative target that was a distractor on the previous trial, but this is absent in depressed patients (Joormann 2004, 2006).

Memoization refers to the reuse of results from previous cognitive efforts. Rather than recomputing the solution to a problem previously faced, individuals tend to reuse the previous solutions (Huys et al. 2015c). This provides a way for the results of cognitive processes to ingrain themselves very rapidly.

Both pruning and memoization have yet to be shown to be directly implicated in depression. Nevertheless, they provide potential algorithms for the implementation of theories of depression that rely on the interaction between two types of processes: one high-level cognitive factor and another more low-level attentive or more reactive factor. For instance, the two-factor model of relapse (Farb et al. 2015) and the cognitive neuropsychological model of depression (Roiser et al. 2012) both suggest that depression involves the integration of two components: (a) negative affective biases are present in terms of attention, memory, and perception, and are risk factors for the development of a first episode; (b) changes in these affective biases precede treatment response (e.g., Wells et al. 2014). The theory suggests that these biases are, over time, consolidated into negative schemata and rumination, but it does not explain how this process might occur. The impact of Pavlovian reflexes on metareasoning is one path by which affective biases might influence cognitive processes, and memoization might provide a rapid way for stamping in the results.

Valuation might thus provide a normative framework for emotions, and investigations into model-based computation and metareasoning might help to integrate cognitive components of depression. This may potentially account for stress sensitization, helplessness, rumination, and emotion regulation.

Clinical Interventions for Recurrence

As stated, both psychopharmacological and psychotherapeutic interventions have proven useful in preventing relapses and recurrences of depression. Here

I briefly review features of ADMs and psychotherapy which aid in the management of recurrent depression.

Antidepressant Medications

ADMs have shown convincing efficacy in reducing the risk of relapse or recurrence of depression. Several meta-analyses (Viguera et al. 1998; Geddes et al. 2003; Kaymaz et al. 2008; Glue et al. 2010) estimate the reduction in the odds of a relapse due to continuation of antidepressants at around 70%, with a relative risk reduction at around 50% and an absolute risk reduction at 20% (from 40%); the number needed to treat (NNT) to avoid one additional relapse is thus 5 over one year (Geddes et al. 2003). These findings have led to recommendations of relatively long treatment periods (e.g., Anderson and Pilling 2010; Bauer et al. 2013), informed by variables describing the course of the illness, mainly severity, the number of prior episodes, and the duration of treatment (Härter et al. 2009; Anderson and Pilling 2010; Bauer et al. 2013). As second and third generation ADMs are tolerated well (Anderson and Tomenson 1995; Cipriani et al. 2009), chronic treatment is a feasible possibility.

Though none of these outweigh the strong current evidence in favor of treatment, it is important to remember that there are only very few trials that have looked at medication on a timeframe of more than two years, and that the impact of the number of past episodes may hide a variety of other variables (e.g., lifetime stress exposure, ADM treatment history, and residual symptoms; see Appendix). The fact that relapse rates after discontinuation of ADMs show a very early excess that slowly drops off (Viguera et al. 1998; El-Mallakh and Briscoe 2012; see Appendix) has fuelled arguments that the antidepressant effect might either wear out (Rothschild et al. 2009) or result in long-term withdrawal syndrome (Fava et al. 2015). There are also reports that long-term treatment could lead to the development of adaptations that render future episodes more likely, particularly after discontinuation. For instance, there are meta-analytic reports that the relapse rate is higher among patients who have responded to antidepressants, than those who have had a spontaneous remission. In fact, the stronger the *in vitro* perturbational effect of the antidepressant, the stronger the effect, even when controlling for the number of prior episodes, treatment duration, and the stringency of recovery definition (Andrews et al. 2011). Hence, it cannot be excluded that adaptations to the antidepressants themselves contribute to the excess relapses seen very early after discontinuation (see Appendix). Such phenomena raise questions about the practice of chronic treatment. Furthermore, antidepressants are also not devoid of other adverse side effects, which range from impairments of sexual function to increased risks of hemorrhage and cardiac arrhythmias. They are discontinued at extremely high rates, with a half-life as low as 20 days; up to 75% of patients discontinue ADMs within the shortest recommended treatment duration (after a first episode) of six months (Olfson et al. 2006; Lee and Lee 2011). Finally, an NNT of 5 is clinically attractive, but chronic long-term medication does place a nonnegligible burden on four out of the five patients who will not benefit from the medication.

Hence, improving the targeting of ADMs to those who are most likely to benefit from continuation or prophylactic treatment would be extremely valuable. Those who will not benefit can be offered alternative preventions, or, if they are at low risk, followed up without intervention. Better indications for treatment might increase the concordance with treatment among those prescribed an ADM, and reduce the number of those treated in vain.

ADMs prominently impact both serotonergic and noradrenergic neuromodulation, and both have been the target of computational models of the type described here. Serotonin has been suggested to relate to automatic behavioral inhibition (Soubrie 1986; Deakin and Graeff 1991; Dayan and Huys 2009; Crockett et al. 2012). However, recent advances in optogenetics have raised doubts about such accounts (Cohen et al. 2015; Dayan and Huys 2015).

Psychotherapy

Psychotherapy is approximately as effective as ADMs in the acute treatment of depression, resulting in remission in 50–60% of patients (DeRubeis et al. 1999). The relapse rates after treatment (approximately 50% over two years; Vittengl et al. 2007) are also high. When compared directly to treatment with ADM, however, psychotherapy appears to have a longer-lasting effect: relapse rates after cognitive therapy, mindfulness-based cognitive therapy (MBCT), or behavioral activation therapy are similar or non-inferior to those under continued ADM treatment (Blackburn et al. 1986; Evans et al. 1992; Teasdale et al. 2002; Fava et al. 2004; Hollon et al. 2005).

Continuation cognitive therapy after remission under either cognitive therapy or antidepressants also reduces relapse rates compared to switchover to placebo (Jarrett et al. 2001; Segal et al. 2010) and effects are similar to those of continuation ADM therapy: approximately 20% absolute risk reduction (Geddes et al. 2003; Piet and Hougaard 2011). Unlike acute-phase psychotherapy, continuation psychotherapy does not seem to have a longer-lasting effect after termination (Jarrett et al. 2013). Adding cognitive therapy to flexible pharmacotherapy may help to achieve recovery, defined as a longer-lasting remission (Hollon et al. 2014). Whether relapse rates after recovery with psychotherapy are higher than after spontaneous recovery, as is true for antidepressants (Andrews et al. 2011), does not appear to have been examined.

As with ADMs (Viguera et al. 1998), there are indications that patients who experience multiple past episodes benefit more from psychotherapy. MBCT, for instance, has a stronger impact on patients with ≥ 3 episodes than those who experience at most 2 episodes. It also appears to be more efficient for treating recurrences that are not initiated by life events (Ma and Teasdale 2004). In fact, MBCT was specifically designed with relapses in mind (Teasdale et al. 2002).

Computational Psychiatry Approaches to Prevent Recurrence

Both psychotherapy and ADMs have well-proven utility in the prevention of relapse and recurrence. However, both appear to have to be administered chronically and both have NNTs around 5, meaning that a substantial portion of those given chronic treatment have little to no benefit. Furthermore, there are drawbacks to both therapies, and individuals discontinue, in particular, ADM therapy at a very high rate. Better allocation to different interventions should improve outcomes and reduce the development and establishment of recurrent depression.

Descriptive Atheoretical Approaches

A first important task for computational psychiatrists is to amalgamate the extant information and use it to derive individual risk scores. What is the relapse risk for a patient who wishes to discontinue medication six months after his second episode, who responded to the first medication, and who has had only some sleeping problems for the past few weeks? The aim in this would not be to necessarily identify particular risk factors, but rather to use agnostic statistical tools to amalgamate a variety of variables into a robust prediction. The literature provides very rich hints as to which variables are most likely to prove important and informative. However, what is as yet not well understood is which variables provide incremental predictive power in which combinations. By identifying individual variables that modulate relapse risk in individual patients, such predictions might be able to guide treatment (which, of course, would have to be assessed specifically). Multiple predictors exist for response to a particular treatment mode, ranging from clinical descriptors (DeRubeis et al. 2014) and EEG measures (Pizzagalli et al. 2001; Mulert et al. 2007; Korb et al. 2009; Leuchter et al. 2009), emotional blunting (Peeters et al. 2010), subgenual resting state activity and response to aversive stimulation (DeRubeis et al. 2008; Roiser et al. 2012; Siegle et al. 2012; Fu et al. 2013) and others. These techniques have not yet entered clinical practice, possibly because they do not really address the actual clinical problem (i.e., treatment allocation). In the setting of relapse prevention, they may prove useful to assess continuance of a specific treatment. However as yet, it is entirely unclear whether the same measures that predict initial treatment response have any value in predicting maintained response.

To begin, purely descriptive, atheoretical machine-learning approaches should be applied to the existing features, to combine them and produce better predictors of an individual's relapse risk (DeRubeis et al. 2014). A next step would be to apply this approach to the differential prediction of relapse under pharmacological versus psychotherapeutic management, and then to the differential relapse risk with different specific treatments. To the extent that such data exists, one could even imagine extending this approach to other interventions which psychiatrists deploy, such as supported living, employment, or psychoeducational interventions.

Mechanistic Approaches

It is an open question to what extent mechanistic approaches will be helpful, in particular, in the valuation account of emotion. Nevertheless, two areas stand out as potential applications. The first is the differential indication for psychotherapy. Psychotherapies are built around specific conceptualizations. For instance, MBCT has extracted a particular component of cognitive therapy, which is the mindful internal distancing from emotions. There is good evidence that the therapy results in changing mindfulness and that this mediates improvement (Teasdale et al. 2002; Ma and Teasdale 2004; Kuyken et al. 2010), but whether a dysfunction (low mindfulness) is a differential indicator is, to my knowledge, not yet settled. In fact, to the best of my knowledge there are no validated measures that predict differential response to different types of psychotherapy. As I argue here, computational notions of valuation provide an integrative framework for emotions and might accommodate several processes that play salient roles in psychotherapy. It might thus be fruitful to build detailed computational models of the active ingredients of psychotherapy. To the extent to which this turns out to be feasible, it might also allow targeted measurements to support differential treatment allocation.

The second area concerns the examination of how dynamic interactions between valuation factors establish and stabilize high-risk states for relapse (i.e., how a single episode is being transformed into a chronic disabling condition). This includes (a) the interaction between helpless cognitions and guided exploration and exploitation, (b) the impact of aversive Pavlovian thought inhibition on mood regulation and on the ability to recognize worthwhile longer-term effort investments, and (c) the impact of memoization on the establishment of automatic thought patterns and more. Building models to explain how these processes result in the establishment of recurrent or chronic disease states might help us identify key variables as well as design efficient measures to manage depression.

Discussion

Psychiatry is a field of medicine that encompasses the extremely complex phenomena of subjective suffering in society and the high-level functions of the human brain. Suffering and social function are hard outcomes to measure reliably or in an appropriately objective manner, and this is arguably at the heart of difficulties in providing hard definitions of illness in psychiatry. Nevertheless, the aim of computational approaches to psychiatry must ultimately be to improve patient outcomes, and hence to improve subjective suffering or functioning in society. Taking a step back, this is most likely to be achieved if the approach addresses specific clinical issues and measures quantities that are directly relevant to care decisions. This could arise directly through better targeting of existing interventions or the development of novel ones. Alternatively, computational approaches might lead to improved diagnoses, in the sense that the diagnoses would encapsulate an improved understanding of the problem at hand and thus lead to a better allocation or development of interventions.

Broadly speaking, mental health practitioners avail themselves of at least five types of interventions:

1. Social support, which defines and validates illness for various purposes (e.g., sick leave, defense arguments in legal issues), provides housing and financial as well as provision and maintenance of support networks
2. Educational efforts, which informs patients, their caregivers, and support networks about the illness and its potential consequences as well as manages stigma and aids understanding of available treatment options; other aspects (e.g., emotional regulation skills, social skills training) may also be relevant
3. Psychotherapy
4. Psychopharmacology
5. Interventions such as electroconvulsive therapy or surgery, which are primarily aimed at patients whose symptoms do not respond to other treatment options

As they seek to improve the conditions of patients, public health workers could benefit from more effective means of assessment and decision making Although the computational tools reviewed here may not directly improve social support measures or psychoeducational approaches, or yield novel molecular targets for medications, they could be crucial in targeting these resources.

Computational psychiatry encompasses two broadly distinguishable approaches. One approach is through mechanistic techniques, which aim to understand the system at hand. Limits to this approach include a dependence on costly data acquisition and difficult learning tasks. Although such tasks can powerfully probe internal processes and mechanisms of cognition, they are often not widely deployed in the clinic because of their temporal costs; they require very substantial attentional resources and cooperation and are poorly adapted to the severely ill patient. They also require substantial expertise to conduct. The other approach involves purely descriptive discovery. This involves agnostic statistical or machine-learning techniques to discover patterns or structure in informative data sets. In terms of the three ways computational techniques could be clinically helpful, it appears that diagnostic refinement and possibly the development of novel interventions might rely more on mechanistic advances, whereas treatment allocation (particularly among treatments with similar rationale) might benefit more imminently from an atheoretical approach.

There are many glaring omissions in this chapter (e.g., the absence of a discussion on neuromodulators). Reinforcement-learning models have provided deep insights into the function of neuromodulators (Dayan 2012) and might provide useful measures on the state of neuromodulatory systems, which could help in the differential allocation of treatment. This is, however, a very tall order, as models are not refined enough at present to be able to distinguish between subtle differences in current pharmacological treatments (though see Maia and Cano-Colino 2015). Here, descriptive methods might guide future advancement (DeBattista et al. 2011).

Conclusion

In this chapter, I have set forth a formal framework for emotions and described how some aspects of dysfunction can be conceptualized using it as well as how it might provide mechanisms for inferring internal variables that determine emotional experience. Descriptive methods are most likely to provide tools to target treatments more efficiently in the near future. The hope is that mechanistic models might yield results by capturing specific processes in a more focused manner, thereby allowing features that are critical to treatment and disease progression to be measured efficiently.

Acknowledgments

I would like to thank the organizers for the opportunity to write this paper. I gratefully acknowledge extensive discussions with Isabel Berwian, Peter Dayan and Nathaniel Daw on these and related issues. This work was partly funded by a Swiss National Science Foundation grant (SNF 320030L_153449).

Appendix

This supplement presents further details on relapse after antidepressant medication discontinuation. The number of previous episodes has a very profound impact. Figure 15.1a, b show the survival curves after medication discontinuation and continuation, respectively, as a function of the number of prior episodes from an early meta-analysis (Viguera et al. 1998), echoed by more recent ones (Kaymaz et al. 2008). However, the number of previous episodes hides substantial individual variability.

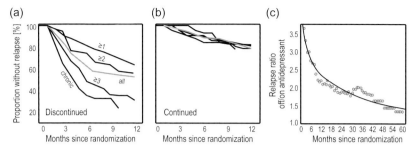

Figure 15.1   Relapse after antidepressant discontinuation. (a) Survival curves over one year after after randomization to placebo as a function of the number of previous episodes. Patients with more previous episodes relapse at a higher rate. (b) Survival curves over one year after continuation of antidepressant as a function of previous episodes. Relapse rate no longer depends on previous history. (c) Five year ratio of relapse risk as a function time since antidepressant discontinuation, compared to relapse risk when continuing medication. The risk is relatively high early on, and over a period of several years slowly falls toward one. Note though that the right tail of the function relies on few studies with small sample sizes. Figure adapted from Viguera et al. (1998).

First is the stage of the depressive disorder at the time of randomization. The probability of treatment response worsens with chronicity (Rush et al. 2006a), symptom severity, and difficult treatment (e.g., the number of treatment steps required to achieve remission) (Kennedy et al. 2003; McGrath et al. 2006; Rush et al. 2006a; Burcusa and Iacono 2007; Patten et al. 2012). Hence these results likely rely on a progressively more selected subset of patients who have responded despite increasingly recurrent disease. Importantly, residual subthreshold symptoms themselves predict early relapses, both in placebo-controlled trials (Andrews et al. 2011; Dunlop et al. 2012) and naturalistically, even after the first episode (Judd et al. 2000); they also can be a stronger risk factor than the number of past episodes (Judd et al. 1998). As there is no accepted definition of remission or recovery (Monroe and Harkness 2011), meta-analyses struggle to control fully for the extent of residual symptoms. Longer and deeper recovery predicts lower recurrence rates, but one important question is whether those with better or worse recovery benefit more from continued treatment (Segal et al. 2010; Dunlop et al. 2012). Importantly, the length of treatment itself does not appear to affect relapse rate, remaining essentially unaltered over courses of medication from one to two months to one year (Geddes et al. 2003; Kaymaz et al. 2008; Glue et al. 2010; Andrews et al. 2011).

However, patients with longer histories of depression typically have had more extensive, longer exposure to antidepressants, possibly leading to long-term alterations which might increase risk after antidepressant discontinuation. Figure 15.1c shows that the relative risk for a relapse occurs very early on and slowly approaches unity over five years. The temporal proximity of the excess risk to discontinuation has raised questions (El-Mallakh and Briscoe 2012) as to its origin and possible confounds, such as partial unblinding (with subjects noticing that the drug might have been been discontinued), a "wearing off" or tachyphylaxis of antidepressant function (Rothschild et al. 2009), or delayed withdrawal effects though drug half-life and discontinuation mode have somewhat unreliable effects on relapse rates (Viguera et al. 1998; Kaymaz et al. 2008; Andrews et al. 2011; Fava et al. 2015). For instance, in the PREVENT trial, which compared venlafaxine to fluoxetine for relapse prevention, continued treatment appeared to have an increased beneficial impact after two years compared to one year (Keller et al. 2007; Kocsis et al. 2007). Such findings speak to the notion of oppositional tolerance, whereby antidepressant discontinuation leads to a relapse rate that is heightened above the "natural" rate. Andrews et al. (2011) found that the relapse rate, after achieving remission on placebo, is lower than after achieving remission on antidepressant and discontinuing it. Importantly, these results hold when controlling for depression history and for recovery definitions. They also found that the extent to which the antidepressant discontinuation relapse rate is increased is proportional to how strongly each medication influenced neuromodulatory systems in animal models, arguing that adaptive processes may increase the risk of relapse independently of any depression disease processes.

16

Clinical Heterogeneity Arising from Categorical and Dimensional Features of the Neurobiology of Psychiatric Diagnoses

Insights from Neuroimaging and Computational Neuroscience

John H. Krystal, Alan Anticevic, John D. Murray, David Glahn,
Naomi Driesen, Genevieve Yang, and Xiao-Jing Wang

Abstract

Clinical heterogeneity presents important challenges to optimizing psychiatric diagnoses and treatments. Patients clustered within current diagnostic schema vary widely on many features of their illness, including their responses to treatments. As outlined by the American Psychiatric Association Diagnostic and Statistical Manual (DSM), psychiatric diagnoses have been refined since DSM was introduced in 1952. These diagnoses serve as the targets for current treatments and supported the emergence of psychiatric genomics. However, the Research Domain Criteria highlight DSM's shortcomings, including its limited ability to encompass dimensional features linking patients across diagnoses. This chapter considers elements of the dimensional and categorical features of psychiatric diagnoses, with a particular focus on schizophrenia. It highlights ways that computational neuroscience approaches have shed light on both dimensional and categorical features of the biology of schizophrenia. It also considers opportunities and challenges associated with attempts to reduce clinical heterogeneity through categorical and dimensional approaches to clustering patients. Finally, discussion will consider ways that one might work with both approaches in parallel or sequentially, as well as diagnostic schema that might integrate both perspectives.

Introduction

> There are known knowns; there are things we know we know. We also know there
> are known unknowns; that is to say we know there are some things we do not
> know. But there are also unknown unknowns – the ones we don't know we don't
> know. —Former U.S. Secretary of Defense, Donald Rumsfeld (February, 2002)

The complexity of the neurobiology of psychiatric clinical conditions is sufficiently great, or our knowledge base sufficiently shallow, as to make it impossible to determine with certainty how close we are coming to a precise understanding of any particular symptom or disorder (Wang and Krystal 2014).
We lack even a precise understanding of how the brain generates typical adaptive human behaviors, and this undermines our ability to establish reliable and
valid psychiatric diagnoses, as reflected in the poor reliability of some psychiatric diagnoses in the initial DSM-5 field trials (Freedman et al. 2013). Yet
at the same time, mental health professions are obliged to do all that they can
to alleviate the suffering, disability, and mortality associated with psychiatric disorders and to identify new treatments that are safer and more effective
than those currently available. Thus, investigators studying the neurobiology
of psychiatric disorders in the service of ultimately identifying new treatments
are inherently working with an incomplete understanding of the neural processes they are studying or how to fix the relevant aspects of their targeted
pathophysiology. To address the limitations in our knowledge, one strategy is
to rely on the power of large-scale exploratory or descriptive research in order
to map the relevant universe of information, such as sequencing the genomes,
epigenomes, connectomes, and microbiomes. These extremely powerful and
informative approaches have transformed our understanding of psychiatry
(Krystal and State 2014), but they have not yet led to a new treatment. The
alternative approach involves informed risk taking through hypothesis building and testing; that is, the development of simplified hypotheses that can be
iteratively refined through experimentation. Computational neuroscience may
be helpful in this process, as this field of research endeavors to transform conceptual hypotheses about the brain into quantitative models (Sejnowski et al.
1988). In turn, computational psychiatry aims to develop and refine quantitative models to explain the features of psychiatric disorders (Montague et al.
2012; Friston et al. 2014; Wang and Krystal 2014).

The task set before my colleagues and I was to address the question of how
computational psychiatry might help to provide insights into the complexity
and heterogeneity of psychiatric disorders. This is a wonderful but overwhelm-
ing challenge, as there is an enormous body of data but very little insight. In
this chapter, we will suggest that biophysically informed computational models can assist in the building of bridges between basic and clinical neuroscience
and ultimately shed light on dimensional features within psychiatric disorders,
transdiagnostic dimensional characteristics, and categorical features of psychiatric diagnoses. Our discussion focuses on schizophrenia, highlighting both

dimensional and categorical features of its biology. Lastly, we will consider opportunities and challenges associated with categorical and dimensional approaches to psychiatric diagnosis, particularly with respect to the prospect of developing novel treatments for psychiatric disorders.

Neurodevelopment as a Source of Clinical Heterogeneity: Possible Implications for Illness Phase-Related Aspects of the Neurobiology and Treatment of Schizophrenia

Stable and Evolving Features of Schizophrenia

The presence of a categorical diagnostic system could be viewed as promoting the assumption that the underlying neurobiology of the disorder does not change fundamentally across the course of illness. In the case of schizophrenia, the fact that the same medications are prescribed to patients regardless of their phase of illness would be consistent with this view. However, neurodevelopment has long been thought to play a fundamental role in the neurobiology of schizophrenia, with genetic or environmental etiologic factors early in life giving rise to a complex and evolving illness pathophysiology (Weinberger 1987; Insel 2010; Tebbenkamp et al. 2014; Volk and Lewis 2014). Further, clinical studies have described illness phases that broadly inform current thinking about schizophrenia, including the prodrome, when individuals display subsyndromal features of the illness; the first episode, where full syndromal features of the illness are expressed; the early course of schizophrenia, which is associated frequently with progressive functional decline; and chronic illness, where patients are thought to plateau clinically with episodic exacerbations (Davidson and McGlashan 1997; Lencz et al. 2001; Keshavan et al. 2005; Agius et al. 2010; Insel 2010). Careful study of the chronic phase of illness, particularly in the elderly, reveals that functional impairments and negative symptoms may progress in a subgroup of these patients, associated with reductions in positive symptoms and formal thought disorder (Davidson et al. 1995; Harvey et al. 1997; Harvey 2014). The changes in symptom profiles and functional impairments are thought to have neurobiological underpinnings which might be targeted by novel treatments that might attenuate or even reverse aspects of the underlying biological changes (Breier et al. 1992; Lieberman et al. 2001; Insel 2010). Although there is still a very superficial understanding of the evolving biology of schizophrenia, recent advances suggest some general principles that might inform future studies.

Broadly speaking, schizophrenia appears to be associated with some biological features that do not substantially change with development and others which do. One relatively stable feature, for example, is a disturbance in the functional connectivity of the thalamus. In "high-risk," first episode, and chronic patients, schizophrenia is associated with reduced functional connectivity of the thalamus with association cortices and overconnectivity of the thalamus with sensorimotor cortices (Woodward et al. 2012; Anticevic et al. 2014a; Cetin et al. 2014; Klingner et al. 2014; Anticevic et al. 2015b; Tu et al. 2015). Patients with bipolar disorder with psychosis appear to have more disruptions in thalamic functional connectivity than bipolar disorder without this symptom, placing their biology at the boundary of schizophrenia and bipolar disorder (Anticevic et al. 2014b). Figure 16.1 illustrates the dimensional relationship between healthy subjects, people with bipolar disorder, and patients diagnosed with schizophrenia.

In contrast to the relatively stable disturbances in thalamic functional connectivity, schizophrenia appears to be associated with many features of illness that evolve over time (Salisbury et al. 2007; Olabi et al. 2011). For example, various findings suggest the balance of excitatory and inhibitory (E/I) connectivity evolves across the life span. Genetic studies strongly implicate genes associated with the development and function of glutamate synapses in the heritable risk for schizophrenia (Walsh et al. 2008; Malhotra et al. 2011; Gulsuner et al. 2013; Timms et al. 2013). Schizophrenia risk genes appear to be particularly expressed prenatally (Gulsuner et al. 2013). These genetic risk mechanisms may contribute to the deficits in glutamate synapses described in postmortem studies (Black et al. 2004; Glausier and Lewis 2013; Datta et al. 2015; MacDonald et al. 2015; Shelton et al. 2015). Development deficits in NMDA glutamate receptor signaling may stimulate neuroadaptations within pyramidal neurons that restore E/I balance in cortical networks by increasing the intrinsic excitability of pyramidal neurons, such as reductions in GIRK2 (Tatard-Leitman et al. 2015). However, this study suggests that increased basal gamma oscillation power may compromise network function by reducing signal-to-noise balance. Further, animal studies of conditional knockout of the NR1 subunit of the NMDA receptor on forebrain interneurons indicate that early-life deficits in NMDA receptor signaling distort cortical development in ways that result in many neurobehavioral stigmata associated with schizophrenia, whereas the same intervention in adults does not reproduce the same profile of effects (Belforte et al. 2010). Recent studies from the Lewis laboratory further support the hypothesis that GABAergic deficits associated with schizophrenia develop as a consequence of glutamate-signaling deficits and may serve to compensate for deficiency in excitatory signaling (Volk and Lewis 2013; Glausier et al. 2014; Hoftman et al. 2015; Kimoto et al. 2015). It is possible that the developmental proliferation of glutamate synapses throughout childhood (Huttenlocher 1979; Petanjek et al. 2011) also serves to restore, to some degree, E/I balance.

However, deficits in GABA signaling may also render cortical networks hyperexcitable (Lazarus et al. 2015) and vulnerable to dysfunction (Krystal et al. 2003; Gonzalez-Burgos et al. 2015), suggesting that the reductions in E/I imbalance are allostatic rather than homeostatic. By allostatic, we mean that the compensation for disturbances in synaptic connectivity may serve to

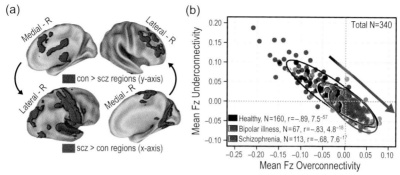

Figure 16.1 Relationship between thalamic over- and underconnectivity across subjects with schizophrenia. (a) Regions showing reduced (blue, top panel) and increased (red, bottom panel) thalamic connectivity for the original discovery sample ($N = 90$) compared to results from healthy comparison subjects. (b) The *x*-axis depicts the extent of functional connectivity of the thalamus and sensorimotor regions where increases are seen in schizophrenia (Fz overconnectivity) and the *y*-axis depicts the functional connectivity of the thalamus with regions in the executive control network where reductions are seen in schizophrenia (Fz underconnectivity). Along each axis, a positive value indicates a positive covariance of thalamic activity and the relevant target regions (*x*-axis: sensorimotor cortex; *y*-axis: prefrontal, cerebellum), whereas negative values reflect negative covariance. A significant negative relationship evident across all healthy controls (gray-black data points, $N = 160$; $r = -0.89$, $P < 7.5^{-57}$) collapsing across three samples (discovery, replication, and healthy subjects matched to patients with bipolar disorder). The same pattern was evident for bipolar patients (blue data points, $N = 67$; $r = -0.83$, $P < 4.8^{-18}$), whereas an attenuated and shifted correlation was found for schizophrenia patients (red data points, $N = 113$; $r = -0.68$, $P < 7.6^{-17}$, collapsing across both discovery and replication samples). Vertical/horizontal green dotted lines mark the zero points. Schizophrenia patients showed a "shift" across the zero lines, indicative of weaker prefrontal-cerebellar-thalamic coupling, but stronger sensorimotor-thalamic coupling. Bipolar patients showed an intermediate degree of disruption, suggesting a "gradient" (inset arrow for qualitative illustration). Ellipses for each group mark the 95% confidence interval. From Anticevic et al. (2014a), reprinted with permission.

restore E/I balance in some contexts, but it renders the network vulnerable to dysfunction, much as adaptations to stress may be successful in the short term but render the organism vulnerable to disease (McEwen and Stellar 1993). We do not mean to imply that GABA deficits necessarily overshoot the extent of glutamate synaptic dysfunction. Instead, we hypothesize that GABA reductions partially compensate for glutamatergic-signaling deficits, but that this adaptation renders the network vulnerable to dysfunction when the network is activated by extrinsic inputs, particularly inputs that are themselves disinhibited and therefore might otherwise be of lower intensity or filtered out (i.e., now are inappropriate to functional context). These extrinsic inputs could be corticocortical (Anticevic et al. 2015a, c) or thalamocortical (Ferrarelli and Tononi 2011; Lisman 2012; Duan et al. 2015). Nonetheless, there are emerging signs of intrinsic abnormalities in GABA systems that might make GABA deficits overshoot excitatory deficits, including copy number variants in GABA genes (Pocklington et al. 2015), and deficits in tonic GABA signaling which might augment phasic-signaling reductions (Maldonado-Aviles et al. 2009).

GABA signaling performs essential functions beyond the regulation of cortical excitability, including the optimization of cortical activity to enable precise spatial working memory (Rao et al. 2000) or olfactory memory (Lin et al. 2014). Computational models support the hypothesis that reduced glutamatergic drive to interneurons could impair the ability to suppress task-irrelevant cortical activity (noise) during working memory (Anticevic et al. 2012b; Murray et al. 2014) and compromise the functional antagonism between the executive and default mode networks at rest (Anticevic et al. 2012b). This work is summarized in Figure 16.2. From another perspective, the loss of adequate noise suppression within local networks could serve to impair memory precision and capacity by compromising the sparse coding of information within local networks (Lin et al. 2014). Network disinhibition in the prefrontal cortex, arising as a consequence of a primary glutamatergic-signaling deficit, could also have important downstream consequences for schizophrenia, such as activating dopamine neurons at the level of the midbrain (Lodge and Grace 2011b; Kim et al. 2015) or perhaps by activating dorsal striatal dopamine terminals directly (de la Fuente-Sandoval et al. 2011). The possibility of increased excitatory drive directly to associative striatum might explain why this region alone shows increased dopamine release in schizophrenia, unlike ventral striatum, cortical regions, limbic regions, and midbrain (Kegeles et al. 2010; Kambeitz et al. 2014; Slifstein et al. 2015; A. Abi-Dargham, pers. comm.).

Evidence for GABA-related pathophysiology in schizophrenia comes from many sources. Increased cortical excitability has been described in the form of short-interval intracortical inhibition in individuals at increased risk for schizophrenia, first episode patients, and patients with chronic illness (Rogasch et al. 2014). Other signs of increased cortical excitability also appear to evolve with the progression of illness. For example, cortical glutamate levels measured with spectroscopy are elevated during the schizophrenia prodrome (Stone et al. 2009) or early in the course of schizophrenia, but decline with illness progression (Marsman et al. 2013). In addition, resting functional connectivity as measured with fMRI appears to be increased early in the course of illness but shows some regional decreases with progression of illness across groups of patients, and perhaps even within patients during treatment (Anticevic et al. 2015a, c). It is possible that the "hyperconnectivity" associated with the early course of schizophrenia arises, at least in part, from deficits in a specific role that subpopulations of GABA neurons play in gating or "filtering" inputs to pyramidal neurons. In particular, somatostatin interneurons, which are compromised in schizophrenia and schizoaffective disorder (Lewis et al. 2008b; Morris et al. 2008a), gate the excitability of distal dendrites of cortical pyramidal neurons in an input-specific manner and may serve to shift the balance between long-term potentiation and depression at dendritic spines

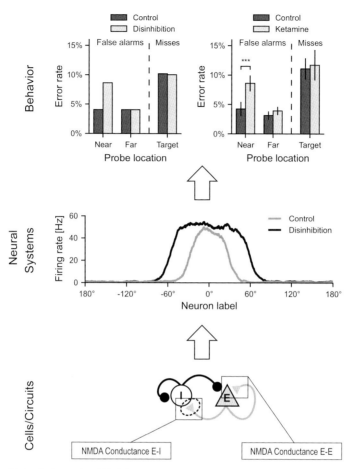

Figure 16.2 Schematic of highlighted findings from a recent computational modeling investigation. The bottom panel shows the manipulation of the NMDA receptor conductance on excitatory (E) and inhibitory (I) cells within a biophysically realistic computational model of working memory (for complete details, see Murray et al. 2014). Because the model is able to capture effects at the microcircuit level (i.e., via firing traces), it generates a specific set of predictions that can be tested at the level of regions or neural systems. As shown in the middle panel, the specific prediction is a broadening of the working memory profile after decreased inhibitory drive onto E cells. This prediction could be tested with electrophysiology (Wang et al. 2013) or BOLD functional MRI at the level of neural systems (Anticevic et al. 2012a, b). The top panel shows how the model generates a behavioral readout such that a specific profile of errors is predicted (top left); this can then be tested with carefully optimized behavioral experiments (top right). Collectively, this approach has the potential to inform across-level understanding disturbances in schizophrenia from receptor to behavior. Nonetheless, this approach is limited because at present it can only be extended to a few well-characterized computational and behavioral processes, such as working memory (see text for further discussion). Asterisks indicate significance (***$p < .001$). From Anticevic et al. (2015d), reprinted with permission.

(Chiu et al. 2013; Higley 2014; Stokes et al. 2014; Sturgill and Isaacson 2015). Breakdown in this function would be expected to allow a much greater range of cross-talk among pyramidal neurons, contributing to "noise" in cortical activity. Schizophrenia also may be associated with increased resting high-frequency cortical activity (Spencer 2011; Gandal et al. 2012; Hirano et al. 2015), as might occur if parvalbumin-containing basket cells were released from inhibition by somatostatin-containing interneurons (Cottam et al. 2013; Pfeffer et al. 2013) or perhaps vasoactive intestinal polypeptide interneurons (Hioki et al. 2013; Pfeffer et al. 2013), which are not yet well characterized in schizophrenia. This increase in resting gamma activity is sometimes referred to as "background noise" in the EEG. The increase in "noise" is thought to compromise cortical signal processing, defined as the ratio of evoked gamma signal (gamma activity evoked by cognitive tasks or 40 Hz auditory clicks) over the resting gamma power.

A central unresolved question in schizophrenia is whether any of the various forms of E/I imbalance described above trigger their own allostatic responses. Over the past twenty years, scientists described homeostatic mechanisms that are engaged by cortical hyperexcitability and downregulate both the presynaptic (Davis 2006a) and postsynaptic (Turrigiano et al. 1994; Lambo and Turrigiano 2013) compartments of glutamate synapses and upregulate GABA synaptic efficacy. There is growing evidence that synaptic homeostatic mechanisms involve proteins implicated in schizophrenia, such as dysbindin (Dickman and Davis 2009). Consistent with this notion, in cross-sectional or longitudinal studies, schizophrenia is associated with an accelerated age-related reduction in both gray and white matter measured with MRI or DTI (Thompson et al. 2001; Vidal et al. 2006; Mori et al. 2007; Nugent et al. 2007; Andreasen et al. 2011; Zhang et al. 2014), cortical glutamate levels measured with MRS (Aoyama et al. 2011; Marsman et al. 2013), and functional connectivity as measured with resting state fMRI (Anticevic et al. 2015a, c). One study directly linked network disinhibition to reduced structural connectivity. Here, patients at high risk for psychosis with hippocampal hypermetabolism on FDG-PET scans showed hippocampal atrophy with longitudinal follow-up (Schobel et al. 2013). In parallel, the study showed in mice that repeated doses of NMDA receptor antagonists also produced hypermetabolism followed by hippocampal atrophy. The authors implicated GABA neuronal deficits in the network disinhibition produced by NMDA receptor antagonist administration by showing that interneuron precursor transplants attenuated the hippocampal hypermetabolism and related physiologic and behavioral phenotypes (Gilani et al. 2014).

Illness Progression and Compounded Neural Allostatic Adaptations

The preceding would suggest that the biology of schizophrenia may evolve throughout the life span and that illness phases may be distinguished by the recruitment of successive allostatic mechanisms that are expressed concurrently with earlier pathological processes (see Figure 16.3). In a gross oversimplification of the complexity of the heterogeneity and progression of schizophrenia, one could build on a prior schema (Insel 2010) and imagine a provisional framework of developmental phases through which schizophrenia might progress. For the purposes of this discussion, we will focus on the interplay of glutamate, GABA, and dopamine signaling.

1. *Predrome*: an early developmental phase where no gross phenotypic alterations arise from the expression of antenatal genetic and environmental risk factors to compromise glutamate synaptic connectivity. This does not preclude the possibility of subtle cognitive or neurophysiologic deficits, as early-life adaptations are not expected to be completely successful.

2. *Prodrome*: the most severe aspects of cognitive and behavioral expression of glutamate synaptic deficits are still largely compensated for by the proliferation of glutamate synapses in childhood combined with deficits in GABA signaling that serve to restore E/I balance. Allostatic adaptation comes at the cost of phenotypic consequences of GABA signaling deficiencies, such as the progressive emergence of cognitive impairments and symptoms. As noted above, dopaminergic activation may be one negative consequence of glutamatergic disinhibition, consistent with recent data in high-risk populations (Bonoldi and Howes 2013). Also, the GABA signaling deficits early in this phase of illness

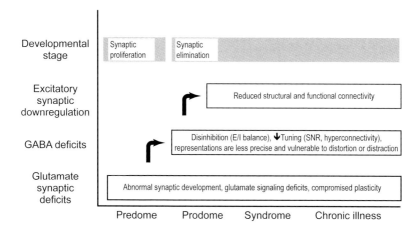

Figure 16.3   A simplified model showing how sequential allostatic neuroadaptations triggered by genetic abnormalities in glutamate synaptic function might interact with programmed synaptic proliferation and elimination, and give rise to an evolving neurobiology of schizophrenia across the life span. SNR: signal-to-noise ratio; see text for description of stages.

may be partially compensated for by an adolescent phase of GABA neuronal proliferation and maturation.

3. *Syndrome*: the full spectrum of symptoms and cognitive dysfunctions emerge as neurodevelopmental trajectories shift from synaptic proliferation to synapse elimination. It signals the end of the production of large numbers of new mature GABA neurons (Kilb 2012; Caballero et al. 2014) and the emergence of a new allostatic adaptation, that is, the downregulation of glutamate synapses in response to altered and hyperconnectivity. The functional downregulation of glutamate synapses during adolescence would be expected to augment the impact of the earlier preprogrammed process of synapse elimination. Neuroimaging studies suggest that the rate of cortical volume loss is greatest early in the course of schizophrenia (Thompson et al. 2001; Andreasen et al. 2011), which may be consistent with the notion that allostatic reductions in functional connectivity may be greatest early in the course of schizophrenia in association with the peak levels of E/I imbalance.

4. *Chronic illness*: deficits in structural and functional connectivity progress throughout the syndromal phase, with current treatments having only limited impact. During this phase, age-related deficits in glutamate synaptic connectivity are expected to be compounded by the intrinsic heritable synaptic dysfunction, the failure to tune synaptic activity optimally, and the compensatory downregulation of structural and functional connectivity.

This model makes predictions related to sources of heterogeneity in schizophrenia. For example, it predicts that the interaction of synaptic proliferation in childhood and the hyperconnectivity arising from GABA deficits, presumably in childhood and early adolescence, serve to delay the expression of schizophrenia symptoms from childhood, when abnormalities in glutamate synapses are already present, to adolescence, when tuning deficits may be more severe and glutamate synaptic deficits are compounded by programmed synaptic elimination. Since some of the same synaptic genes are implicated in the risk for autism and schizophrenia, it is possible that these disorders are distinguished, in part, by the relative success of the neural allostatic adaptations in schizophrenia relative to autism. Further, the model suggests that childhood onset of schizophrenia could be distinguished from typical schizophrenia by the greater severity of the initial glutamatergic synaptic dysfunction, failure in allostatic adaptations, or disturbances in synaptic proliferation or elimination. It supports the observation that females might have a later onset (Szymanski et al. 1995; Lindamer et al. 1997), better treatment response (Szymanski et al. 1995), or perhaps lower incidence of schizophrenia (Kendler and Walsh 1995) than males, perhaps as a consequence of the synaptogenic effects of estrogen (Woolley and McEwen 1994). It is also consistent with the increased risk for schizophrenia by prenatal environmental factors that disturb normal synaptic development (Insel 2010). Further, it suggests that drugs which compromise the integrity of GABA neuronal function, such as CB1 agonists (Hajos et al. 2008; Eggan et al. 2010; Volk and Lewis 2015), might worsen symptoms (D'Souza et al. 2004; 2005) and perhaps even promote the transition from subclinical to clinical symptoms of psychosis (Wilkinson et al. 2014), although this remains to be clearly demonstrated (van der Meer et al. 2012a).

Toward Illness Phase-Specific Treatments for Schizophrenia

If the pathophysiology of schizophrenia progresses through predictable illness phases, then treatments which target distinct neurobiological mechanisms may have their greatest impact at particular phases of illness. Dopamine D2 receptor antagonists, for example, would be corrective for psychosis in illness phases where dopamine release is increased (Frankle et al. 2004), but they would not be expected to be effective in hyperglutamatergic patients who do not have dopaminergic hyperactivity (Demjaha et al. 2014). These drugs would not be expected to prevent the onset of the schizophrenia prodrome nor reverse the consequences of synaptic pruning in chronic schizophrenia, although they might blunt the severity of symptoms and, in so doing, delay the point where patients met diagnostic criteria for schizophrenia (van der Gaag et al. 2013).

The efficacy of drugs targeting features of glutamate signaling may be particularly affected by the evolving biology of schizophrenia. For example, drugs enhancing the stimulation of the glycine/D-serine co-agonist site of the NMDA receptor might be expected to treat symptoms of schizophrenia attributable to deficits in glutamate synaptic signaling. Indeed, meta-analyses suggest that in chronic schizophrenia, glycine and D-serine may have some modest adjunctive therapeutic value (Tuominen et al. 2005, 2006; Singh and Singh 2011), although this efficacy has been questioned (Iwata et al. 2015). Perhaps, though, these medications are more effective early in the course of schizophrenia, when glutamatergic synaptic elimination has not been fully expressed or perhaps GABA neuronal function may be rescued by enhanced glutamatergic input. Consistent with this view, tantalizing preliminary data raise the possibility that glycine, D-serine, and glycine transporter-1 inhibitors may be effective as a monotherapy in prodromal or patients early in their course of illness (Lane et al. 2008; Woods et al. 2013). These studies need definitive rigorous replication.

Metabotropic glutamate receptor 2 (mGluR2) agonists may also show illness phase-specific efficacy for the treatment of schizophrenia. Preclinical and clinical studies suggest that mGluR2/3 agonist drugs reduced the physiologic and behavioral consequences of acute NMDA receptor antagonist effects by reducing glutamate hyperactivity (Moghaddam and Adams 1998; Cartmell et al. 1999; Krystal et al. 2005), thus raising the possibility that this class of medications could treat symptoms of schizophrenia arising from disinhibition of cortical networks (Krystal et al. 2003). The initial mGluR2/3 agonist trial in schizophrenia was positive (Patil et al. 2007); however, efforts to replicate this result as monotherapy or adjunctive therapy did not yield positive results (Stauffer et al. 2013; Adams et al. 2014; Downing et al. 2014). Based on the illness phase model presented above, one might hypothesize that patients early in the course of schizophrenia might benefit from an mGluR2/3 agonist because it would reduce cortical network disinhibition. However, in patients with chronic illness, one might hypothesize that the benefits of modest levels of cortical inhibition would be to halt illness progression, and that higher doses would exacerbate the negative impact of growing deficits in synaptic connectivity. This hypothesis was explored in a secondary analysis performed by investigators at Lilly Research Laboratories in schizophrenia clinical trials of their mGluR2/3 agonist prodrug, LY2140023 monohydrate (Kinon et al. 2015). Their analysis reported that a low dose, 40 mg, of LY2140023 monohydrate was more effective than an 80 mg dose of this drug and comparably effective to risperidone in early course schizophrenia patients. However, in patients with long-standing illness, the 40 mg dose was no better than placebo and the 80 mg dose significantly worsened symptoms relative to placebo (Figure 16.4).

Other medications could be tested for illness phase-specific effects in schizophrenia. mGluR2 agonists were first tested in schizophrenia because they reduced the effects of NMDA receptor antagonists on glutamate release and cognition in animals and humans (Moghaddam and Adams 1998; Krystal et al. 2005). However, other drugs that reduce cortical excitability have been shown to reduce ketamine effects in animals or humans, including AMPA receptor antagonists (Moghaddam et al. 1997), lamotrigine (a drug that blocks several voltage-gated ion channels) (Anand et al. 2000a), alpha7 nicotine receptor agonists (Castner et al. 2011), dopamine-1 receptor agonists, glycine transporter-1 receptor antagonists (Castner et al. 2014), AMPAkines (Roberts et al. 2010), and subtype-selective GABA$_A$ receptor facilitators (Castner et al. 2010). It remains to be seen whether any of these drugs or mechanisms exhibits illness phase-dependent efficacy.

Phase-specific pharmacotherapies may create opportunities for disease-modifying treatments. For example, to the extent that glutamate synaptic deficits contribute to deficient maturation of GABA neurons, early remediation of these signaling deficits might promote normal GABA neuronal development and prevent the emergence of hyperglutamatergic states, and in turn might attenuate the subsequent decline in synaptic connectivity. Reductions in cortical E/I balance might not salvage early synaptic connectivity or the emergence of GABA neuronal deficits, but they may attenuate allostatic loss in synaptic connectivity in response to glutamatergic hyperactivity.

However, it is evident that we do not yet understand how to restore deficiencies in glutamate synaptic connectivity, which evidently are more complicated than deficits in stimulation of the glycine/D-serine site of NMDA receptors. It is tempting to think that drugs which enhance synaptic excitability—such as AMPAkines, low- (alpha7 subunit containing) and high-affinity nicotine receptor agonists or positive allosteric modulators, muscarinic cholinergic

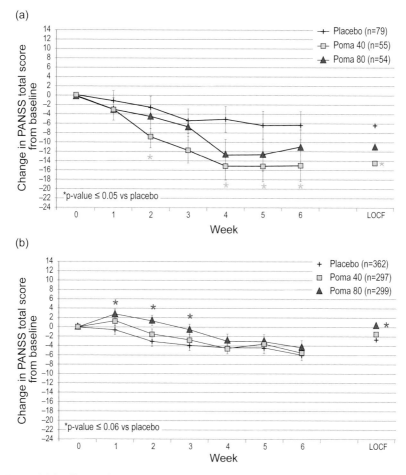

Figure 16.4   Change in PANSS total score with pomaglumetad in Eli Lilly study HBBM illustrates the possibility that mGluR2/3 agonism might treat symptoms in early course schizophrenia patients but worsen them in patients with chronic illness. From Kinon et al. (2015), reprinted with permission. Duration of illness for early-in-disease patients is ≤3 years (a) and for late-in-disease patients is ≥10 years (b). PANSS: positive and negative syndrome scale; Poma: pomaglumetad, an mGluR2/3 agonist pro-drug; RIS: risperidone; SE: standard error.

agonists, and estrogen (in relation to its capacity to stimulate dendritic spine growth) (Woolley and McEwen 1994)—or drugs that enhance neurotrophic factor signaling might play a role in this phase of illness. However, few of these mechanisms have been adequately tested, and evidence is limited to support the efficacy of those mechanisms that have been tested and is generally suggestive of modest effects. It will be important to determine whether these treatments enhance cortical signal processing or merely increase cortical noise. Perhaps one value of viewing the pharmacotherapy of schizophrenia in an illness phase-dependent manner is that it may help to focus attention on important gaps in treatments that might get targeted by new treatments.

Summary

Developmental changes in the neurobiology of schizophrenia and other CNS disorders are an important source of clinical heterogeneity that is not addressed adequately in current diagnostic schema. Better understanding of this evolving neurobiology may help guide the development of more effective treatments aimed at addressing aspects of the pathophysiology that give rise to symptoms and functional impairments at the various stages of illness.

Dimensional Properties of Psychiatric Diagnoses and Clinical Heterogeneity

Another source of clinical heterogeneity arises from a "known unknown"; that is, our inability to classify patients on the basis of etiology or pathophysiology. Present diagnostic schemas (DSM, ICD) are based on symptom profiles that are likely to lump together patients with markedly different pathophysiologies and to divide patient groups that share common elements of both etiology and pathophysiology (Wiecki et al. 2015).

There is a high degree of overlap in the common gene variants contributing to the risk for schizophrenia and bipolar disorder (Cardno and Owen 2014; Maier et al. 2015). However, large numbers of both rare and common gene variants contribute to the risk for each disorder. Thus one source of neurobiological heterogeneity may be genetic heterogeneity within a diagnostic category. Further, some neurobiological properties might constitute a dimension of neural dysfunction that could span several diagnoses. This dimensional perspective would be particularly important if an array of distinct abnormalities at molecular and cellular levels produced common disturbances in the function of cortical microcircuits or macrocircuits. Informed by this type of thinking, the Research Domain Criteria (RDoC) incorporate some of the advantages that result from approaching psychiatric pathophysiology from a transdiagnostic dimensional perspective (Cuthbert 2014b; Insel and Cuthbert 2015).

There is some preclinical evidence to support this approach. One recent example of this type of convergence comes from animal models of deficits in NMDA receptor function, where mice with selective knockouts of GluN1 NMDA receptors on forebrain cortical pyramidal neurons and mice with selective knockouts of GluN1 subunits in parvalbumin-containing interneurons resulted in some common cortical electrophysiological and behavioral alterations (Billingslea et al. 2014; Krystal 2015; Tatard-Leitman et al. 2015).

One exemplar where a dimensional approach to diagnosis may yield important insights is in the relationship between schizophrenia and bipolar disorder.

In many domains of clinical status, including symptom severity, cognitive dysfunction, and functional impairment, patients diagnosed with schizophrenia appear to have similar but more severe problems than patients with bipolar disorder (Badcock et al. 2005; Green 2006; Sanchez-Morla et al. 2009; Brosey and Woodward 2015). A growing number of studies of circuit-based imaging studies have highlighted dimensional relationships between these two disorders (Qiu et al. 2007; Sui et al. 2011; Argyelan et al. 2014; Lui et al. 2015). For example, disturbances in thalamic functional connectivity may be a dimensional trait shared by schizophrenia and bipolar disorder; that is, where schizophrenia appears to express qualitatively similar but quantitatively more severe version of a disturbance found in patients diagnosed with bipolar disorder. As noted earlier, a growing number of studies of schizophrenia have described reduced thalamic functional connectivity with executive control circuits and increased functional connectivity with sensorimotor regions. Thalamic overconnectivity was associated with symptoms of schizophrenia, whereas deficits in association cortex and thalamus were linked to executive cognitive dysfunction. This pattern of alterations is also observed in bipolar disorder but to a lesser degree. The dimensional relationship between the two groups of patients is evident when data from both patient groups are plotted on the same graph (see Figure 16.1).

Therapeutic insights may emerge from dimensional approaches to diagnoses. For example, in the case of the disturbances in thalamic connectivity, one might wonder whether depotentiating (1 Hz) TMS stimulation over sensorimotor cortex (Hoffman et al. 2005) or potentiating (10 Hz) TMS stimulation over executive cortices (Shi et al. 2014; Wolwer et al. 2014; Wobrock et al. 2015) would reduce symptoms or cognitive impairments, perhaps by correcting illness-related alterations in thalamocortical connectivity. Similarly tDCS over association cortices might activate the underlying regions and enhance cortical plasticity (Hoy et al. 2015; Tarur Padinjareveettil et al. 2015). These approaches need to be explored with some care to ensure that increasing cortical excitability did not simply exacerbate cortical E/I imbalances and increase "noise" rather than "signal." It will be important to determine whether neurostimulation treatments, like some pharmacologic ones, show illness phase-dependent efficacy.

Categorical Features of Psychiatric Diagnoses

Despite the genetic complexity of psychiatric disorders, the extensive progress made in identifying genes associated with traditional categorical diagnoses suggests that these diagnostic entities are linked in ways that are still unclear regarding etiology and pathophysiology (Krystal and State 2014). A critical question is whether dimensional "transdiagnostic" approaches to the neurobiology of psychiatric disorders will replace previous categorical diagnostic systems or whether the current emphasis on dimensional approaches will lead to a synthesis of categorical and dimensional approaches to psychiatric diagnosis.

Supporting the importance of categorical distinctions, recent neuroimaging studies provide strong evidence for qualitative differences in the neurobiology of schizophrenia and bipolar disorder. For example, to date increased dopamine release in the associative striatum has been widely demonstrated in schizophrenia (Laruelle et al. 1996; Kegeles et al. 2010) but not documented in bipolar disorder (Anand et al. 2000b). While this difference may reflect inadequate study of the neurobiology of bipolar disorder, other important differences have emerged. Postmortem dorsolateral prefrontal cortex cellular deficits in schizophrenia and bipolar disorder show evidence of qualitative and quantitative differences (Selemon and Rajkowska 2003). Also, as presented in Figure 16.5, schizophrenia appears to be associated with increased signal power in the low-frequency range and increased signal variance in resting-state fMRI data, whereas bipolar disorder is not associated with either of these properties (Yang et al. 2014). Two features of the increase in variance were demonstrated: an increase in the voxel-wise variance across the brain and an increase in the global signal. It would be important to know whether these features, which distinguish schizophrenia from bipolar disorder, contribute to the progressive and persisting cortical volume in schizophrenia. Bipolar disorder, which does not appear to be prominently associated with increased signal power in the low-frequency range or increased signal variance, also does not appear to be prominently associated with persisting enhancements in cortical volume loss after early adulthood (Woods et al. 1990; Blumberg et al. 2006).

To explore mechanisms that might account for the categorical differences between schizophrenia and bipolar disorder, a biophysically based computational model of resting state fluctuations was applied (Deco et al. 2013). In this approach BOLD signals are simulated using mean-field dynamics (Wong and Wang 2006) for each of 66 neural nodes that are coupled following a structure based on diffusion-weighted imaging studies in humans (Hagmann et al. 2008). Key parameters in the model include the strength of recurrent "self-coupling" (w) within nodes and long-range or "global" coupling (G) between nodes. Each of these parameters represents the combination of excitatory and inhibitory connectivity. In this model (Figure 16.6), the local variance of each node increased with increasing values of w and G. Together these data suggest that the functional hyperconnectivity observed in rs-fMRI data might contribute to the increase in the voxel-wise signal variance observed in schizophrenia. This finding provides support for the hypothesis that the increased variance observed in schizophrenia, but not bipolar disorder, has a neural rather than artifactual origin. Further, the similarity in the increases in functional connectivity produced by the NMDA glutamate receptor antagonist, ketamine, in healthy human subjects (Driesen et al. 2013) to the hyperconnectivity documented in association with schizophrenia may suggest that the increases in functional

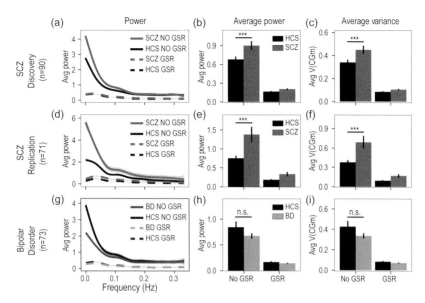

Figure 16.5 Illustration of diagnosis-related disturbances in cortical activity where schizophrenia and bipolar disorder exhibit qualitative differences in relation to healthy subjects; from Yang et al. (2014), reprinted with permission. Power (proportion of the signal power falling within discrete frequency bins) and variance of the cortical gray matter signal in schizophrenia (SCZ) and bipolar disorder (BD). (a) Power of the cortical gray matter signal in 90 SCZ patients (dark gray) relative to 90 healthy comparison subjects (HCS) (black). (b) Mean power across all frequencies before and after global signal reduction (GSR) indicating an increase in SCZ [$F(1, 178) = 7.42, P < 0.01$], and attenuation by GSR [$F(1, 178) = 5.37, P < 0.025$]. (c) Cortical gray matter variance also showed increases in SCZ [$F(1, 178) = 7.25, P < 0.01$] and GSR-induced reduction in SCZ [$F(1, 178) = 5.25, P < 0.025$]. (d)–(f) Independent SCZ samples confirming increased cortical gray matter power [$F(1, 143) = 9.2, P < 0.01$] and variance [$F(1, 143) = 9.25, P < 0.01$] effects, but also the attenuating impact of GSR on power [$F(1, 143) = 7.75, P < 0.01$] and variance [$F(1, 143) = 8.1, P < 0.01$]. (g)–(i) Results for BD patients ($n = 73$) relative to matched HCS did not reveal GSR effects observed in SCZ samples [$F(1, 127) = 2.89, P = 0.092$, n.s.] and no evidence for increase in cortical gray matter power or variance. All effects remained when examining all gray matter voxels. Error bars mark ± 1 SEM. ***$P < 0.001$ level of significance; n.s., not significant.

connectivity observed in schizophrenia have their origins in deficits in NMDA glutamate receptor signaling, hypothesized to contribute to network disinhibition in schizophrenia (Olney and Farber 1995; Grunze et al. 1996; Moghaddam et al. 1997; Krystal et al. 2003).

The implications of deficits in NMDA glutamate receptor signaling and secondary GABA signaling in connectivity disturbances in schizophrenia return the therapeutic focus to the issues raised earlier. Restoration of normal glutamatergic signaling capacity as early as possible in the life span might reduce the emergence or expression of GABAergic deficits and other sources of

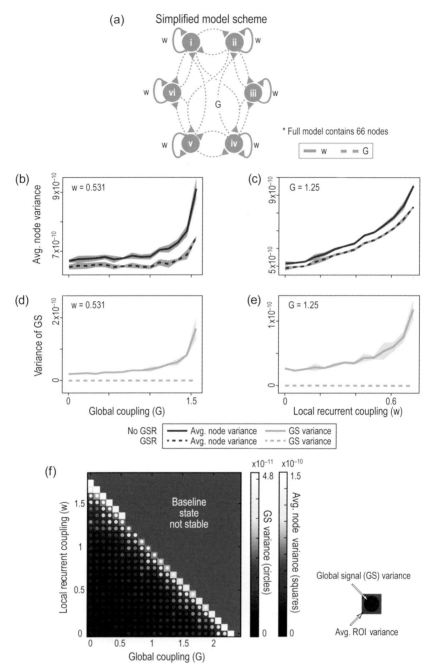

Figure 16.6 Computational modeling simulation of BOLD signal variance illustrates a biologically grounded hypothetical mechanism for increased global and local variance; from Yang et al. (2014), reprinted with permission.

cortical functional hyperconnectivity and increased signal variance (i.e., "cortical noise") associated with schizophrenia. Similarly, suppression of cortical hyperconnectivity would be expected to reduce the expression of noisy cortical activity, a benefit that must be balanced against the cost of suppressing cortical signals with the progression of illness. The striking prediction of the finding that hyperconnectivity and increased cortical signal variance are features of schizophrenia but not bipolar disorder suggests that diagnosis-specific treatment strategies may emerge from careful study of the categorical features of the neurobiology of psychiatric disorders.

Toward Synthesis: Balancing Categorical and Dimensional Approaches

The preceding discussion suggests that tensions between categorical and dimensional psychiatric diagnoses are unavoidable and that they reflect important contributors to clinical heterogeneity. Thus a critical question at the moment is whether either approach on its own—that is, the categorical approach espoused in DSM-5 (APA 2013) and ICD-10 (1992) or the dimensional approach presented in RDoC (Insel and Cuthbert 2015)—has sufficient explanatory power to guide treatment and stimulate research, so as to serve as the framework for future diagnostic schema. Perhaps, future diagnostic schema will adopt one or the other approach as the backbone for classification and then integrate the other perspective within a hierarchical or matrix model.

As in all models, there may be elements of the organizational structure that arise arbitrarily or from convention. In other words, the same data could be clustered using either dimensional or categorical approaches. For example, as discussed above, schizophrenia and bipolar disorder are associated with some

Figure 16.6 (continued) (a) A biophysically based computational model of resting-state BOLD signals is used to explore parameters that could reflect empirical observations in schizophrenia. The two key parameters are the strength of local, recurrent self-coupling (w) within nodes (solid lines) and the strength of long-range, global coupling (G) between 66 nodes in total (dashed lines). (b) and (c) Simulations indicate increased variance of local BOLD signals originating from each node, in response to increased w or G. (d)–(e) The global signal (GS), computed as the spatial average across all nodes, also showed increased variance by elevating w or G. Shading represents the standard deviation at each value of w or G computed across four realizations with different starting noise, illustrating model stability. Dotted lines indicate effects after *in silico* GS regression. (f) Two-dimensional parameter space, capturing the positive relationship between w/G and variance of the BOLD signal at the local node level (squares, far right color bar) and the GS level (circles in each square, the adjacent color bar). The blue area marks regimes where the model baseline is associated with unrealistically elevated firing rates of simulated neurons. Model simulations illustrate how alterations in biophysically based parameters (rather than physiological noise) can increase GS and local variance observed empirically in schizophrenia. Of note in (b)–(e), when w is modulated, $G = 1.25$. Conversely, when G is modulated, $w = 0.531$.

qualitative differences in functional connectivity and voxel-based signal variance. These data would tend to support categorical diagnostic approaches, exemplified by DSM-5 (APA 2013). However, having defined their clinical significance, these traits could be used within a dimensional or transdiagnostic approach to group patients based on their biology.

Keeping an open mind with respect to categorical and dimensional features of psychiatric disorders over the course of illness may be important because our current understanding of disorders like schizophrenia is based on data from medicated patients, and it is possible that drug and illness effects are somewhat confounded. Some postmortem studies of schizophrenia have attempted to address these confounding effects by comparing data from patients to medication-treated nonhuman primates (Volk et al. 2013; Georgiev et al. 2014). Careful retrospective comparisons in patients combined with prospective studies in primates suggest that antipsychotic treatment affects glial populations and reduces cortical volume (Dorph-Petersen et al. 2005; Konopaske et al. 2008), among other effects. One could imagine that the differential prescription of antipsychotics and mood-stabilizing medications might have contributed to categorical differences in the neurobiology of schizophrenia and bipolar disorder. However, mood stabilizers are widely prescribed to people with schizophrenia and antipsychotic-resistant symptoms (Meltzer 1992; Wolkowitz 1993), and second-generation antipsychotic medications have emerged as first line treatments for bipolar disorder. Thus, it may be the case that these medications now contribute to dimensional features of the neurobiology of these disorders.

Further, even when diagnostic groups are distinguished by qualitative differences in their underlying neurobiology, as highlighted in Figure 16.5, there may still be dimensional relationships across these diagnoses (i.e., substantial overlap in the biology of individuals across diagnoses). To illustrate this point, we conducted a secondary analysis of data from a subgroup of individuals from the data presented in Figure 16.5. As can be seen in Figure 16.7, even though the groups differ qualitatively in their average power and average variance, there is still overlap of individuals with schizophrenia and bipolar disorder. While it is possible that this overlap simply reflects a failure of symptom-based categorization to adequately separate patients with schizophrenia and bipolar disorder, it is also possible that these disorders intrinsically overlap in their biology even in dimensions where they qualitatively differ as groups.

Categorizing patients on the basis of particular biological traits is likely to reduce some sources of clinical heterogencity while increasing other aspects (see Table 16.1). For example, one might cluster schizophrenia and bipolar disorder on the basis of their shared common gene variants, but this would ignore the significant differences between these disorders with respect to the impact of rare gene variants (Cardno and Owen 2014) and fail to capture links between schizophrenia and autism with regards to these rare gene variants (Sebat et al. 2009). There may be a fundamental tension between unidimensional and multidimensional clustering strategies. DSM-5 embraces clinical

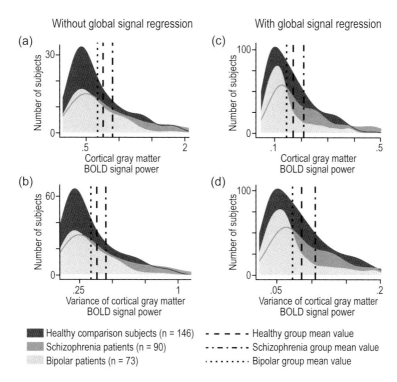

Figure 16.7   Power and variance of cortical gray matter signal in schizophrenia and bipolar disorder. The distribution of the average signal variance and average power of the gray matter signal of individual subjects from a subgroup of the sample presented in Figure 16.5: healthy subjects, n = 146; schizophrenia, n = 90; bipolar disorder, n = 73. Data are presented without and with global signal regression. (a) Group distributions of BOLD signal power averaged across the cortical gray matter for each subject, computed without removing the global mean signal. (b) Group distributions for the variance of BOLD signal averaged across cortical gray matter, computed without removing the global mean signal. (c) Same as in (a), except computed on the BOLD signal after removing the global mean signal through regression. (d) Same as in (b), except computed on the BOLD signal after removing the global mean signal through regression. This figure shows that even though schizophrenia patients increased on these dimensions compared to healthy subjects and bipolar patients were numerically reduced on these dimensions compared to healthy subjects, there was overlap in the values from all three groups.

heterogeneity on any particular dimension, in theory, to reduce the clinical heterogeneity of the overall treatment and clinical course. In contrast, RDoC allows for a greater degree of heterogeneity in clinical presentation in the pursuit of greater homogeneity, in particular prioritized biological or behavioral traits. The benefit of the latter strategy is that it might serve to guide the development of treatments and, more importantly, increase the likelihood of detecting clinical benefits of treatments that targeted that particular dimension of illness. This approach actually validates clinical practice, as clinicians frequently

Table 16.1   Clinical heterogeneity associated with categorical and dimensional categorization exemplified by DSM-5 and RDoC, respectively.

	DSM-5	RDoC
Type	Categorical	Dimensional
Objective	Guide clinical practice and reimbursement	Advance translational neuroscience and discovery of novel treatments
Strategy for reducing heterogeneity	Optimizing statistical association of clinical features within diagnoses	Clustering patients on the basis of well-defined traits that have emerged from translational neuroscience
Sources of heterogeneity	Syndromes defined very broadly Most syndromes defined without basis in etiology or neural mechanism	Clustering patients with evident heterogeneity in clinical presentation

target specific treatment nonresponsive symptoms using a dimensional approach (i.e., treating residual anxiety in schizophrenia with an anxiolytic rather than antipsychotic). There is, however, no guarantee that this patient clustering strategy will recapitulate the degree of homogeneity of multidimensional clustering strategies with regards to genetics or other dimensions of neurobiology, as endophenotypes (Gottesman and Gould 2003) appear to be associated with many of the challenges in genetic association as traditional psychiatric diagnoses (Krystal and State 2014).

Can computational approaches yield evidence-based schema that would more effectively categorize patients (see Chapters 1 and 2, this volume)? Machine-learning approaches, for example, may make it possible to cluster patients empirically in ways that are both more homogeneous and better predictors of treatment response (Wiecki et al. 2015) than current categorical schema. Further, these approaches may be applied to many types of clinical data so that symptoms; cognitive function; structural, functional, and chemical neuroimaging data; social and vocational function; course of illness; patterns of comorbidity; and other data can be more meaningfully integrated. Computational approaches that employ Bayesian approaches might enable the integration of a wide array of information within a nosological framework and the updating of this framework on an empirical basis when important new data are generated (see Mathys as well as Flagel et al., this volume). This strategy may reduce the appearance of tension between dimensional and categorical features of nosology by incorporating these perspectives in a single schema. The Bayesian approach to diagnosis, which considers multiple potential diagnoses concurrently and updates the prioritization of these diagnoses as new information emerges, is not alien to medical thinking: it is at the core of medical practice.

Efforts to reduce clinical heterogeneity in the diagnostic schema are irrevocably linked to the search for novel treatment mechanisms. For example, RDoC was introduced to facilitate closer links between neural processes and psychiatric medication development within a translational neuroscience framework (Insel and Cuthbert 2015). The anticipated outcome is that new treatment mechanisms will be identified that "fix" disturbances in these neural processes. When that occurs, it may then become important to assess these neural processes, which may be measured using genomic, epigenomic, neuroimaging, or other approaches in routine clinical practice. This is the natural progression that has emerged in all other areas of medicine. Measuring blood pressure became important when it was appreciated that treating elevated blood pressure reduced risk for cardiovascular and cerebrovascular disease. Measuring the presence of Her2 in breast tumors became important clinically when Herceptin was introduced (Mukerjee 2010). When these measurements become clinically useful, we can expect them to be incorporated, whether as dimensional or categorical traits, within treatment algorithms and diagnostic schema.

Conclusions

In this chapter we have considered sources of clinical heterogeneity related to the effort to understand psychiatric disorders from categorical and dimensional perspectives. Examples from the neurobiology of schizophrenia were used to illustrate several important points. Computational neuroscience approaches help frame the interpretation of neuroimaging findings to illustrate where the biology of schizophrenia appears to be similar but more severe than bipolar disorder (or were qualitatively different from findings in bipolar disorder), highlighting dimensional and categorical diagnostic properties of the neurobiology of schizophrenia.

Important sources of clinical heterogeneity include the evolving neurobiology of psychiatric disorders with development, which may have important treatment implications. The presence of categorical features of neuropsychiatric disorders supports the maintenance of some elements of categorical diagnoses even though current diagnostic schema give rise to clusters of patients who vary widely on any particular clinical or neurobiological dimension. Data supporting transdiagnostic dimensional features of psychiatric disorders reinforce an approach consistent with the RDoC approach of clustering patients, even though it may not conform to traditional diagnostic schema and may thus produce greater levels of clinical heterogeneity in clinical or neurobiological dimensions. Moving fluidly between these two approaches may enable both clinical practice and research to address specific clinical or research challenges. It is possible that categorical and dimensional approaches could be integrated, better than they are currently in DSM or ICD, within future approaches to the classification of patients.

Acknowledgments

The preparation of this commentary was supported by the National Center for Advancing Translational Science (1UH2TR000960-01), the Department of Veterans Affairs (VA National Center for PTSD), the National Institute on Alcohol Abuse and Alcoholism (P50AA12870, M01RR00125), and the Yale Center for Clinical Investigation (UL1 RR024139). The authors have no financial interests related to the development of diagnostic tests in psychiatry.

Conclusion

17

From Psychiatry to Computation and Back Again

A. David Redish and Joshua A. Gordon

In the opening chapters of this volume, we outlined a series of challenges facing psychiatry, as well as a description of its various promises, and suggested that taking a computational perspective could potentially illuminate a way forward. In this concluding chapter, we revisit these challenges and promises, in the context of what transpired at this Ernst Strüngmann Forum, to highlight the connections between the various themes raised. In particular, we will bring out the points of agreement and disagreement between the discussion groups and the chapters that arose from those discussions. We conclude with a description of the efforts, current and ongoing, to bring the potential synergy between psychiatry and computational neuroscience emphasized in this volume to a reality in the scientific and clinical arenas.

The Challenges of Psychiatry

The principal task for psychiatry is clear: using what we know of how the mind arises from interactions between the physical brain and its environmental and social milieu, how do we define and treat psychiatric disorders? In the opening chapter to this book, we identified three challenges that psychiatry currently faces which, if we could address, would go a long way toward that goal of defining and treating psychiatric disorders:

1. We need a *diagnostic nosology* that is better suited to access the current knowledge in psychology and neuroscience.
2. We need new *biomarkers* capable of assisting with diagnosis and prediction.
3. We need to develop improved *treatments*.

Furthermore, we emphasized that the path from genetics to behavior is complex and nonlinear, and that it depends on neural circuits. Finally, we pointed out that the end goal was personalized medicine, uniquely identified to be what was best for a specific patient.

Many of these challenges were echoed and expanded throughout this volume. However, several complexities were also delineated which make these challenges particularly difficult. For example, psychiatric disorders are *heterogeneous*, both at the *observational level*—each person has a very individual reaction to their neuropsychiatric, environmental, and social situation (Totah et al., this volume)—and the *etiological level*—many of the current psychiatric diagnoses are actually observations built around symptoms, not causal entities in themselves (Totah et al. and Flagel et al., this volume). As discussed by MacDonald and colleagues in Chapter 9, psychiatric dysfunction is *multisourced* (also known as *multifinal*, i.e., a single cause can have divergent outcomes) and *multipotential* (also known as *equifinal*, i.e., multiple causes can lead to observationally similar outcomes). Thus, an important fourth challenge relates to how one can develop a nosology of a system that is marked by a kaleidoscope of causes, each of which can be expressed as a pleiotropy of symptoms.

The *dynamics* of psychiatric disease add an additional layer of complexity. As noted by Totah et al. (see also Flagel et al., Barch, and Krystal et al., this volume), dysfunction proceeds through phases. Thus, patients can present quite differently at the various phases. There are several elements to this complexity. The first and most obvious is that diseases have a time course (episodic, chronic, or progressive) which helps define them. Although psychiatric syndromes often have canonical time courses associated with them, here too there is tremendous heterogeneity within the categories (Totah et al., Barch, and Krystal et al., this volume). Next, one must consider dynamic differences that occur because the patient is developing: a dysfunction in childhood may well manifest differently in adolescence or adulthood (Rutter et al. 2006; Totah et al., this volume). Finally, dysfunction itself can evolve over time, because the dysfunction itself may be progressing, because processes in the brain are attempting to compensate for the dysfunction (Krystal et al., this volume), or as a result of interactions with the environment (Totah et al., this volume; Borsboom et al. 2011; Borsboom and Cramer 2013). This reveals a fifth challenge: It is necessary to take into account the dynamics of illness while remembering that the key to treatment is to change the patient's trajectory (Flagel et al. and Friston, this volume).

The Promise of Computation

The computational perspective uses formal methods to relate how processes at one level can explain processing at other levels (e.g., how information processing in neural circuits can drive mental processes and behavior) and, in particular, how changes at one level can explain effects at other levels (e.g., how genetic differences can change the function of neural circuits) (Chapter 2). Computational neuroscience provides a diverse toolkit of models and theories that can be used to provide that explanatory power (Kurth-Nelson et al., this volume). The most appropriate theory or model to use depends on the levels involved in one's questions.

The computational perspective has three effects which we suggest hold promise to address these challenges. First, the necessary *formalism* of computation forces one to be more complete and often reveals obscure consequences. All four discussion groups found themselves independently using a similar formalism to link causal dysfunction with observational effects that will have large implications for nosology (see Totah et al., Kurth-Nelson et al., Moran et al., and Flagel et al., this volume, and discussion below).

Second, because computation in neural circuits is fundamentally about *information processing* and how that information processing drives behavior, the questions one begins to ask about psychiatric dysfunction changes. For instance, when addressing surprising actions—such as continued use of addictive drugs (Redish 2004; Redish et al. 2008; Moran et al., this volume), misstated logic in semantic dementia (McClelland and Rogers 2003; Moran et al., this volume), unreasonable actions taken in schizophrenia (see chapters by Barch and Krystal et al., this volume), or a lack of action in depression (Huys, this volume)—the computational perspective changes the question from merely identifying *what* the subject is doing differently to identifying *how* the subject is recognizing information and processing that information differently. In particular, physical changes in neural circuits can have profound effects on how information is stored and processed in that neural circuit.

Third, this suggests that *treatment* can be aimed at (a) repairing the physical dysfunction in the neural circuit, (b) changing the environmental or social milieu in which that dysfunction occurs, or (c) changing other neural circuits to accommodate or replace the function of the dysfunctional circuit.

Synthesis from the Four Groups

As noted in the Chapters 1 and 2, and can be seen in this book's structure, Forum participants were divided into four working groups, each of which was tasked with a key topic relating to the question of how the computational perspective changes psychiatry, both in theory and practice.

Totah et al.: The Complexity and Heterogeneity of Psychiatry Disorders

This discussion group was tasked with examining what it was that computation needed to address. A central theme of this chapter is that complexity and heterogeneity are not noise to be abstracted away, but rather critical factors that need to be included in any computational theory. Using three case studies, they point out the variability from patient to patient and exhort us not to forget that patients are individuals with specific life stories. They recommend embracing this heterogeneity. Moreover, they remind us that psychiatric dysfunctions change over time, whether through development, through progression of the dysfunction itself, or through interactions with the environment.

An issue raised by the accompanying chapter (Barch, this volume) is that a single dysfunction ("something is going on with dopamine in schizophrenia") can have many consequences throughout neural function, making the relationship between biological dysfunction and psychiatric observations even more complex. An important factor raised in this section is that we need to find a way to connect the biological and environmental factors that are dysfunctional to the observed psychiatric categories. It is suggested that this connection is going to be both multisourced and multipotential.

Kurth-Nelson et al.: Computational Approaches

The next group was tasked with examining what computational models were available and how they could be applied. Their discussion raised the very important issue that there is not a single computational model. Computational models and the computational perspective are very broad and wide ranging; however, they have in common the ability to link levels, particularly in non-intuitive and complex ways. Importantly, not all computational models have to be mechanistic; they can formally describe interactions between processes without specifying mechanism. Furthermore, the group pointed out that our goal in judging the success of computational approaches should not be to replicate current diagnoses of psychiatry. Success needs to be measured against more fundamental outcomes, reminding us that the goal is treatment and improvement in patient prognoses.

In an accompanying chapter, Frank (this volume) reviewed computational neuroscience approaches across specific levels and suggested that the computational framework characterizes mental illness in terms of difficulties in balancing trade-offs. Here again, we need to think of psychiatry in terms of how an individual with a specific biological brain interacts with the environmental and social milieu.

Mathys (this volume) reviewed a specific computational framework that was explicitly designed to handle multisourced and multipotential connections: Bayesian inference (Pearl 1988, 2009b; Jaynes 2003). This framework can formalize the relationship between a dimensional model of underlying potential causes and the pleiotropy of observed behaviors. Importantly, this framework allows one to formalize the inverse logic that allows reasoning from observations to potential dimensional causes. The key to this framework is to see both causes and observations as probabilistic rather than certainties. In setting out the principles of Bayesian inference, Mathys introduced a key method to develop frameworks for integrating computation and psychiatry in novel ways that was used by both Flagel et al. and Moran et al. (this volume).

Flagel et al.: A New Framework

This group was tasked with identifying how the computational perspective can be used to improve nosology. Supporting chapters by First, MacDonald et al. and Mathys (this volume) framed the issues for the group. First (this volume) reminds us of the importance of nosology, noting that the original intention of the Diagnostic and Statistical Manual of Mental Disorders (DSM) and International Classification of Diseases (ICD) was to maintain an atheoretical categorical perspective. Of course, no perspective is atheoretical, and the categories in the DSM and ICD have become the definitions of named syndromes. The new, highly theoretical Research Domain Criteria (RDoC) project, implemented by the U.S. National Institute of Mental Health (Cuthbert and Insel 2010; Insel et al. 2010), has taken a wholly new dimensional approach. It has been extremely difficult, however, to meld the dimensional descriptions in RDoC with the categorical descriptions of DSM and ICD (First, this volume). This is partially due to issues raised by MacDonald et al. (this volume)—the relationship between causes and consequences are both multifinal (multisourced) and equifinal (multipotential). They suggest that causal networks such as those used in reliability engineering could be a way to link these multisourced and multipotential relationships. These relationships are directly formalizable in the Bayesian inference perspective raised by Mathys (this volume).

This discussion group came to the conclusion that if one takes the novel perspective that *psychiatric diagnoses are observations*, not causes, then a natural nosology appears in which there are biological and environmental causes that probabilistically lead to observations (Flagel et al., this volume). The biological and environmental causes can be highly dimensional and highly theoretical (like RDoC). Psychiatric diagnoses remain an important part of clinical practice, but they become cues to the underlying hypothesized causes rather than syndromes themselves. Importantly, observations can also include other measurements, such as biological measurements, clinical instruments, or cognitive tasks. Furthermore, because Bayesian inference allows reasoning in both directions (from observations to hypothesized causes and from probability distributions over causes to predicted observations), prognoses can be thought of as yet another observation: one that unfolds over future time.

In his accompanying chapter, Friston (this volume) uses a computational simulation to show how this logic might work. Importantly, as noted by Totah et al. (this volume), a patient's experience is a trajectory, and the goal must be to shift that trajectory. This means that the prognosis needs to be seen as a probability distribution across trajectories, and that the goal of treatment is to change that probability distribution. Therefore, this framework naturally includes the dynamic nature of psychiatric illness as well as the heterogeneity inherent in these dynamics, but with the capacity to formally analyze and quantify these dynamics, as well as the effects of treatment on disease course.

Moran et al.: Candidate Examples

The fourth group was tasked with finding targets, dysfunctions, syndromes, and disorders that could be used as canonical examples of the computational perspective in action. Moran et al. sketched out several examples of specific models that have had an impact on the field, including dopamine models in schizophrenia and addiction, computational analyses of neuroimaging and EEG measurements, and a model of treatment in changing allostasis along the amygdala-HPA axis. They began, however, with a generative Bayesian inference model in which biological parameters are linked to symptoms, biomarkers, and diagnoses through computational parameters. This model is identical to the new nosology suggested by Flagel et al. (compare Figure 12.1 with Figure 10.3). In the general framework by Moran et al. (identified as a "generative model"), the biological parameters are the underlying structure (putative causes and hidden physiological states of Flagel et al.), the computational parameters are the theoretical structure (the latent constructs of Flagel et al.), and the symptoms, biomarkers, and diagnoses are the observations. Importantly, Moran et al. view this generative model as a trajectory that needs to be shifted by treatment. In the companion chapters, Montague, Paulus et al., Huys, and Krystal et al. (all this volume) lay out examples of these relationships, looking specifically at the relationship between computation and dysfunction in risk-sensitivity, value and addiction, depression, and schizophrenia.

Common Themes and New Breakthroughs

Remarkably, all four groups converged independently on a similar breakthrough: that the key to connecting the fundamental science with clinical practice is a multipotential and multisourced computational perspective which links psychiatric observations with underlying dysfunction, but which breaks the one-to-one assumptions currently underlying clinical practice. This key breakthrough can be summarized by a simple statement: DSM diagnoses are symptoms, not syndromes. This realization led us to propose a new system in which fundamental science identifies *neuropsychological processes* and *failure modes* within those processes. These processes can be computational theoretic models (e.g., reinforcement-learning algorithms; Montague, this volume) or psychological constructs, such as are used to define the RDoC matrix. These processes probabilistically produce *outcomes*, which can be either observations (e.g., scales on a psychiatric instrumental questionnaire), measurements (e.g., scores on a task), or DSM diagnoses. This new perspective unifies the fundamental science processes, such as RDoC, with psychiatric categorizations, such as used in the DSM or ICD-10.

Early discussions of computational models talked in terms of "vulnerabilities" or "failure modes" (Redish 2004; Huys 2007; Rangel et al. 2008; Redish et al. 2008; Huys et al. 2015a), in which one derived observed dysfunction from specific errors in hypothesized underlying processes. However, as became clear in the discussions, the multisourced and multipotential nature of psychiatric dysfunction is going to depend on a more complex path from dysfunction to symptom (see Totah et al., MacDonald et al., Flagel et al., Krystal et al., this volume), taking into account genetic and environmental causes, heterogeneity, and dynamics.

Models of fundamental dysfunctions are inherently dimensional, based on errors in specific parameters and processes (e.g., RDoC), whereas models of clinical practice are inherently categorical (e.g., DSM, ICD). The relationship between fundamental science and clinical practice is more than a simple threshold on dimensionality. To accommodate this complexity, we propose a Bayesian causal model of failure modes in underlying fundamental neuropsychological processes leading to observations (diagnoses) with specific probabilities. This new system (exemplified by the figures in Flagel et al., this volume) opens up an entirely new perspective on psychiatry, providing a way to connect clinical observation with underlying neuropsychological causes. Moreover, this probabilistic framework has heterogeneity built into it: a given dysfunction in a fundamental process might lead to psychosis with some (non-zero) probability or mood instability with a different (non-zero) probability. Thus the same failure in the same neuropsychological process could lead to two different diagnoses from the perspective of the clinician. The framework also explains comorbidity: some fraction of people with that dysfunction may find themselves with both diagnoses.

Importantly, these Bayesian inferences can be applied to trajectories, both through the past and into the future. Bayesian inferences, taking into account past trajectories, allow the hypothesized dysfunction to explain pathophysiological processes and symptoms that change and develop over time. Bayesian inferences about future trajectories are *prognoses*. As pointed out by Flagel et al. (see also Friston, this volume), a prognosis is simply a predicted observation about the future trajectory.

This new nosology solves the heterogeneity and comorbidity problems completely, while unifying RDoC (and other neuropsychological hypothesis-based schemes) with clinical categorizations (such as are used in DSM and ICD). This proposal unifies observations, clinical categorizations (DSM syndromes), measurements (e.g., clinical questionnaire instruments, performance on cognitive tasks), and prognoses as observations linked through Bayesian inference to a scientific hypothesis of latent constructs.

An important open question that we still must face is whether the hypothesized latent constructs are the correct taxonomy. Of course, science is always progressing by refining and replacing theoretical constructs. Aristotle's theory of gravity was replaced by Newton's theory, and Newton's equations were replaced by Einstein's. Current observations of galactic motion are hinting at the possibility of further refinement in Einstein's equations. Nevertheless, at each stage, scientific theory was able to provide practical consequences. Do we have the right fundamental science dimensions and the right experimental tests to identify patient treatments? One of the major advantages of the Bayesian perspective is that it provides an *ongoing process* rather than a direct answer to this open question. Latent constructs which do not inform prognoses will be excluded mathematically because they are unpredictive, whereas those that provide for better description of the observations and better prognoses will be preserved.

The Way Forward

Establishing a Computational Nosology

Our breakthrough is a proposal: it describes a methodology for translational psychiatry, for a way to connect fundamental (basic) science research to clinical practice. Actually implementing this methodology, however, is going to require buy-in from the current stakeholders in these processes.

Clinicians might be daunted by the complexity of Bayesian inference, or the notion of carrying out such analyses for every patient mathematically. However, it would not be hard to implement the algorithm in a computer app or online website. Essentially, a clinician would enter a series of observations (e.g., diagnoses, clinical instrument test scores, cognitive task results, perhaps even genetic tests) and would receive a probability distribution over potential treatments, explanations, and outcomes.

This is actually not that different from what current practice is supposed to be. The clinician determines the presence or absence of symptoms through a clinical interview; consults the diagnostic "algorithms" in the DSM (which define categories from lists of symptoms), and determines the appropriate diagnoses. A treatment is then selected based on the likelihood of response for that diagnosis. In reality, of course, both diagnoses and treatments are selected based much more on experience than on an algorithm. This reality, however, is based on a lack of specific information, which if present would enable decisions to be made with greater confidence. In this sense, what we are proposing is an improved DSM: one that might be complicated enough to require a computer, but one which would be much more useful than current systems. It supplements the clinician's judgment by providing explicit probabilities over latent causes, treatments, diagnoses, and prognoses that the clinician can use to communicate with the patient and to help make decisions with the patient about treatment. This requires communicating probabilities to patients, but medical fields are already communicating probabilities for other complex diseases (e.g., cancer), in terms of the probability distribution of survival curves and future prognosis trajectories based on different treatments.

A final consideration for the clinician is the notion that the algorithmic recommendations are personalized and personalizeable. One inputs patient-specific information and the outputs are probabilities that a given treatment will work in a given way. Thus, there is not only room but rather a requirement for clinical judgment in adopting these recommendations.

Acceptance of the model will require cooperation from scientists as well. Fundamental (basic) scientists will determine the latent constructs (scientific hypotheses) that underlie neuropsychological function, including how those processes interact with each other as well as with environmental and sociological factors. Translational scientists will determine how observations and measurements predict (and are predicted by) those underlying constructs. A team of experts will convene and codify those relationships. Presumably that team of experts will meet regularly to update that codification. There is no reason not to expect that team of experts to have computational help to run complex calculations measuring specific relationships. Once the experts have codified the relationships, they can be compiled into an algorithm that can be accessed by any clinician anywhere, even without the level of expertise needed to codify the relationships. This algorithm would be updated periodically, not unlike the iterative editions of the DSM, but more frequently (and with mathematical rigor applied to any changes, which would have to have documented efficacy).

Of course, as with any scientific endeavor, this is a cyclical process by which new fundamental science hypotheses are derived from clinical observations, which will lead to new discoveries, requiring new codifications, and (hopefully) improved patient outcomes (Figure 17.1).

Challenges and Promises

We began with a list of challenges (a new diagnostic nosology, new biomarkers, and improved treatments), complexities (the heterogeneity of patients, the dynamics through which psychiatric dysfunction progresses, that the path from genetics to behavior goes through neural circuits), and promises (the hope of personalized medicine). We believe that this Forum succeeded in addressing these issues far beyond what we could have hoped. The new perspective integrating dimensional science (RDoC, fundamental/basic hypotheses) with observations (DSM/ICD syndromes and categories, clinical instruments, tests, etc.) is explicitly a new diagnostic nosology that builds on the progress made over the last half-century in both the clinical and fundamental sciences. It provides a direct way to incorporate new biomarkers as they are discovered and provides a new way to identify the best treatments (as probabilities over prognoses). It directly addresses the complexities, including taking into account nonlinearities and the heterogeneity of patients. Because our proposed diagnostic nosology reflects trajectories in the past (history) and future (prognoses), it takes into account the dynamics of psychiatric dysfunction. Because it depends on latent constructs, one does not have to go directly from genetics to

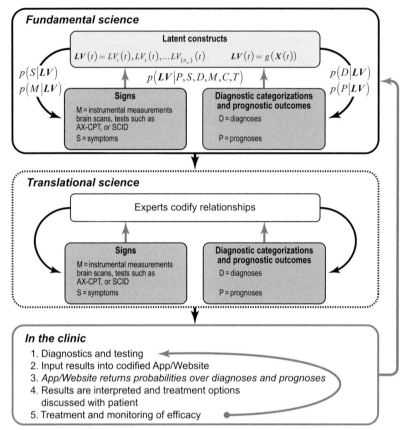

Figure 17.1   The proposed process to translate fundamental science to clinical practice. Fundamental (basic) scientists determine the most appropriate latent constructs
and the relationship between those latent constructs and signs and outcomes. A committee of experts codifies these relationships so that one can integrate the two steps
of Bayesian inference from signs and observations to diagnostic categorizations and
prognostic outcomes, maintaining the probabilities. A clinician then uses this codified
relationship to translate from signs (observations) and diagnoses to a probability distribution over diagnostic categorizations and prognoses. Treatments can be presented
in how they would change the prognoses (probabilistically). Because continued observations, including the consequences of treatments, provide additional information to
the probabilities of diagnoses and prognoses, the clinician will keep returning to the
codification for refinement of diagnosis, treatment, and prognosis. Of course, this path
from fundamental science to the clinic is a continual cycle of scientific progress and
refinement, with fundamental science learning from continued clinical observations.

behavior or psychiatric categories; instead, one can go (probabilistically) from
genetics to latent constructs and then (probabilistically) from latent constructs
to behavior. Finally, each person will have a unique set of genetics, history, and
symptoms, which probabilistically interact to produce a unique diagnosis and
prognosis, directly giving us personalized medicine.

New Structures

The field of "computational psychiatry" is exploding with a new ongoing workshop group that has been formed, new departments forming, a new journal, and several books being developed, including this one. A community bringing together experienced psychiatrists, experienced computational neuroscientists, and newly trained students with expertise in both fields is appearing. This community is developing experimental tests which can be used to identify parameters in underlying dysfunctions, such as the two-step decision task capable of differentiating model-based and model-free decision-making processes, which predicts a broad spectrum of obsessive behaviors (Gillan et al. 2016). New studies suggest that these multidimensional (genetic and task-based) categorizations produce more reliable categorizations than traditional DSM categories (Clementz et al. 2016). More work is clearly needed to translate the tasks developed for measuring fundamental science dimensions (in both human and nonhuman models) into clinically relevant measures. We believe that the field is ready to have a direct effect on the practice of psychiatry.

Bibliography

Note: Numbers in square brackets denote the chapter in which an entry is cited.

Aarts, E., M. van Holstein, and R. Cools. 2011. Striatal Dopamine and the Interface between Motivation and Cognition. *Front. Psychol.* **2**:163. [4]

Abbeel, P., D. Dolgov, A. Y. Ng, and S. Thrun. 2008. Apprenticeship Learning for Motion Planning with Application to Parking Lot Navigation. RSJ Intl. Conf. on Intelligent Robots and Systems. IROS 2008, pp. 1083–1090. Piscataway, NJ: IEEE Press. [14]

Abbeel, P., and A. Ng. 2004. Apprenticeship Learning via Inverse Reinforcement Learning. Proc. 21st Intl. Conf. on Machine Learning, vol. 69, p. 8. Banff: ACM Series. [14]

Abi-Dargham, A., L. S. Kegeles, Y. Zea-Ponce, et al. 2004. Striatal Amphetamine-Induced Dopamine Release in Patients with Schizotypal Personality Disorder Studied with Single Photon Emission Computed Tomography and [123i]Iodobenzamide. *Biol. Psychiatry* **55**:1001–1006. [4]

Abi-Dargham, A., O. Mawlawi, I. Lombardo, et al. 2002. Prefrontal Dopamine D1 Receptors and Working Memory in Schizophrenia. *J. Neurosci.* **22**:3708–3719. [5]

Abi-Dargham, A., E. van de Giessen, M. Slifstein, L. S. Kegeles, and M. Laruelle. 2009. Baseline and Amphetamine-Stimulated Dopamine Activity Are Related in Drug-Naive Schizophrenic Subjects. *Biol. Psychiatry* **65**:1091–1093. [4]

Abrahams, B. S., and D. H. Geschwind. 2008. Advances in Autism Genetics: on the Threshold of a New Neurobiology. *Nat. Rev. Genet.* **9**:341–355. [3]

Abramowitz, J. S. 2013. The Practice of Exposure Therapy: Relevance of Cognitive-Behavioral Theory and Extinction Theory. *Behav. Ther.* **44**:548–558. [5]

Abramson, L. Y., G. I. Metalsky, and L. B. Alloy. 1989. Hopelessness Depression: A Theory-Based Subtype of Depression. *Psychol. Rev.* **96**:358–372. [15]

Adams, D. H., L. Zhang, B. A. Millen, B. J. Kinon, and J. C. Gomez. 2014. Pomaglumetad Methionil (LY2140023 Monohydrate) and Aripiprazole in Patients with Schizophrenia: A Phase 3, Multicenter, Double-Blind Comparison. *Schizophr. Res. Treatment* **2014**:758212. [16]

Adams, R. A., K. E. Stephan, H. R. Brown, and K. J. Friston. 2013. The Computational Anatomy of Psychosis. *Front. Psychiatry* **4**:47. [7, 12]

Addington, J., I. Epstein, L. Liu, et al. 2011. A Randomized Controlled Trial of Cognitive Behavioral Therapy for Individuals at Clinical High Risk of Psychosis. *Schizophr. Res.* **125**:54–61. [3]

Adler, C. M., P. McDonough-Ryan, K. W. Sax, et al. 2000. fMRI of Neuronal Activation with Symptom Provocation in Unmedicated Patients with Obsessive Compulsive Disorder. *J. Psychiatr. Res.* **34**:317–324. [12]

Adriano, F., C. Caltagirone, and G. Spalletta. 2012. Hippocampal Volume Reduction in First-Episode and Chronic Schizophrenia: A Review and Meta-Analysis. *Neuroscientist* **18**:180–200. [1, 4]

Adriano, F., I. Spoletini, C. Caltagirone, and G. Spalletta. 2010. Updated Meta-Analyses Reveal Thalamus Volume Reduction in Patients with First-Episode and Chronic Schizophrenia. *Schizophr. Res.* **123**:1–14. [4]

Afraz, A., D. L. Yamins, and J. J. DiCarlo. 2014. Neural Mechanisms Underlying Visual Object Recognition. *Cold Spring Harb. Symp. Quant. Biol.* **79**:99–107. [5]

Agid, O., L. Schulze, T. Arenovich, et al. 2013. Antipsychotic Response in First-Episode Schizophrenia: Efficacy of High Doses and Switching. *Eur. Neuropsychopharmacology* **23**:1017–1022. [12]

Agius, M., C. Goh, S. Ulhaq, and P. McGorry. 2010. The Staging Model in Schizophrenia, and Its Clinical Implications. *Psychiatria Danubina* **22**:211–220. [16]

Agnew-Blais, J., and L. J. Seidman. 2013. Neurocognition in Youth and Young Adults under Age 30 at Familial Risk for Schizophrenia: A Quantitative and Qualitative Review. *Cogn. Neuropsychiatry* **18**:44–82. [4]

Ahmari, S. E., T. Spellman, N. L. Douglass, et al. 2013. Repeated Cortico-Striatal Stimulation Generates Persistent OCD-Like Behavior. *Science* **340**:1234–1239. [10]

Ainslie, G. 1992. Picoeconomics. Cambridge: Cambridge Univ. Press. [2]

———. 2001. Breakdown of Will. Cambridge Cambridge Univ. Press. [2]

Akil, H., S. Brenner, E. Kandel, et al. 2010. Medicine. The Future of Psychiatric Research: Genomes and Neural Circuits. *Science* **327**:1580–1581. [1]

Albin, R. L., A. B. Young, and J. B. Penney. 1989. The Functional Anatomy of Basal Ganglia Disorders. *Trends Neurosci.* **12**:366–374. [2]

Allen, P., C. A. Chaddock, O. D. Howes, et al. 2012. Abnormal Relationship between Medial Temporal Lobe and Subcortical Dopamine Function in People with an Ultra High Risk for Psychosis. *Schizophr. Bull.* **38**:1040–1049. [4]

Alloy, L. B., L. Y. Abramson, M. E. Hogan, et al. 2000. The Temple-Wisconsin Cognitive Vulnerability to Depression Project: Lifetime History of Axis I Psychopathology in Individuals at High and Low Cognitive Risk for Depression. *J. Abnorm. Psychol.* **109**:403–418. [15]

Alloy, L. B., L. Y. Abramson, W. G. Whitehouse, et al. 1999. Depressogenic Cognitive Styles: Predictive Validity, Information Processing and Personality Characteristics, and Developmental Origins. *Behav. Res. Ther.* **37**:503–531. [15]

Alptekin, K., B. Degirmenci, B. Kivircik, et al. 2001. Tc-99m HMPAO Brain Perfusion SPECT in Drug-Free Obsessive-Compulsive Patients without Depression. *Psychiatry Res.* **107**:51–56. [10]

Amemori, K.-I., and A. M. Graybiel. 2012. Localized Microstimulation of Primate Pregenual Cingulate Cortex Induces Negative Decision-Making. *Nat. Neurosci.* **15**: 776–785. [15]

Amsel, L., and J. J. Mann. 2001. Suicide Risk Assessment and the Suicide Process Approach. In: Understanding Suicidal Behaviour: The Suicidal Process Approach to Research, Treatment and Prevention, ed. K. van Heeringen, pp. 163–181. Chichester: John Wiley. [3]

Anand, A., D. S. Charney, D. A. Oren, et al. 2000a. Attenuation of the Neuropsychiatric Effects of Ketamine with Lamotrigine: Support for Hyperglutamatergic Effects of N-Methyl-D-Aspartate Receptor Antagonists. *Arch. Gen. Psychiatry* **57**:270–276. [16]

Anand, A., P. Verhoeff, N. Seneca, et al. 2000b. Brain SPECT Imaging of Amphetamine-Induced Dopamine Release in Euthymic Bipolar Disorder Patients. *Am. J. Psychiatry* **157**:1108–1114. [16]

Anda, R. F., V. J. Felitti, J. D. Bremner, et al. 2006. The Enduring Effects of Abuse and Related Adverse Experiences in Childhood. A Convergence of Evidence from Neurobiology and Epidemiology. *Eur. Arch. Psychiatry Clin. Neurosci.* **256**:174–186. [3]

Anders, S. L., C. K. Peterson, L. M. James, et al. 2015. Neural Communication in Posttraumatic Growth. *Exp. Brain Res.* **233**:2013–2020. [10]

Anderson, I. M., and S. Pilling. 2010. Depression: The Treatment and Management of Depression in Adults (Updated edition). Leicester: British Psychological Society. [15]

Anderson, I. M., and B. M. Tomenson. 1995. Treatment Discontinuation with Selective Serotonin Reuptake Inhibitors Compared with Tricyclic Antidepressants: A Meta-Analysis. *BMJ* **310**:1433–1438. [15]

Anderson, M. L., and T. Oates. 2007. A Review of Recent Research in Metareasoning and Metalearning. *AI Mag.* **28**:12. [15]

Andreasen, N. C., P. Nopoulos, V. Magnotta, et al. 2011. Progressive Brain Change in Schizophrenia: A Prospective Longitudinal Study of First-Episode Schizophrenia. *Biol. Psychiatry* **70**:672–679. [16]

Andrews, A., M. Knapp, P. McCrone, and M. Parsonage. 2012. Effective Interventions in Schizophrenia the Economic Case: A Report Prepared for the Schizophrenia Commission. London: Personal Social Services Research Unit, London School of Economics and Political Science. [3]

Andrews, G., T. Brugha, M. E. Thase, et al. 2007. Dimensionality and the Category of Major Depressive Episode. *Int. J. Methods Psychiatr. Res.* **16** 541–551. [8]

Andrews, G., D. S. Charney, P. J. Sirovatka, and D. A. Regier. 2009. Stress-Induced and Fear Circuitry Disorders Refining the Research Agenda for DSM-V. Arlington, VA: American Psychiatric Association. [8]

Andrews, P. W., S. G. Kornstein, L. J. Halberstadt, C. O. Gardner, and M. C. Neale. 2011. Blue Again: Perturbational Effects of Antidepressants Suggest Monoaminergic Homeostasis in Major Depression. *Front. Psychol.* **2**:159. [15]

Angst, J. 1992. Epidemiology of Depression. *Psychopharmacology (Berl.)* **106**:S71–S74. [15]

Anticevic, A., M. W. Cole, J. D. Murray, et al. 2012a. The Role of Default Network Deactivation in Cognition and Disease. *Trends Cogn. Sci.* **16**:584–592. [16]

Anticevic, A., M. W. Cole, G. Repovs, et al. 2014a. Characterizing Thalamo-Cortical Disturbances in Schizophrenia and Bipolar Illness. *Cereb. Cortex* **24**:3116–3130. [16]

Anticevic, A., P. R. Corlett, M. W. Cole, et al. 2015a. N-Methyl-D-Aspartate Receptor Antagonist Effects on Prefrontal Cortical Connectivity Better Model Early Than Chronic Schizophrenia. *Biol. Psychiatry* **77**:569–580. [16]

Anticevic, A., M. Gancsos, J. D. Murray, et al. 2012b. NMDA Receptor Function in Large-Scale Anticorrelated Neural Systems with Implications for Cognition and Schizophrenia. *PNAS* **109**:16720–16725. [16]

Anticevic, A., K. Haut, J. D. Murray, et al. 2015b. Association of Thalamic Dysconnectivity and Conversion to Psychosis in Youth and Young Adults at Elevated Clinical Risk. *JAMA Psychiatry* **72**:882–891. [16]

Anticevic, A., X. Hu, Y. Xiao, et al. 2015c. Early-Course Unmedicated Schizophrenia Patients Exhibit Elevated Prefrontal Connectivity Associated with Longitudinal Change. *J. Neurosci.* **35**:267–286. [16]

Anticevic, A., J. D. Murray, and D. M. Barch. 2015d. Bridging Levels of Understanding in Schizophrenia through Computational Modeling. *Clin. Psychol. Sci.* **3**:433–459. [16]

Anticevic, A., G. Yang, A. Savic, et al. 2014b. Mediodorsal and Visual Thalamic Connectivity Differ in Schizophrenia and Bipolar Disorder with and without Psychosis History. *Schizophr. Bull.* **40**:1227–1243. [16]

Aoyama, N., J. Theberge, D. J. Drost, et al. 2011. Grey Matter and Social Functioning Correlates of Glutamatergic Metabolite Loss in Schizophrenia. *Br. J. Psychiatry* **198**:448–456. [16]

APA. 1980. Diagnostic and Statistical Manual of Mental Disorders, DSM-III. Arlington: American Psychiatric Association Publishing. [8, 10]

———. 2000. Diagnostic and Statistical Manual of Mental Disorders (DSM-IV-TR), 4th Edition. Arlington, VA: American Psychiatric Association Publishing. [2, 8, 15]

APA. 2013. Diagnostic and Statistical Manual of Mental Disorders (DSM-5). Arlington: American Psychiatric Association Publishing. [1, 2, 8, 10, 14, 16]

APA Task Force on Laboratory Tests in Psychiatry. 1987. The Dexamethasone Suppression Test: An Overview of Its Current Status in Psychiatry. *Am. J. Psychiatry* **144**:1253–1262. [1]

Argyelan, M., T. Ikuta, P. DeRosse, et al. 2014. Resting-State fMRI Connectivity Impairment in Schizophrenia and Bipolar Disorder. *Schizophr. Bull.* **40**:100–110. [16]

Arnsten, A. F. T. 2011. Catecholamine Influences on Dorsolateral Prefrontal Cortical Networks. *Biol. Psychiatry* **69**:e89–e99. [3]

Autism Genome Project Consortium. 2007. Mapping Autism Risk Loci Using Genetic Linkage and Chromosomal Rearrangements. *Nat. Genet.* **39**:319–328. [3]

Badcock, J. C., P. T. Michiel, and D. Rock. 2005. Spatial Working Memory and Planning Ability: Contrasts between Schizophrenia and Bipolar I Disorder. *Cortex* **41**:753–763. [16]

Badre, D., B. B. Doll, N. M. Long, and M. J. Frank. 2012. Rostrolateral Prefrontal Cortex and Individual Differences in Uncertainty-Driven Exploration. *Neuron* **73**: 595–607. [6]

Barabási, A.-L., N. Gulbahce, and J. Loscalzo. 2011. Network Medicine: A Network-Based Approach to Human Disease. *Nat. Rev. Genet.* **12**:56–68. [3]

Barch, D. M. 2004. Pharmacological Manipulation of Human Working Memory. *Psychopharmacology* **174**:126–135. [4]

———. 2005. The Cognitive Neuroscience of Schizophrenia. In: Annual Review of Clinical Psychology, ed. T. Cannon and S. Mineka, pp. 321–353, vol. 1. Washington, D.C.: American Psychological Association. [4]

Barch, D. M., and T. S. Braver. 2007. Cognitive Control in Schizophrenia: Psychological and Neural Mechanisms. In: Cognitive Limitations in Aging and Psychopathology, ed. R. W. Engle et al., pp. 122–159. Cambridge: Cambridge Univ. Press. [4]

Barch, D. M., T. S. Braver, C. S. Carter, R. A. Poldrack, and T. W. Robbins. 2009. CNTRICS Final Task Selection: Executive Control. *Schizophr. Bull.* **35**:115–135. [4]

Barch, D. M., and C. S. Carter. 2005. Amphetamine Improves Cognitive Function in Medicated Individuals with Schizophrenia and in Healthy Volunteers. *Schizophr. Res.* **77**:43–58. [4]

Barch, D. M., C. S. Carter, T. S. Braver, et al. 2001. Selective Deficits in Prefrontal Cortex Regions in Medication Naive Schizophrenia Patients. *Arch. Gen. Psychiatry* **50**:280–288. [4]

Barch, D. M., and A. E. Ceaser. 2012. Cognition in Schizophrenia: Core Psychological and Neural Mechanisms. *Trends Cogn. Sci.* **16**:27–34. [4]

Barch, D. M., and J. M. Sheffield. 2014. Cognitive Impairments in Psychotic Disorders: Common Mechanisms and Measurement. *World Psychiatry* **13**:224–232. [4]

———. 2016. Cognitive Control in Schizophrenia: Psychological and Neural Mechanisms. In: Handbook of Cognitive Control, ed. T. Egner. Chichester: John Wiley & Sons, in press. [4]

Barch, D. M., M. T. Treadway, and N. Schoen. 2014. Effort, Anhedonia, and Function in Schizophrenia: Reduced Effort Allocation Predicts Amotivation and Functional Impairment. *J. Abnorm. Psychol.* **123**:387–397. [4]

Barker, J. M., M. M. Torregrossa, A. P. Arnold, and J. R. Taylor. 2010. Dissociation of Genetic and Hormonal Influences on Sex Differences in Alcoholism-Related Behaviors. *J. Neurosci.* **30**:9140–9144. [3]

Barto, A. G. 1995. Adaptive Critics and the Basal Ganglia. In: Models of Information Processing in the Basal Ganglia, ed. J. C. Houk et al., pp. 215–232. Cambridge, MA: MIT Press. [5]

Barton, J. J., M. V. Cherkasova, K. Lindgren, et al. 2002. Antisaccades and Task Switching: Studies of Control Processes in Saccadic Function in Normal Subjects and Schizophrenic Patients. *Ann. NY Acad. Sci.* **956**:250–263. [4]

Bartos, M., I. Vida, and P. Jonas. 2007. Synaptic Mechanisms of Synchronized Gamma Oscillations in Inhibitory Interneuron Networks. *Nat. Rev. Neurosci.* **8**:45–56. [4]

Bastos, A. M., W. Usrey, R. A. Adams, et al. 2012. Canonical Microcircuits for Predictive Coding. *Neuron* **76**:695–711. [7, 12]

Bauer, M., A. Pfennig, E. Severus, et al. 2013. World Federation of Societies of Biological Psychiatry (WFSBP) Guidelines for Biological Treatment of Unipolar Depressive Disorders, Part 1: Update 2013 on the Acute and Continuation Treatment of Unipolar Depressive Disorders. *World J. Biol. Psychiatry* **14**:334–385. [15]

Baxter, L. R., Jr., J. M. Schwartz, K. S. Bergman, et al. 1992. Caudate Glucose Metabolic Rate Changes with Both Drug and Behavior Therapy for Obsessive-Compulsive Disorder. *Arch. Gen. Psychiatry* **49**:681–689. [1]

Baxter, L. R., Jr., J. M. Schwartz, J. C. Mazziotta, et al. 1988. Cerebral Glucose Metabolic Rates in Nondepressed Patients with Obsessive-Compulsive Disorder. *Am. J. Psychiatry* **145**:1560–1563. [10]

Bayes, T., and R. Price. 1763. An Essay Towards Solving a Problem in the Doctrine of Chances. By the Late Rev. Mr. Bayes, F. R. S. Communicated by Mr. Price, in a Letter to John Canton. *Phi. Trans.* **53**:370–418. [7]

Beck, A. T. 1967. Depression: Clinical, Experimental and Theoretical Aspects. New York: Harper & Row. [15]

Bedi, G., F. Carrillo, G. A. Cecchi, et al. 2015. Automated Analysis of Free Speech Predicts Psychosis Onset in High-Risk Youths. *NPJ Schizophr.* **1**:15030. [3]

Beeler, J. A., M. J. Frank, J. McDaid, et al. 2012. A Role for Dopamine-Mediated Learning in the Pathophysiology and Treatment of Parkinson's Disease. *Cell Rep.* **2**:1747–1761. [3, 5]

Beesdo, K., S. Knappe, and D. Pine. 2009. Anxiety and Anxiety Disorders in Children and Adolescents: Developmental Issues and Implications for DSM-V. *Psychiatr. Clin. North Am.* **32**:483–524. [3]

Behrens, T. E., M. W. Woolrich, M. E. Walton, and M. F. Rushworth. 2007. Learning the Value of Information in an Uncertain World. *Nat. Neurosci.* **10**:1214–1221. [6, 14]

Belforte, J. E., V. Zsiros, E. R. Sklar, et al. 2010. Postnatal NMDA Receptor Ablation in Corticolimbic Interneurons Confers Schizophrenia-Like Phenotypes. *Nat. Neurosci.* **13**:76–83. [16]

Belin, D., A. C. Mar, J. W. Dalley, T. W. Robbins, and B. J. Everitt. 2008. High Impulsivity Predicts the Switch to Compulsive Cocaine-Taking. *Science* **320**:1352–1355. [2]

Bellman, R. 1957. Dynamic Proramming. Princeton: Princeton Univ. Press. [13]

Beltramo, R., G. D'Urso, M. Dal Maschio, et al. 2013. Layer-Specific Excitatory Circuits Differentially Control Recurrent Network Dynamics in the Neocortex. *Nat. Neurosci.* **16**:227–234. [3]

Beninger, R. J., J. Wasserman, K. Zanibbi, et al. 2003. Typical and Atypical Antipsychotic Medications Differentially Affect Two Nondeclarative Memory Tasks in Schizophrenic Patients: A Double Dissociation. *Schizophr. Res.* **61**:281–292. [4]

Bernard, J. A., J. M. Orr, and V. A. Mittal. 2015. Abnormal Hippocampal-Thalamic White Matter Tract Development and Positive Symptom Course in Individuals at Ultra-High Risk for Psychosis. *NPJ Schizophr.* **1**:15009. [3]

Berridge, K. C. 2006. The Debate over Dopamine's Role in Reward: The Case for Incentive Salience. *Psychopharmacology* **191**:391–431. [3]

Bickel, W. K., R. D. Landes, Z. Kurth-Nelson, and A. D. Redish. 2014. A Quantitative Signature of Self-Control Repair Rate-Dependent Effects of Successful Addiction Treatment. *Clin. Psychol. Sci.* **2**:685–695. [2]

Bickel, W. K., A. J. Quisenberry, L. Moody, and A. G. Wilson. 2015. Therapeutic Opportunities for Self-Control Repair in Addiction and Related Disorders: Change and the Limits of Change in Trans-Disease Processes. *Clin. Psychol. Sci.* **3**:140–153. [2]

Bickel, W. K., R. Yi, R. D. Landes, P. F. Hill, and C. Baxter. 2011. Remember the Future: Working Memory Training Decreases Delay Discounting among Stimulant Addicts. *Biol. Psychiatry* **69**:260–265. [2]

Billingslea, E. N., V. M. Tatard-Leitman, J. Anguiano, et al. 2014. Parvalbumin Cell Ablation of NMDA-R1 Causes Increased Resting Network Excitability with Associated Social and Self-Care Deficits. *Neuropsychopharmacology* **39**:1603–1613. [16]

Binder, E. B., R. G. Bradley, W. Liu, et al. 2008. Association of FKBP5 Polymorphisms and Childhood Abuse with Risk of Posttraumatic Stress Disorder Symptoms in Adults. *JAMA* **299**:1291–1305. [3]

Black, J. E., I. M. Kodish, A. W. Grossman, et al. 2004. Pathology of Layer V Pyramidal Neurons in the Prefrontal Cortex of Patients with Schizophrenia. *Am. J. Psychiatry* **161**:742–744. [16]

Blackburn, I. M., K. M. Eunson, and S. Bishop. 1986. A Two-Year Naturalistic Follow-up of Depressed Patients Treated with Cognitive Therapy, Pharmacotherapy and a Combination of Both. *J. Affect. Disord.* **10**:67–75. [15]

Blackwood, D. H., A. Fordyce, M. T. Walker, et al. 2001. Schizophrenia and Affective Disorders: Cosegregation with a Translocation at Chromosome 1q42 That Directly Disrupts Brain-Expressed Genes: Clinical and P300 Findings in a Family. *Am. J. Hum. Genet.* **69**:428–433. [3]

Bloch, M. H., C. A. Bartley, L. Zipperer, et al. 2014. Meta-Analysis: Hoarding Symptoms Associated with Poor Treatment Outcome in Obsessive-Compulsive Disorder. *Mol. Psychiatry* **19**:1025–1030. [10]

Bloch, M. H., A. Landeros-Weisenberger, B. Kelmendi, et al. 2006. A Systematic Review: Antipsychotic Augmentation with Treatment Refractory Obsessive-Compulsive Disorder. *Mol. Psychiatry* **11**:622–632. [10]

Bloch, M. H., and C. Pittenger. 2010. The Genetics of Obsessive-Compulsive Disorder. *Curr. Psychiatry Rev.* **6**:91–103. [10]

Blumberg, H. P., J. H. Krystal, R. Bansal, et al. 2006. Age, Rapid-Cycling, and Pharmacotherapy Effects on Ventral Prefrontal Cortex in Bipolar Disorder: A Cross-Sectional Study. *Biol. Psychiatry* **59**:611–618. [16]

Boccaletti, S., V. Latora, Y. Moreno, M. Chavez, and D. Hwang. 2006. Complex Networks: Structure and Dynamics. *Phys. Rep.* **424**:175–308. [3]

Bogacz, R., and K. Gurney. 2007. The Basal Ganglia and Cortex Implement Optimal Decision Making between Alternative Actions. *Neural Comput.* **19**:442–477. [5, 6]

Bois, C., L. Levita, I. Ripp, et al. 2015. Hippocampal, Amygdala and Nucleus Accumbens Volume in First-Episode Schizophrenia Patients and Individuals at High Familial Risk: A Cross-Sectional Comparison. *Schizophr. Res.* **165**:45–51. [3]

Boland, R. J., and M. B. Keller. 2002. The Course of Depression. In: Neuropsychopharmacology: The Fifth Generation of Progress, ed. K. L. Davis et al., pp. 1009–1015. Philadelphia: Lippincott Williams and Wilkins. [1]

Boly, M., M. I. Garrido, O. Gosseries, et al. 2011. Preserved Feedforward but Impaired Top-Down Processes in the Vegetative State. *Science* **332**:858–862. [10]

Bondi, C. O., A. Y. Taha, J. L. Tock, et al. 2014. Adolescent Behavior and Dopamine Availability Are Uniquely Sensitive to Dietary Omega-3 Fatty Acid Deficiency. *Biol. Psychiatry* **75**:38–46. [3]

Bonoldi, I., and O. D. Howes. 2013. The Enduring Centrality of Dopamine in the Pathophysiology of Schizophrenia: *In Vivo* Evidence from the Prodrome to the First Psychotic Episode. *Adv. Pharmacol.* **68**:199–220. [4, 16]

Bora, E., A. Fornito, J. Radua, et al. 2011. Neuroanatomical Abnormalities in Schizophrenia: A Multimodal Voxelwise Meta-Analysis and Meta-Regression Analysis. *Schizophr. Res.* **127**:46–57. [4]

Bora, E., and R. M. Murray. 2014. Meta-Analysis of Cognitive Deficits in Ultra-High Risk to Psychosis and First-Episode Psychosis: Do the Cognitive Deficits Progress over, or after, the Onset of Psychosis? *Schizophr. Bull.* **40**:744–755. [4]

Borsboom, D., and A. O. J. Cramer. 2013. Network Analysis: An Integrative Approach to the Structure of Psychopathology. *Annu. Rev. Clin. Psychol.* **9**:91–121. [5, 17]

Borsboom, D., A. O. J. Cramer, V. D. Schmittmann, S. Epskamp, and W. L. J. 2011. The Small World of Psychopathology. *PLoS One* **6**:e27407. [5, 17]

Bosl, W., A. Tierney, H. Tager-Flusberg, and C. Nelson. 2011. EEG Complexity as a Biomarker for Autism Spectrum Disorder Risk. *BMC Med.* **9**:18. [3]

Botvinick, M. M., T. S. Braver, D. M. Barch, C. S. Carter, and J. D. Cohen. 2001. Conflict Monitoring and Cognitive Control. *Psychol. Rev.* **108**:624. [13]

Botvinick, M. M., Y. Niv, and A. C. Barto. 2009. Hierarchically Organized Behavior and Its Neural Foundations: A Reinforcement Learning Perspective. *Cognition* **113**: 262–280. [6, 12, 13]

Botvinick, M. M., L. E. Nystrom, K. Fissell, C. S. Carter, and J. D. Cohen. 1999. Conflict Monitoring versus Selection-for-Action in Anterior Cingulate Cortex. *Nature* **402**:179–181. [13]

Botvinick, M. M., and A. Weinstein. 2014. Model-Based Hierarchical Reinforcement Learning and Human Action Control. *Phil. Trans. R. Soc. B* **369**:20130480. [12]

Boureau, Y. L., and P. Dayan. 2011. Opponency Revisited: Competition and Cooperation between Dopamine and Serotonin. *Neuropsychopharmacology* **36**:74–97. [6]

Brandon, N. J., and A. Sawa. 2011. Linking Neurodevelopmental and Synaptic Theories of Mental Illness through DISC1. *Nat. Rev. Neurosci.* **12**:707–722. [1]

Braver, T. S. 1997. Mechanisms of Cognitive Control: A Neurocomputational Model. Ph.D. Thesis, Psychology Department, Carnegie Mellon University, Pittsburgh, PA. [4]

Braver, T. S., and D. M. Barch. 2002. A Theory of Cognitive Control, Aging Cognition, and Neuromodulation. *Neurosci. Biobehav. Rev.* **26**:809–817. [4]

Braver, T. S., D. M. Barch, and J. D. Cohen. 1999. Cognition and Control in Schizophrenia: A Computational Model of Dopamine and Prefrontal Function. *Biol. Psychiatry* **46**:312–328. [4]

Braver, T. S., J. R. Gray, and G. C. Burgess. 2007. Explaining the Many Varieties of Working Memory Variation: Dual Mechanisms of Cognitive Control, Part 1. In: Variation in Working Memory, ed. A. R. Conway et al. Oxford: Oxford Univ. Press. [4]

Braver, T. S., J. L. Paxton, H. S. Locke, and D. M. Barch. 2009. Flexible Neural Mechanisms of Cognitive Control within Human Prefrontal Cortex. *PNAS* **106**:7351–7356. [4]

Breakspear, M., G. Roberts, M. J. Green, et al. 2015. Network Dysfunction of Emotional and Cognitive Processes in Those at Genetic Risk of Bipolar Disorder. *Brain* **138**:3427–3439. [12]

Breakspear, M., J. A. Roberts, J. R. Terry, et al. 2006. A Unifying Explanation of Primary Generalized Seizures through Nonlinear Brain Modeling and Bifurcation Analysis. *Cereb. Cortex* **16**:1296–1313. [12]

Breier, A., J. L. Schreiber, J. Dyer, and D. Pickar. 1992. Course of Illness and Predictors of Outcome in Chronic Schizophrenia: Implications for Pathophysiology. *Br. J. Psychiatry Suppl.*38–43. [16]

Breiman, L. 2001. Random Forests. *Mach. Learn.* **45**:5–32. [14]

Brennan, M. D. 2014. Pharmacogenetics of Second-Generation Antipsychotics. *Pharmacogenomics* **15**:869–884. [11]

Bressloff, P. C., J. D. Cowan, P. J. Golubitsky, and M. C. Wiener. 2002. What Geometric Visual Hallucinations Tell Us About the Visual Cortex. *Neural Comput.* **14**:473–491. [2]

Brian, A. J., C. Roncadin, E. Duku, et al. 2014. Emerging Cognitive Profiles in High-Risk Infants with and without Autism Spectrum Disorder. *Res. Autism Spectr. Disord.* **8**:1557–1566. [3]

Brodersen, K. H., L. Deserno, F. Schlagenhauf, et al. 2014. Dissecting Psychiatric Spectrum Disorders by Generative Embedding. *NeuroImage Clin.* **4**:98–111. [12]

Brodersen, K. H., T. M. Schofield, A. P. Leff, et al. 2011. Generative Embedding for Model-Based Classification of fMRI Data. *PLoS Comput. Biol.* **7**:e1002079. [12]

Brosey, E., and N. D. Woodward. 2015. Schizotypy and Clinical Symptoms, Cognitive Function, and Quality of Life in Individuals with a Psychotic Disorder. *Schizophr. Res.* **166**:92–97. [16]

Brown, A. S. 2011. The Environment and Susceptibility to Schizophrenia. *Prog. Neurobiol.* **93**:23–58. [4]

Brown, A. S., and E. J. Derkits. 2010. Prenatal Infection and Schizophrenia: A Review of Epidemiologic and Translational Studies. *Am. J. Psychiatry* **167**:261–280. [3, 4]

Brown, H. R., R. A. Adams, I. Parees, M. Edwards, and K. J. Friston. 2013. Active Inference, Sensory Attenuation and Illusions. *Cogn. Process.* **14**:411–427. [7]

Brown, R., and J. Kulik. 1977. Flashbulb Memories. *Cognition* **5**:73–99. [10]

Browning, M., T. E. Behrens, G. Jocham, J. X. O'Reilly, and S. J. Bishop. 2015. Anxious Individuals Have Difficulty Learning the Causal Statistics of Aversive Environments. *Nat. Neurosci.* **18**:590–596. [6]

Brunel, N., and X. J. Wang. 2001. Effects of Neuromodulation in a Cortical Network Model of Object Working Memory Dominated by Recurrent Inhibition. *J. Comput. Neurosci.* **11**:63–85. [6]

Buchanan, A. 1999. Risk and Dangerousness. *Psychol. Med.* **29**:465–473. [5]

Buckley, P. F., B. J. Miller, D. S. Lehrer, and D. J. Castle. 2009. Psychiatric Comorbidities and Schizophrenia. *Schizophr. Bull.* **35**:383–402. [9]

Bullmore, E., and O. Sporns. 2009. Complex Brain Networks: Graph Theoretical Analysis of Structural and Functional Systems. *Nat. Rev. Neurosci.* **10**:186–198. [3]

Burcusa, S. L., and W. G. Iacono. 2007. Risk for Recurrence in Depression. *Clin. Psychol. Rev.* **27**:959–985. [15]

Bush, R. R., and F. Mosteller. 1951a. A Mathematical Model for Simple Learning. *Psychol. Rev.* **58**:313–323. [13]

————. 1951b. A Model for Stimulus Generalization and Discrimination. *Psychol. Rev.* **58**:413–423. [13]

————. 1953. A Stochastic Model with Applications to Learning. *Ann. Math. Stat.* **24**:559–585. [13]

————. 1955. Stochastic Models for Learning. New York: Wiley. [13]

Buzsáki, G., and K. Mizuseki. 2014. The Log-Dynamic Brain: How Skewed Distributions Affect Network Operations. *Nat. Rev. Neurosci.* **15**:264–278. [3]

Caballero, A., E. Flores-Barrera, D. K. Cass, and K. Y. Tseng. 2014. Differential Regulation of Parvalbumin and Calretinin Interneurons in the Prefrontal Cortex During Adolescence. *Brain Struct. Funct.* **219**:395–406. [16]

Cai, X., G. D. Evrony, H. S. Lehmann, et al. 2014. Single-Cell, Genome-Wide Sequencing Identifies Clonal Somatic Copy-Number Variation in the Human Brain. *Cell Rep.* **8**:1280–1289. [3]

Calkins, M. E., and W. G. Iacono. 2000. Eye Movement Dysfunction in Schizophrenia: A Heritable Characteristic for Enhancing Phenotype Definition. *Am. J. Med. Genet.* **97**:72–76. [1]

Camchong, J., A. W. I. MacDonald, B. A. Mueller, et al. 2014. Changes in Resting Functional Connectivity During Abstinence in Stimulant Use Disorder: A Preliminary Comparison of Relapsers and Abstainers. *Drug Alcohol Depend.* **139**:145–151. [2]

Camperi, M., and X. J. Wang. 1998. A Model of Visuospatial Working Memory in Prefrontal Cortex: Recurrent Network and Cellular Bistability. *J. Comput. Neurosci.* **5**:383–405. [4]

Cannon, M., P. B. Jones, and R. M. Murray. 2002. Obstetric Complications and Schizophrenia: Historical and Meta-Analytic Review. *Am. J. Psychiatry* **159**:1080–1092. [4]

Cannon, T. D., and M. C. Keller. 2006. Endophenotypes in the Genetic Analyses of Mental Disorders. *Annu. Rev. Clin. Psych.* **2**:267–290. [5]

Cantor, C. 2005. Evolution and Posttraumatic Stress: Disorders of Vigilance and Defence. London: Routledge. [10]

Cardin, J. A., M. Carlén, K. Meletis, et al. 2009. Driving Fast-Spiking Cells Induces Gamma Rhythm and Controls Sensory Responses. *Nature* **459**:663–667. [3]

Cardno, A. G., and M. J. Owen. 2014. Genetic Relationships between Schizophrenia, Bipolar Disorder, and Schizoaffective Disorder. *Schizophr. Bull.* **40**:504–515. [16]

Carr, J. 1981. Applications of Centre Manifold Theory. Berlin: Springer-Verlag. [11]

Carroll, L. S., and M. J. Owen. 2009. Genetic Overlap between Autism, Schizophrenia and Bipolar Disorder. *Genome Med.* **1**: [3]

Carter, C. S., T. S. Braver, D. M. Barch, et al. 1998. Anterior Cingulate Cortex, Error Detection, and the Online Monitoring of Performance. *Science* **280**:747–749. [13]

Cartmell, J., J. A. Monn, and D. D. Schoepp. 1999. The Metabotropic Glutamate 2/3 Receptor Agonists LY354740 and LY379268 Selectively Attenuate Phencyclidine versus D-Amphetamine Motor Behaviors in Rats. *J. Pharmacol. Exp. Ther.* **291**:161–170. [16]

Caspi, A., A. R. Hariri, A. Holmes, R. Uher, and T. E. Moffitt. 2010. Genetic Sensitivity to the Environment: The Case of the Serotonin Transporter Gene and Its Implications for Studying Complex Diseases and Traits. *Am. J. Psychiatry* **167**:509–527. [3]

Caspi, A., K. Sugden, T. E. Moffitt, et al. 2003. Influence of Life Stress on Depression: Moderation by a Polymorphism in the 5-HTT Gene. *Science* **301**:386–389. [3]

Castner, S. A., J. L. Arriza, J. C. Roberts, et al. 2010. Reversal of Ketamine-Induced Working Memory Impairments by the GABAAalpha2/3 Agonist TPA023. *Biol. Psychiatry* **67**:998–1001. [16]

Castner, S. A., N. V. Murthy, K. Ridler, et al. 2014. Relationship between Glycine Transporter 1 Inhibition as Measured with Positron Emission Tomography and Changes in Cognitive Performances in Nonhuman Primates. *Neuropsychopharmacology* **39**:2742–2749. [16]

Castner, S. A., G. N. Smagin, T. M. Piser, et al. 2011. Immediate and Sustained Improvements in Working Memory after Selective Stimulation of Alpha7 Nicotinic Acetylcholine Receptors. *Biol. Psychiatry* **69**:12–18. [16]

Cavanagh, J. F., T. V. Wiecki, M. X. Cohen, et al. 2011. Subthalamic Nucleus Stimulation Reverses Mediofrontal Influence over Decision Threshold. *Nat. Neurosci.* **14**:1462–1467. [5, 6]

CDC. 2014. Fatal Injury Data. Web-Based Injury Statistics Query and Reporting System. http://www.cdc.gov/injury/wisqars/fatal_injury_reports.html (accessed March 2, 2016). [3]

Ceaser, A. E., T. E. Goldberg, M. F. Egan, et al. 2008. Set-Shifting Ability and Schizophrenia: A Marker of Clinical Illness or an Intermediate Phenotype? *Biol. Psychiatry* **64**:782–788. [4]

Cetin, M. S., F. Christensen, C. C. Abbott, et al. 2014. Thalamus and Posterior Temporal Lobe Show Greater Inter-Network Connectivity at Rest and across Sensory Paradigms in Schizophrenia. *NeuroImage* **97**:117–126. [16]

Cha, C. B., S. Najmi, J. M. Park, C. T. Finn, and M. K. Nock. 2010. Attentional Bias toward Suicide-Related Stimuli Predicts Suicidal Behavior. *J. Abnorm. Psychol.* **119**:616–622. [3]

Chan, M. K., M.-O. Krebs, D. Cox, et al. 2015. Development of a Blood-Based Molecular Biomarker Test for Identification of Schizophrenia before Disease Onset. *Transl. Psychiatry* **5**:e601. [3]

Charney, D. S., D. H. Barlow, K. Botteron, et al. 2002. Neuroscience Research Agenda to Guide Development of a Pathophysiologically Based Classification System. In: A Research Agenda for DSM-V, ed. D. J. Kupfer et al., pp. 31–84. Washington, D.C.: American Psychiatric Association. [8]

Chekroud, A. M., and J. H. Krystal. 2015. Personalised Pharmacotherapy: An Interim Solution for Antidepressant Treatment? *BMJ* **350**:h2502. [11]

Chen, C., T. Takahashi, S. Nakagawa, T. Inoue, and I. Kusumi. 2015. Reinforcement Learning in Depression: A Review of Computational Research. *Neurosci. Biobehav. Rev.* **55**:247–267. [3]

Cheney, D. L., and R. M. Seyfarth. 1990. How Monkeys See the World. Chicago: Univ. of Chicago Press. [2]

Cheng, G. L., J. C. Tang, F. W. Li, E. Y. Lau, and T. M. Lee. 2012. Schizophrenia and Risk-Taking: Impaired Reward but Preserved Punishment Processing. *Schizophr. Res.* **136**:122–127. [4]

Chiu, C. Q., G. Lur, T. M. Morse, et al. 2013. Compartmentalization of Gabaergic Inhibition by Dendritic Spines. *Science* **340**:759–762. [16]

Chiu, P. H., T. M. Lohrenz, and P. R. Montague. 2008. Smokers' Brains Compute, but Ignore, a Fictive Error Signal in a Sequential Investment Task. *Nat. Neurosci.* **11**:514–520. [2]

Cho, K. K., R. Hoch, A. T. Lee, et al. 2015. Gamma Rhythms Link Prefrontal Interneuron Dysfunction with Cognitive Inflexibility in Dlx5/6(+/-) Mice. *Neuron* **85**:1332–1343. [1]

Cho, R. Y., R. O. Konecky, and C. S. Carter. 2006. Impairments in Frontal Cortical Gamma Synchrony and Cognitive Control in Schizophrenia. *PNAS* **103**:19878–19883. [4]

Chu, C. M., S. D. Thomas, J. R. Ogloff, and M. Daffern. 2013. The Short- to Medium-Term Predictive Accuracy of Static and Dynamic Risk Assessment Measures in a Secure Forensic Hospital. *Assessment* **20**:230–241. [5]

Chubb, J. E., N. J. Bradshaw, D. C. Soares, D. J. Porteous, and J. K. Millar. 2008. The DISC Locus in Psychiatric Illness. *Mol. Psychiatry* **13**:36–64. [3]

Chudasama, Y., and T. W. Robbins. 2004. Psychopharmacological Approaches to Modulating Attention in the Five-Choice Serial Reaction Time Task: Implications for Schizophrenia. *Psychopharmacology (Berl.)* **174**:86–98. [4]

Churchland, P., and T. J. Sejnowski. 1994. The Computational Brain. Cambridge, MA: MIT Press. [2]

Cicchetti, D., and F. Rogosch. 1996. Equifinality and Multifinality in Developmental Pscyhopathology. *Dev. Psychopathol.* **8**:597–600. [10]

Cicero, D. C., E. A. Martin, T. M. Becker, and J. G. Kerns. 2014. Reinforcement Learning Deficits in People with Schizophrenia Persist after Extended Trials. *Psychiatry Res.* **220**:760–764. [4]

Cipriani, A., T. A. Furukawa, G. Salanti, et al. 2009. Comparative Efficacy and Acceptability of 12 New-Generation Antidepressants: A Multiple-Treatments Meta-Analysis. *Lancet* **373**:746–758. [15]

Clark, L. 2010. Decision-Making During Gambling: An Integration of Cognitive and Psychobiological Approaches. *Phil. Trans. R. Soc. B* **365**:319–330. [12]

Clark, L. A. 2005. Temperament as a Unifying Concept in the Study of Personality and Psychopathology. *J. Abnorm. Psychol.* **114**:505–521. [8]

Clark, L. A., D. Watson, and S. Reynolds. 1995. Diagnosis and Classification of Psychopathology: Challenges to the Current System and Future Directions. *Annu. Rev. Psychol.* **46**:121–153. [8]

Clarke, D. E., W. E. Narrow, D. A. Regier, et al. 2013. DSM-5 Field Trials in the United States and Canada, Part I: Study Design, Sampling Strategy, Implementation, and Analytic Approaches. *Am. J. Psychiatry* **170**:43–58. [7]

Clementz, B. A., J. A. Sweeney, J. P. Hamm, et al. 2016. Identification of Distinct Psychosis Biotypes Using Brain-Based Biomarkers. *Am. J. Psychiatry* **173**:373–384. [5, 17]

Cloninger, C. R. 1998. A New Conceptual Paradigm from Genetics and Psychobiology for the Science of Mental Health. *Aust. NZ J. Psychiatry* **33**:174–186. [8]

Coan, J. A., and J. J. Allen. 2007. Handbook of Emotion Elicitation and Assessment: Oxford Univ. Press. [15]

Cohen, J. 1983. The Cost of Dichotomization. *Appl. Psychol. Measurement* **7**:249–253. [8]

Cohen, J. D., D. M. Barch, C. Carter, and D. Servan-Schreiber. 1999. Context-Processing Deficits in Schizophrenia: Converging Evidence from Three Theoretically Motivated Cognitive Tasks. *J. Abnorm. Psychol.* **108**:120–133. [4]

Cohen, J. D., T. S. Braver, and J. W. Brown. 2002. Computational Perspectives on Dopamine Function in Prefrontal Cortex. *Curr. Opin. Neurobiol.* **12**:223–229. [5, 6]

Cohen, J. D., S. M. McClure, and A. J. Yu. 2007. Should I Stay or Should I Go? How the Human Brain Manages the Trade-Off between Exploitation and Exploration. *Phil. Trans. R. Soc. B* **362**:933–942. [6]

Cohen, J. D., and D. Servan-Schreiber. 1992. Context, Cortex, and Dopamine: A Connectionist Approach to Behavior and Biology in Schizophrenia. *Psychol. Rev.* **99**:45–77. [5]

Cohen, J. Y., M. W. Amoroso, and N. Uchida. 2015. Serotonergic Neurons Signal Reward and Punishment on Multiple Timescales. *eLife* **4**: [15]

Collins, A. G. E., J. K. Brown, J. M. Gold, J. A. Waltz, and M. J. Frank. 2014. Working Memory Contributions to Reinforcement Learning Impairments in Schizophrenia. *J. Neurosci.* **34**:13747–13756. [4]

Collins, A. G. E., and M. J. Frank. 2012. How Much of Reinforcement Learning Is Working Memory, Not Reinforcement Learning? A Behavioral, Computational, and Neurogenetic Analysis. *Eur. J. Neurosci.* **35**:1024–1035. [6]

———. 2013. Cognitive Control over Learning: Creating, Clustering and Generalizing Task-Set Structure. *Psychol. Rev.* **120**:190–229. [6, 12]

———. 2014. Opponent Actor Learning (OpAL): Modeling Interactive Effects of Striatal Dopamine on Reinforcement Learning and Choice Incentive. *Psychol. Rev.* **121**:337–366. [3, 5, 6]

Colzato, L. S., W. P. M. Van Den Wildenberg, and B. Hommel. 2007. Impaired Inhibitory Control in Recreational Cocaine Users. *PLoS One* **2**:e1143. [14]

Conant, R. C., and R. W. Ashby. 1970. Every Good Regulator of a System Must Be a Model of That System. *Int. J. Syst. Sci.* **1**:89–97. [7]

Conklin, C. A., and S. T. Tiffany. 2002. Applying Extinction Research and Theory to Cue-Exposure Addiction Treatments. *Addiction* **97**:155–167. [5]

CONVERGE Consortium. 2015. Sparse Whole-Genome Sequencing Identifies Two Loci for Major Depressive Disorder. *Nature* **523**:588–591. [3]

Cook, N. R. 2007. Use and Misuse of the Receiver Operating Characteristic Curve in Risk Prediction. *Circulation* **115**:928–935. [14]

Cools, R. 2011. Dopaminergic Control of the Striatum for High-Level Cognition. *Curr. Opin. Neurobiol.* **21**:402–407. [4]

Cools, R., and M. D'Esposito. 2011. Inverted-U-Shaped Dopamine Actions on Human Working Memory and Cognitive Control. *Biol. Psychiatry* **69**:e113–125. [4, 6]

Cools, R., K. Nakamura, and N. D. Daw. 2011. Serotonin and Dopamine: Unifying Affective, Activational, and Decision Functions. *Neuropsychopharmacology* **36**:98–113. [15]

Cooray, G. K., B. Sengupta, P. Douglas, et al. 2015. Characterising Seizures in Anti-NMDA-Receptor Encephalitis with Dynamic Causal Modelling. *NeuroImage* **118**:508–519. [12]

Corlett, P. R., G. D. Honey, J. H. Krystal, and P. C. Fletcher. 2011. Glutamatergic Model Psychoses: Prediction Error, Learning, and Inference. *Neuropsychopharmacology* **36**:294–315. [12]

Cortes, C., and V. Vapnik. 1995. Support-Vector Networks. *Mach. Learn.* **20**:273–297. [5]

Cottam, J. C., S. L. Smith, and M. Hausser. 2013. Target-Specific Effects of Somatostatin-Expressing Interneurons on Neocortical Visual Processing. *J. Neurosci.* **33**:19567–19578. [16]

Courtet, P., I. I. Gottesman, F. Jollant, and T. D. Gould. 2011. The Neuroscience of Suicidal Behaviors: What Can We Expect from Endophenotype Strategies? *Transl. Psychiatry* **1**:e7. [3]

Cox, R. T. 1946. Probability, Frequency and Reasonable Expectation. *Am. J. Phys.* **14**:1–13. [7, 10]

Cox, S. M. L., M. J. Frank, K. Larcher, et al. 2015. Striatal D1 and D2 Signaling Differentially Predict Learning from Positive and Negative Outcomes. *NeuroImage* **109**:95–101. [12]

Craddock, N., and M. J. Owen. 2010. The Kraepelinian Dichotomy: Going, Going...But Still Not Gone. *Br. J. Psychiatry* **196**:92–95. [7]

Craske, M. G., K. Kircanski, M. Zelikowsky, et al. 2008. Optimizing Inhibitory Learning During Exposure Therapy. *Behav. Res. Ther.* **46**:5–27. [5]

Critchley, H., and A. Seth. 2012. Will Studies of Macaque Insula Reveal the Neural Mechanisms of Self-Awareness? *Neuron* **74**:423–426. [12]

Crockett, M. J., L. Clark, A. M. Apergis-Schoute, S. Morein-Zamir, and T. W. Robbins. 2012. Serotonin Modulates the Effects of Pavlovian Aversive Predictions on Response Vigor. *Neuropsychopharmacology* **37**:2244–2252. [15]

Cross-Disorder Group of the Psychiatric Genomics Consortium, S. H. Lee, S. Ripke, et al. 2013. Genetic Relationship between Five Psychiatric Disorders Estimated from Genome-Wide Snps. *Nat. Genet.* **45**:984–994. [3]

Curtis, D., A. E. Vine, A. McQuillin, et al. 2011. Case-Case Genome-Wide Association Analysis Shows Markers Differentially Associated with Schizophrenia and Bipolar Disorder and Implicates Calcium Channel Genes. *Psychiatr. Genet.* **21**:1–4. [1]

Cuthbert, B. N. 2014a. Research Domain Criteria: Toward Future Psychiatric Nosology. *Asian J. Psychiatry* **7**:4–5. [4]

———. 2014b. Translating Intermediate Phenotypes to Psychopathology: The NIMH Research Domain Criteria. *Psychophysiology* **51**:1205–1206. [16]

Cuthbert, B. N., and T. R. Insel. 2010. The Data of Diagnosis: New Approaches to Psychiatric Classification. *Psychiatry* **73**:311–314. [8, 17]

Cuthbert, B. N., and M. J. Kozak. 2013. Constructing Constructs for Psychopathology: The NIMH Research Domain Criteria. *J. Abnorm. Psychol.* **122**:928–937. [4, 9]

Dahle, K. P. 2006. Strengths and Limitations of Actuarial Prediction of Criminal Reoffence in a German Prison Sample: A Comparative Study of LSI-R, HCR-20 and PCL-R. *Int. J. Law Psychiatry* **29**:431–442. [5]

Dahlem, M. A., and E. P. Chronicle. 2004. A Computational Perspective on Migraine Aura. *Prog. Neurobiol.* **74**:351–361. [2]

Dalley, J. W., R. N. Cardinal, and T. W. Robbins. 2004. Prefrontal Executive and Cognitive Functions in Rodents: Neural and Neurochemical Substrates. *Neurosci. Biobehav. Rev.* **28**:771–784. [2]

Dapretto, M., M. S. Davies, and J. H. Pfeifer. 2006. Understanding Emotions in Others: Mirror Neuron Dysfunction in Children with Autism Spectrum Disorders. *Nat. Neurosci.* **9**:28–30. [2]

Datta, A., D. Truong, P. Minhas, L. C. Parra, and M. Bikson. 2012. Inter-Individual Variation During Transcranial Direct Current Stimulation and Normalization of Dose Using MRI-Derived Computational Models. *Front. Psychiatry* **3**:91. [5]

Datta, D., D. Arion, J. P. Corradi, and D. A. Lewis. 2015. Altered Expression of CDC42 Signaling Pathway Components in Cortical Layer 3 Pyramidal Cells in Schizophrenia. *Biol. Psychiatry* **78**:775–785. [16]

Daunizeau, J., O. David, and K. E. Stephan. 2011a. Dynamic Causal Modelling: A Critical Review of the Biophysical and Statistical Foundations. *NeuroImage* **58**:312–322. [11]

Daunizeau, J., H. E. M. den Ouden, M. Pessiglione, et al. 2010. Observing the Observer (II): Deciding When to Decide. *PLoS One* **5**:e15555. [7]

Daunizeau, J., K. Preuschoff, K. J. Friston, and K. E. Stephan. 2011b. Optimizing Experimental Design for Comparing Models of Brain Function. *PLoS Comput. Biol.* **7**:e1002280. [12]

Davidson, L., and T. H. McGlashan. 1997. The Varied Outcomes of Schizophrenia. *Can. J. Psychiatry* **42**:34–43. [16]

Davidson, L., M. J. O'Connell, J. Tondora, M. Lawless, and A. C. Evans. 2005. Recovery in Serious Mental Illness: A New Wine or Just a New Bottle? *Prof. Psychol. Res. Pr.* **36**:480. [12]

Davidson, M., P. D. Harvey, P. Powchik, et al. 1995. Severity of Symptoms in Chronically Institutionalized Geriatric Schizophrenic Patients. *Am. J. Psychiatry* **152**: 197–207. [16]

Davis, G. W. 2006a. Homeostatic Control of Neural Activity: From Phenomenology to Molecular Design. *Annu. Rev. Neurosci.* **29**:307–323. [16]

Davis, M. J. 2006b. Low-Dimensional Manifolds in Reaction-Diffusion Equations. 1. Fundamental Aspects. *J. Phys. Chem. A* **110**:5235–5256. [11]

Daw, N. D., and P. Dayan. 2014. The Algorithmic Anatomy of Model-Based Evaluation. *Phil. Trans. R. Soc. B* **369**: [15]

Daw, N. D., S. J. Gershman, B. Seymour, P. Dayan, and R. J. Dolan. 2011. Model-Based Influences on Humans' Choices and Striatal Prediction Errors. *Neuron* **69**:1204–1215. [4, 6, 15]

Daw, N. D., Y. Niv, and P. Dayan. 2005. Uncertainty-Based Competition between Prefrontal and Dorsolateral Striatal Systems for Behavioral Control. *Nat. Neurosci.* **8**:1704–1711. [6, 15]

Daw, N. D., J. P. O'Doherty, P. Dayan, B. Seymour, and R. J. Dolan. 2006. Cortical Substrates for Exploratory Decisions in Humans. *Nature* **441**:876–879. [6, 15]

Daw, N. D., and D. S. Touretzky. 2002. Long-Term Reward Prediction in TD Models of the Dopamine System. *Neural Comput.* **14**:2567–2583. [14]

Day, L. B., M. Weisend, R. J. Sutherland, and T. Schallert. 1999. The Hippocampus Is Not Necessary for a Place Response but May Be Necessary for Pliancy. *Behav. Neurosci.* **113**:914–924. [2]

Dayan, P. 2009. Dopamine, Reinforcement Learning, and Addiction. *Pharmacopsychiatry* **42(Suppl 1)**:S56–65. [4]

———. 2012. Twenty-Five Lessons from Computational Neuromodulation. *Neuron* **76**:240–256. [13, 15]

Dayan, P., and L. F. Abbot. 2001. Theoretical Neuroscience: Computational and Mathematical Modeling of Neural Systems. Cambridge, MA: MIT Press. [13]

Dayan, P., and B. W. Balleine. 2002. Reward, Motivation, and Reinforcement Learning. *Neuron* **36**:285–298. [10, 15]

Dayan, P., and K. C. Berridge. 2014. Model-Based and Model-Free Pavlovian Reward Learning: Revaluation, Revision, and Revelation. *Cogn. Affect. Behav. Neurosci.* **14**:473–492. [10]

Dayan, P., and N. D. Daw. 2008. Decision Theory, Reinforcement Learning, and the Brain. *Cogn. Affect. Behav. Neurosci.* **8**:429–453. [13]

Dayan, P., G. E. Hinton, R. M. Neal, and R. S. Zemel. 1995. The Helmholtz Machine. *Neural Comput.* **7**:889–904. [7]

Dayan, P., and Q. J. M. Huys. 2008. Serotonin, Inhibition, and Negative Mood. *PLoS Comput. Biol.* **4**:e4. [12, 15]

Dayan, P., and Q. J. M. Huys. 2009. Serotonin in Affective Control. *Annu. Rev. Neurosci.* **32**:95–126. [15]

Dayan, P., and Q. J. M. Huys. 2015. Serotonin's Many Meanings Elude Simple Theories. *eLife* **4**: [12, 15]

Dayan, P., S. Kakade, and P. R. Montague. 2000. Learning and Selective Attention. *Nat. Neurosci.* **3**:1218–1223. [13]

Dayan, P., and Y. Niv. 2008. Reinforcement Learning: The Good, the Bad, and the Ugly. *Curr. Opin. Neurobiol.* **18**:185–196. [13]

Dayan, P., and M. E. Walton. 2012. A Step-by-Step Guide to Dopamine. *Biol. Psychiatry* **71**:842–843. [4]

Deakin, J. F. W., and F. G. Graeff. 1991. 5-Ht and Mechanisms of Defence. *J. Psychopharm.* **5**:305–316. [15]

DeBattista, C., G. Kinrys, D. Hoffman, et al. 2011. The Use of Referenced-EEG (rEEG) in Assisting Medication Selection for the Treatment of Depression. *J. Psychiatr. Res.* **45**:64–75. [15]

de Boer, R. J., and A. S. Perelson. 1991. Size and Connectivity as Emergent Properties of a Developing Immune Network. *J. Theor. Biol.* **149**:381–424. [11]

Decker, H. 2007. How Kraepelian Was Kraepelin? Joe Kraepelinian Are the Neo-Kraepelinians? From Emil Kraepelin to DSM-III. *Hist. Psychiatry* **18**:337–360. [8]

Deco, G., V. K. Jirsa, P. A. Robinson, M. Breakspear, and K. J. Friston. 2008. The Dynamic Brain: From Spiking Neurons to Neural Masses and Cortical Fields. *PLoS Comput. Biol* **4**:e1000092. [12]

Deco, G., and M. L. Kringelbach. 2014. Great Expectations: Using Whole-Brain Computational Connectomics for Understanding Neuropsychiatric Disorders. *Neuron* **84**:892–905. [12]

Deco, G., A. Ponce-Alvarez, D. Mantini, et al. 2013. Resting-State Functional Connectivity Emerges from Structurally and Dynamically Shaped Slow Linear Fluctuations. *J. Neurosci.* **33**:11239–11252. [16]

de Koning, P. P., M. Figee, P. van den Munckhof, P. R. Schuurman, and D. Denys. 2011. Current Status of Deep Brain Stimulation for Obsessive-Compulsive Disorder: A Clinical Review of Different Targets. *Curr. Psychiatry Rep.* **13**:274–282. [10]

de la Fuente-Sandoval, C., P. Leon-Ortiz, R. Favila, et al. 2011. Higher Levels of Glutamate in the Associative-Striatum of Subjects with Prodromal Symptoms of Schizophrenia and Patients with First-Episode Psychosis. *Neuropsychopharmacology* **36**: 1781–1791. [16]

Demjaha, A., A. Egerton, R. M. Murray, et al. 2014. Antipsychotic Treatment Resistance in Schizophrenia Associated with Elevated Glutamate Levels but Normal Dopamine Function. *Biol. Psychiatry* **75**:e11–13. [16]

Demjaha, A., R. M. Murray, P. K. McGuire, S. Kapur, and O. D. Howes. 2012. Dopamine Synthesis Capacity in Patients with Treatment-Resistant Schizophrenia. *Am. J. Psychiatry* **169**:1203–1210. [4]

Deng, Z. D., S. H. Lisanby, and A. V. Peterchev. 2013. Controlling Stimulation Strength and Focality in Electroconvulsive Therapy via Current Amplitude and Electrode Size and Spacing: Comparison with Magnetic Seizure Therapy. *J. ECT* **29**:325–335. [5]

Der-Avakian, A., and A. Markou. 2012. The Neurobiology of Anhedonia and Other Reward-Related Deficits. *Trends Neurosci.* **35**:68–77. [14]

DeRubeis, R. J., Z. D. Cohen, N. R. Forand, et al. 2014. The Personalized Advantage Index: Translating Research on Prediction into Individualized Treatment Recommendations. A Demonstration. *PLoS One* **9**:e83875. [15]

DeRubeis, R. J., L. A. Gelfand, T. Z. Tang, and A. D. Simons. 1999. Medications versus Cognitive Behavior Therapy for Severely Depressed Outpatients: Mega-Analysis of Four Randomized Comparisons. *Am. J. Psychiatry* **156**:1007–1013. [15]

DeRubeis, R. J., G. J. Siegle, and S. D. Hollon. 2008. Cognitive Therapy versus Medication for Depression: Treatment Outcomes and Neural Mechanisms. *Nat. Rev. Neurosci.* **9**:788–796. [15]

Deserno, L., Q. J. M. Huys, R. Boehme, et al. 2015. Ventral Striatal Dopamine Reflects Behavioral and Neural Signatures of Model-Based Control During Sequential Decision Making. *PNAS* **112**:1595–1600. [15]

Deumens, R., A. Blokland, and J. Prickaerts. 2002. Modeling Parkinson's Disease in Rats: An Evaluation of 6-OHDA Lesions of the Nigrostriatal Pathway. *Exp. Neurol.* **175**:303–317. [2]

Dezfouli, A., and B. W. Balleine. 2012. Habits, Action Sequences and Reinforcement Learning. *Eur. J. Neurosci.* **35**:1036–1051. [10]

Diaconescu, A. O., C. Mathys, L. A. E. Weber, et al. 2014. Inferring on the Intentions of Others by Hierarchical Bayesian Learning. *PLoS Comput. Biol.* **10**:e1003810. [7]

Dickman, D. K., and G. W. Davis. 2009. The Schizophrenia Susceptibility Gene Dysbindin Controls Synaptic Homeostasis. *Science* **326**:1127–1130. [16]

Dickson, H., K. R. Laurens, A. E. Cullen, and S. Hodgins. 2012. Meta-Analyses of Cognitive and Motor Function in Youth Aged 16 Years and Younger Who Subsequently Develop Schizophrenia. *Psychol. Med.* **42**:743–755. [10]

Dima, D., D. E. Dietrich, W. Dillo, and H. M. Emrich. 2010. Impaired Top-Down Processes in Schizophrenia: A Dcm Study of ERPs. *NeuroImage* **52**:824–832. [12]

Dima, D., J. P. Roiser, D. E. Dietrich, et al. 2009. Understanding Why Patients with Schizophrenia Do Not Perceive the Hollow-Mask Illusion Using Dynamic Causal Modelling. *NeuroImage* **46**:1180–1186. [12]

Docx, L., J. de la Asuncion, B. Sabbe, et al. 2015. Effort Discounting and Its Association with Negative Symptoms in Schizophrenia. *Cogn. Neuropsychiatry* **20**:172–185. [4]

Dolan, R. J., and P. Dayan. 2013. Goals and Habits in the Brain. *Neuron* **80**:312–325. [10]

Doll, B. B., K. D. Duncan, D. A. Simon, D. Shohamy, and N. D. Daw. 2015. Model-Based Choices Involve Prospective Neural Activity. *Nat. Neurosci.* **18**:767–772. [15]

Doll, B. B., D. A. Simon, and N. D. Daw. 2012. The Ubiquity of Model-Based Reinforcement Learning. *Curr. Opin. Neurobiol.* **22**:1075–1081. [4, 6]

Dorph-Petersen, K. A., J. N. Pierri, J. M. Perel, et al. 2005. The Influence of Chronic Exposure to Antipsychotic Medications on Brain Size before and after Tissue Fixation: A Comparison of Haloperidol and Olanzapine in Macaque Monkeys. *Neuropsychopharmacology* **30**:1649–1661. [16]

Douglas, R. J. 1995. News and Views: The Bee's Needs. *Nature* **377**:683–684. [13]

Douglas, R. J., and K. A. Martin. 1991. A Functional Microcircuit for Cat Visual Cortex. *J. Physiology* **440**:735–769. [7]

Downar, J., J. Geraci, T. V. Salomons, et al. 2014. Anhedonia and Reward-Circuit Connectivity Distinguish Nonresponders from Responders to Dorsomedial Prefrontal Repetitive Transcranial Magnetic Stimulation in Major Depression. *Biol. Psychiatry* **76**:176–185. [3]

Downing, A. M., B. J. Kinon, B. A. Millen, et al. 2014. A Double-Blind, Placebo-Controlled Comparator Study of LY2140023 Monohydrate in Patients with Schizophrenia. *BMC Psychiatry* **14**:351. [16]

Doya, K. 2000. Reinforcement Learning in Continuous Time and Space. *Neural Comput.* **12**:219–245. [6]

Drevets, W. C. 2000. Functional Anatomical Abnormalities in Limbic and Prefrontal Cortical Structures in Major Depression. *Prog. Brain Res.* **126**:413–431. [3]

Driesen, N. R., G. McCarthy, Z. Bhagwagar, et al. 2013. Relationship of Resting Brain Hyperconnectivity and Schizophrenia-Like Symptoms Produced by the NMDA Receptor Antagonist Ketamine in Humans. *Mol. Psychiatry* **18**:1199–1204. [16]

D'Souza, D. C., W. M. Abi-Saab, S. Madonick, et al. 2005. Delta-9-Tetrahydrocannabinol Effects in Schizophrenia: Implications for Cognition, Psychosis, and Addiction. *Biol. Psychiatry* **57**:594–608. [16]

D'Souza, D. C., E. Perry, L. MacDougall, et al. 2004. The Psychotomimetic Effects of Intravenous Delta-9-Tetrahydrocannabinol in Healthy Individuals: Implications for Psychosis. *Neuropsychopharmacology* **29**:1558–1572. [16]

Du, Y., and A. A. Grace. 2013. Peripubertal Diazepam Administration Prevents the Emergence of Dopamine System Hyperresponsivity in the MAM Developmental Disruption Model of Schizophrenia. *Neuropsychopharmacology* **38**:1881–1888. [3]

Duan, A. R., C. Varela, Y. Zhang, et al. 2015. Delta Frequency Optogenetic Stimulation of the Thalamic Nucleus Reuniens Is Sufficient to Produce Working Memory Deficits: Relevance to Schizophrenia. *Biol. Psychiatry* **77**:1098–1107. [16]

Dunlop, B. W., P. Holland, W. Bao, P. T. Ninan, and M. B. Keller. 2012. Recovery and Subsequent Recurrence in Patients with Recurrent Major Depressive Disorder. *J. Psychiatr. Res.* **46**:708–715. [15]

Durstewitz, D., and T. Gabriel. 2007. Dynamical Basis of Irregular Spiking in NMDA-Driven Prefrontal Cortex Neurons. *Cereb. Cortex* **17**:894–908. [5]

Durstewitz, D., and J. K. Seamans. 2002. The Computational Role of Dopamine D1 Receptors in Working Memory. *Neural Netw.* **15**:561–572. [5]

———. 2008. The Dual-State Theory of Prefrontal Cortex Dopamine Function with Relevance to Catechol-O-Methyltransferase Genotypes and Schizophrenia. *Biol. Psychiatry* **64**:739–749. [2, 5, 6]

Durstewitz, D., J. K. Seamans, and T. J. Sejnowski. 2000. Dopamine-Mediated Stabilization of Delay-Period Activity in a Network Model of Prefrontal Cortex. *J. Neurophysiol.* **83**:1733–1750. [5]

Durstewitz, D., N. M. Vittoz, S. B. Floresco, and J. K. Seamans. 2010. Abrupt Transitions between Prefrontal Neural Ensemble States Accompany Behavioral Transitions During Rule Learning. *Neuron* **66**:438–448. [5]

Dykshoorn, K. L. 2014. Trauma-Related Obsessive-Compulsive Disorder: A Review. *Health Psychol. Behav. Med.* **2**:517–528. [10]

Eaton, R. C., ed. 1984. Neural Mechanisms of Startle Behavior. New York: Springer. [2]

Eaton, W. W., R. Thara, B. Federman, B. Melton, and K. Y. Liang. 1995. Structure and Course of Positive and Negative Symptoms in Schizophrenia. *Arch. Gen. Psychiatry* **52**:127–134. [3]

Edwards, B. G., D. M. Barch, and T. S. Braver. 2010. Improving Prefrontal Cortex Function in Schizophrenia through Focused Training of Cognitive Control. *Front. Hum. Neurosci.* **4**:32. [4]

Edwards, M. J., R. A. Adams, H. R. Brown, I. Pareés, and K. J. Friston. 2012. A Bayesian Account of "Hysteria." *Brain* **135**:3495–3512. [7]

Egerton, A., S. Brugger, M. Raffin, et al. 2012. Anterior Cingulate Glutamate Levels Related to Clinical Status Following Treatment in First-Episode Schizophrenia. *Neuropsychopharmacology* **37**:2515–2521. [3]

Egerton, A., C. A. Chaddock, T. T. Winton-Brown, et al. 2013. Presynaptic Striatal Dopamine Dysfunction in People at Ultra-High Risk for Psychosis: Findings in a Second Cohort. *Biol. Psychiatry* **74**:106–112. [4]

Eggan, S. M., D. S. Melchitzky, S. R. Sesack, K. N. Fish, and D. A. Lewis. 2010. Relationship of Cannabinoid CB1 Receptor and Cholecystokinin Immunoreactivity in Monkey Dorsolateral Prefrontal Cortex. *Neuroscience* **169**:1651–1661. [16]

Eichenbaum, H., C. Stewart, and R. G. M. Morris. 1990. Hippocampal Representation in Place Learning. *J. Neurosci.* **10**:3531–3542. [2]

El-Mallakh, R. S., and B. Briscoe. 2012. Studies of Long-Term Use of Antidepressants: How Should the Data from Them Be Interpreted? *CNS Drugs* **26**:97–109. [15]

Elia, J., X. Gai, H. M. Xie, et al. 2010. Rare Structural Variants Found in Attention-Deficit Hyperactivity Disorder Are Preferentially Associated with Neurodevelopmental Genes. *Mol. Psychiatry* **15**:637–646. [3]

Elliott, R., P. J. McKenna, T. W. Robbins, and B. J. Sahakian. 1995. Neuropsychological Evidence for Frontostriatal Dysfunction in Schizophrenia. *Psychol. Med.* **25**:619–630. [4]

Ellison-Wright, I., D. C. Glahn, A. R. Laird, S. M. Thelen, and E. Bullmore. 2008. The Anatomy of First-Episode and Chronic Schizophrenia: An Anatomical Likelihood Estimation Meta-Analysis. *Am. J. Psychiatry* **165**:1015–1023. [4]

Elvevåg, B., J. Duncan, and P. J. McKenna. 2000. The Use of Cognitive Context in Schizophrenia: An Investigation. *Psychol. Med.* **30**:885–897. [4]

Eppinger, B., N. W. Schuck, L. E. Nystrom, and J. D. Cohen. 2013. Reduced Striatal Responses to Reward Prediction Errors in Older Compared with Younger Adults. *J. Neurosci.* **33**:9905–9912. [12]

Eshel, N., and J. P. Roiser. 2010. Reward and Punishment Processing in Depression. *Biol. Psychiatry* **68**:118–124. [14]

Esslinger, C., S. Englisch, D. Inta, et al. 2012. Ventral Striatal Activation During Attribution of Stimulus Saliency and Reward Anticipation Is Correlated in Unmedicated First Episode Schizophrenia Patients. *Schizophr. Res.* **140**:114–121. [4]

Evans, M. D., S. D. Hollon, R. J. DeRubeis, et al. 1992. Differential Relapse Following Cognitive Therapy and Pharmacotherapy for Depression. *Arch. Gen. Psychiatry* **49**:802–808. [15]

Everitt, B. J., and T. W. Robbins. 2005. Neural Systems of Reinforcement for Drug Addiction: From Actions to Habits to Compulsion. *Nat. Neurosci.* **8**:1481–1489. [12]

———. 2013. From the Ventral to the Dorsal Striatum: Devolving Views of Their Roles in Drug Addiction. *Neurosci. Biobehav. Rev.* **37**:1946–1954. [12]

Fani, N., T. Z. King, T. Jovanovic, et al. 2012. White Matter Integrity in Highly Traumatized Adults with and without Post-Traumatic Stress Disorder. *Neuropsychopharmacology* **37**:2740–2746. [3]

Farb, N. A. S., A. K. Anderson, R. T. Bloch, and Z. V. Segal. 2011. Mood-Linked Responses in Medial Prefrontal Cortex Predict Relapse in Patients with Recurrent Unipolar Depression. *Biol. Psychiatry* **70**:366–372. [15]

Farb, N. A. S., J. A. Irving, A. K. Anderson, and Z. V. Segal. 2015. A Two-Factor Model of Relapse/Recurrence Vulnerability in Unipolar Depression. *J. Abnorm. Psychol.* **124**:38–53. [15]

Farrar, A. M., K. N. Segovia, P. A. Randall, et al. 2010. Nucleus Accumbens and Effort-Related Functions: Behavioral and Neural Markers of the Interactions between Adenosine A2A and Dopamine D2 Receptors. *Neuroscience* **166**:1056–1067. [4]

Fatouros-Bergman, H., S. Cervenka, L. Flyckt, G. Edman, and L. Farde. 2014. Meta-Analysis of Cognitive Performance in Drug-Naive Patients with Schizophrenia. *Schizophr. Res.* **158**:156–162. [4]

Fava, G. A., A. Gatti, C. Belaise, J. Guidi, and E. Offidani. 2015. Withdrawal Symptoms after Selective Serotonin Reuptake Inhibitor Discontinuation: A Systematic Review. *Psychother. Psychosom.* **84**:72–81. [15]

Fava, G. A., C. Ruini, C. Rafanelli, et al. 2004. Six-Year Outcome of Cognitive Behavior Therapy for Prevention of Recurrent Depression. *Am. J. Psychiatry* **161**:1872–1876. [15]

Fazzari, P., A. Snellinx, V. Sabanov, et al. 2014. Cell Autonomous Regulation of Hippocampal Circuitry via Aph1b-Γ-Secretase/Neuregulin 1 Signalling. *eLife* **3**:e02196. [3]

Feighner, J. P., E. Robins, S. B. Guze, et al. 1972. Diagnostic Criteria for Use in Psychiatric Research. *Arch. Gen. Psychiatry* **26**:57–63. [9]

Feinberg, I. 1982. Schizophrenia: Caused by a Fault in Programmed Synaptic Elimination During Adolescence? *J. Psychiatr. Res.* **17**:319–334. [3]

Felmingham, K. L., C. Dobson-Stone, P. R. Schofield, G. J. Quirk, and R. A. Bryant. 2013. The Brain-Derived Neurotrophic Factor Val66Met Polymorphism Predicts Response to Exposure Therapy in Posttraumatic Stress Disorder. *Biol. Psychiatry* **73**:1059–1063. [3]

Fergusson, D. M., and L. J. Horwood. 1995. Predictive Validity of Categorically and Dimensionally Scored Measures of Disruptive Childhood Behaviors. *J. Am. Acad. Child Adolesc. Psychiatry* **34**:477–485; discussion 485–477. [10]

Fergusson, D. M., L. J. Horwood, E. M. Ridder, and A. L. Beautrais. 2005. Subthreshold Depression in Adolescence and Mental Health Outcomes in Adulthood. *Arch. Gen. Psychiatry* **62**:66–72. [10]

Ferrarelli, F., and G. Tononi. 2011. The Thalamic Reticular Nucleus and Schizophrenia. *Schizophr. Bull.* **37**:306–315. [16]

Ferreira, M. A. , M. C. O'Donovan, Y. A. Meng, et al. 2008. Collaborative Genome-Wide Association Analysis Supports a Role for ANK3 and CACNA1C in Bipolar Disorder. *Nat. Genet.* **40**:1056–1058. [3]

Fervaha, G., O. Agid, G. Foussias, and G. Remington. 2013a. Impairments in Both Reward and Punishment Guided Reinforcement Learning in Schizophrenia. *Schizophr. Res.* **150**:592–593. [4]

Fervaha, G., A. Graff-Guerrero, K. K. Zakzanis, et al. 2013b. Incentive Motivation Deficits in Schizophrenia Reflect Effort Computation Impairments During Cost-Benefit Decision-Making. *J. Psychiatr. Res.* **47**:1590–1596. [4]

Figee, M., P. de Koning, S. Klaassen, et al. 2014. Deep Brain Stimulation Induces Striatal Dopamine Release in Obsessive-Compulsive Disorder. *Biol. Psychiatry* **75**:647–652. [10]

First, M. B. 2005a. Clinical Utility: A Prerequisite for the Adoption of a Dimensional Approach in DSM. *J. Abnorm. Psychol.* **114**:560–564. [8]

———. 2005b. Mutually Exclusive versus Co-Occurring Diagnostic Categories: The Challenge of Diagnostic Comorbidity. *Psychopathology* **38**:206–210. [8]

———. 2013. Diagnostic and Statistical Manual of Mental Disorders, 5th Edition, and Clinical Utility. *J. Nerv. Ment. Disorder* **201**:727–729. [8]

First, M. B., and D. Westen. 2007. Classification for Clinical Practice: How to Make ICD and DSM Better Able to Serve Clinicians. *Int. Rev. Psychiatry* **19**:473–481. [8]

Foa, E. B., M. R. Liebowitz, M. J. Kozak, et al. 2005. Randomized, Placebo-Controlled Trial of Exposure and Ritual Prevention, Clomipramine, and Their Combination in the Treatment of Obsessive-Compulsive Disorder. *Am. J. Psychiatry* **162**:151–161. [1]

Forbes, N. F., L. A. Carrick, A. M. McIntosh, and S. M. Lawrie. 2009. Working Memory in Schizophrenia: A Meta-Analysis. *Psychol. Med.* **39**:889–905. [4]

Ford, J. M., D. H. Mathalon, S. Whitfield, W. O. Faustman, and W. T. Roth. 2002. Reduced Communication between Frontal and Temporal Lobes During Talking in Schizophrenia. *Biol. Psychiatry* **51**:485–492. [3]

Ford, J. M., S. E. Morris, R. E. Hoffman, et al. 2014. Studying Hallucinations within the NIMH RDoC Framework. *Schizophr. Bull.* **40 (Suppl. 4)**:S295–S304. [9]

Fox, H. C., M. Garcia, K. Kemp, et al. 2006. Gender Differences in Cardiovascular and Corticoadrenal Response to Stress and Drug Cues in Cocaine Dependent Individuals. *Psychopharmacology* **185**:348–357. [3]

Frank, E., R. F. Prien, R. B. Jarrett, et al. 1991. Conceptualization and Rationale for Consensus Definitions of Terms in Major Depressive Disorder. Remission, Recovery, Relapse, and Recurrence. *Arch. Gen. Psychiatry* **48**:851–855. [15]

Frank, G. K. W., J. R. Reynolds, M. E. Shott, and R. C. O'Reilly. 2011. Altered Temporal Difference Learning in Bulimia Nervosa. *Biol. Psychiatry* **70**:728–735. [12]

Frank, M. J. 2005. Dynamic Dopamine Modulation in the Basal Ganglia: A Neurocomputational Account of Cognitive Deficits in Medicated and Nonmedicated Parkinsonism. *J. Cogn. Neurosci.* **17**:51–72. [6, 15]

———. 2006. Hold Your Horses: A Dynamic Computational Role for the Subthalamic Nucleus in Decision Making. *Neural Netw.* **19**:1120–1136. [12]

———. 2011. Computational Models of Motivated Action Selection in Corticostriatal Circuits. *Curr. Opin. Neurol.* **21**:381–386. [2]

———. 2015. Linking across Levels of Computation in Model-Based Cognitive Neuroscience. In: An Introduction to Model-Based Cognitive Neuroscience, ed. B. U. Forstmann and E.-J. Wagenmakers, pp. 159–177. New York: Springer. [5, 6]

Frank, M. J., B. B. Doll, J. Oas-Terpstra, and F. Moreno. 2009. Prefrontal and Striatal Dopaminergic Genes Predict Individual Differences in Exploration and Exploitation. *Nat. Neurosci.* **12**:1062–1068. [6]

Frank, M. J., and J. A. Fossella. 2011. Neurogenetics and Pharmacology of Learning, Motivation, and Cognition. *Neuropsychopharmacology* **36**:133–152. [5]

Frank, M. J., C. Gagne, E. Nyhus, et al. 2015. fMRI and EEG Predictors of Dynamic Decision Parameters During Human Reinforcement Learning. *J. Neurosci.* **35**:484–494. [5]

Frank, M. J., B. Loughry, and R. C. O'Reilly. 2001. Interactions between Frontal Cortex and Basal Ganglia in Working Memory: A Computational Model. *Cogn. Affect. Behav. Neurosci.* **1**:137–160. [6]

Frank, M. J., A. A. Moustafa, H. M. Haughey, T. Curran, and K. E. Hutchison. 2007a. Genetic Triple Dissociation Reveals Multiple Roles for Dopamine in Reinforcement Learning. *PNAS* **104**:16311–16316. [2]

Frank, M. J., J. Samanta, A. A. Moustafa, and S. J. Sherman. 2007b. Hold Your Horses: Impulsivity, Deep Brain Stimulation and Medication in Parkinsonism. *Science* **318**:1309–1312. [6]

Frank, M. J., L. C. Seeberger, and R. C. O'Reilly. 2004. By Carrot or by Stick: Cognitive Reinforcement Learning in Parkinsonism. *Science* **306**:1940–1943. [2, 12, 15]

Frank, T. D. 2004. Nonlinear Fokker-Planck Equations: Fundamentals and Applications. Springer Series in Synergetics. Berlin: Springer. [11]

Franke, C., B. Reuter, L. Schulz, and N. Kathmann. 2007. Schizophrenia Patients Show Impaired Response Switching in Saccade Tasks. *Biological psychology* **76**:91–99. [4]

Frankle, W. G., R. Gil, E. Hackett, et al. 2004. Occupancy of Dopamine D2 Receptors by the Atypical Antipsychotic Drugs Risperidone and Olanzapine: Theoretical Implications. *Psychopharmacology (Berl.)* **175**:473–480. [16]

Franklin, N. T., and M. J. Frank. 2015. A Cholinergic Feedback Circuit to Regulate Striatal Population Uncertainty and Optimize Reinforcement Learning. *eLife* **4**:e12029. [6]

Freedman, D. 2001. False Prediction of Future Dangerousness: Error Rates and Psychopathy Checklist—revised. *J. Am. Acad. Psychiatry Law* **29**:89–95. [5]

Freedman, R., D. A. Lewis, R. Michels, et al. 2013. The Initial Field Trials of DSM-5: New Blooms and Old Thorns. *Am. J. Psychiatry* **170**:1–5. [1, 7, 16]

Freeman, W. J. 1975. Mass Action in the Nervous System. New York: Academic Press. [12]

Friedman, A. K., J. J. Walsh, and B. Juarez. 2014. Enhancing Depression Mechanisms in Midbrain Dopamine Neurons Achieves Homeostatic Resilience. *Science* **344**:313–319. [5]

Friston, K. J. 1998. The Disconnection Hypothesis. *Schizophr. Res.* **30**:115–125. [12]

———. 2005. A Theory of Cortical Responses. *Phil. Trans. R. Soc. B* **360**:815–836. [7]

———. 2008. Hierarchical Models in the Brain. *PLoS Comput. Biol.* **4**:e1000211. [7]

———. 2009. The Free-Energy Principle: A Rough Guide to the Brain? *Trends Cogn. Sci.* **13**:293–301. [12]

Friston, K. J., L. Harrison, and W. D. Penny. 2003. Dynamic Causal Modelling. *Neuro-Image* **19**:1273–1302. [11, 12]

Friston, K. J., J. Kilner, and L. Harrison. 2006. A Free Energy Principle for the Brain. *J. Physiol. (Paris)* **100**:70–87. [12]

Friston, K. J., J. Mattout, N. Trujillo-Barreto, J. Ashburner, and W. Penny. 2007. Variational Free Energy and the Laplace Approximation. *NeuroImage* **34**:220–234. [7]

Friston, K. J., and W. D. Penny. 2011. Post Hoc Bayesian Model Selection. *NeuroImage* **56**:2089–2099. [11]

Friston, K. J., K. E. Stephan, R. Montague, and R. J. Dolan. 2014. Computational Psychiatry: The Brain as a Phantastic Organ. *Lancet Psychiatry* **1**:148–158. [16]

Friston, K. J., N. Trujillo-Barreto, and J. Daunizeau. 2008. Dem: A Variational Treatment of Dynamic Systems. *NeuroImage* **41**:849–885. [11]

Fu, C. H. Y., H. Steiner, and S. G. Costafreda. 2013. Predictive Neural Biomarkers of Clinical Response in Depression: A Meta-Analysis of Functional and Structural Neuroimaging Studies of Pharmacological and Psychological Therapies. *Neurobiol. Dis.* **52**:75–83. [15]

Fusar-Poli, P., A. Bechdolf, M. J. Taylor, et al. 2013a. At Risk for Schizophrenic or Affective Psychoses? A Meta-Analysis of DSM/ICD Diagnostic Outcomes in Individuals at High Clinical Risk. *Schizophr. Bull.* **39**:923–932. [3, 10]

Fusar-Poli, P., I. Bonoldi, A. R. Yung, et al. 2012a. Predicting Psychosis: Meta-Analysis of Transition Outcomes in Individuals at High Clinical Risk. *Arch. Gen. Psychiatry* **69**:220–229. [10]

Fusar-Poli, P., S. Borgwardt, A. Bechdolf, et al. 2013b. The Psychosis High-Risk State: A Comprehensive State-of-the-Art Review. *JAMA Psychiatry* **70**:107–120. [10]

Fusar-Poli, P., W. T. Carpenter, S. W. Woods, and T. H. McGlashan. 2014. Attenuated Psychosis Syndrome: Ready for DSM-5.1? *Annu. Rev. Clin. Psych.* **10**:155–192. [3]

Fusar-Poli, P., G. Deste, R. Smieskova, et al. 2012b. Cognitive Functioning in Prodromal Psychosis: A Meta-Analysis. *Arch. Gen. Psychiatry* **69**:562–571. [10]

Fusar-Poli, P., O. D. Howes, P. Allen, et al. 2010. Abnormal Frontostriatal Interactions in People with Prodromal Signs of Psychosis: A Multimodal Imaging Study. *Arch. Gen. Psychiatry* **67**:683–691. [4]

Fusar-Poli, P., O. D. Howes, P. Allen, et al. 2011. Abnormal Prefrontal Activation Directly Related to Pre-Synaptic Striatal Dopamine Dysfunction in People at Clinical High Risk for Psychosis. *Mol. Psychiatry* **16**:67–75. [4]

Fusar-Poli, P., and A. Meyer-Lindenberg. 2013. Striatal Presynaptic Dopamine in Schizophrenia, Part II: Meta-Analysis of [(18)F/(11)C]-DOPA PET Studies. *Schizophr. Bull.* **39**:33–42. [4]

Galletly, C. A., A. C. McFarlane, and C. R. Clark. 2007. Impaired Updating of Working Memory in Schizophrenia. *Int. J. Psychophysiol.* **63**:265–274. [4]

Gandal, M. J., J. C. Edgar, K. Klook, and S. J. Siegel. 2012. Gamma Synchrony: Towards a Translational Biomarker for the Treatment-Resistant Symptoms of Schizophrenia. *Neuropharmacology* **62**:1504–1518. [16]

Garrido, M. I., J. M. Kilner, K. E. Stephan, and K. J. Friston. 2009. The Mismatch Negativity: A Review of Underlying Mechanisms. *Clin. Neurophysiol.* **120**:453–463. [12]

Gasser, T. 2009. Mendelian Forms of Parkinson's Disease. *Biochim. Biophys. Acta* **1792**:587–596. [2]

Gaugler, T., L. Klei, S. J. Sanders, et al. 2014. Most Genetic Risk for Autism Resides with Common Variation. *Nat. Genet.* **46**:881–885. [3]

Geddes, J. R., S. M. Carney, C. Davies, et al. 2003. Relapse Prevention with Antidepressant Drug Treatment in Depressive Disorders: A Systematic Review. *Lancet* **361**:653–661. [15]

Georgiev, D., D. Arion, J. F. Enwright, et al. 2014. Lower Gene Expression for Kcns3 Potassium Channel Subunit in Parvalbumin-Containing Neurons in the Prefrontal Cortex in Schizophrenia. *Am. J. Psychiatry* **171**:62–71. [16]

Georgopoulos, A. P., H. R. Tan, S. M. Lewis, et al. 2010. The Synchronous Neural Interactions Test as a Functional Neuromarker for Post-Traumatic Stress Disorder (PTSD): A Robust Classification Method Based on the Bootstrap. *J. Neural Eng.* **7**:16011. [10]

Gerds, T. A., T. Cai, and M. Schumacher. 2008. The Performance of Risk Prediction Models. *Biometrical J.* **50**:457–479. [14]

Gershman, S. J., and Y. Niv. 2010. Learning Latent Structure: Carving Nature at Its Joints. *Curr. Opin. Neurol.* **20**:251–256. [2]

Gershman, S. J., B. Pesaran, and N. D. Daw. 2009. Human Reinforcement Learning Subdivides Structured Action Spaces by Learning Effector-Specific Values. *J. Neurosci.* **29**:13524–13531. [13]

Geurts, D. E. M., Q. J. M. Huys, H. E. M. den Ouden, and R. Cools. 2013. Aversive Pavlovian Control of Instrumental Behavior in Humans. *J. Cogn. Neurosci.* **25**:1428–1441. [15]

Ghahramani, Z., and G. E. Hinton. 2000. Variational Learning for Switching State-Space Models. *Neural Comput.* **12**:831–864. [5]

Gibson, J. R., K. M. Huber, and T. C. Südhof. 2009. Neuroligin-2 Deletion Selectively Decreases Inhibitory Synaptic Transmission Originating from Fast-Spiking but Not from Somatostatin-Positive Interneurons. *J. Neurosci.* **29**:13883–13897. [3]

Gilani, A. I., M. O. Chohan, M. Inan, et al. 2014. Interneuron Precursor Transplants in Adult Hippocampus Reverse Psychosis-Relevant Features in a Mouse Model of Hippocampal Disinhibition. *PNAS* **111**:7450–7455. [16]

Gilbertson, M. W., M. E. Shenton, A. Ciszewski, et al. 2002. Smaller Hippocampal Volume Predicts Pathologic Vulnerability to Psychological Trauma. *Nat. Neurosci.* **5**:1242–1247. [2, 10]

Gill, K. M., J. M. Cook, M. M. Poe, and A. A. Grace. 2014. Prior Antipsychotic Drug Treatment Prevents Response to Novel Antipsychotic Agent in the Methylazoxymethanol Acetate Model of Schizophrenia. *Schizophr. Bull.* **40**:341–350. [3]

Gillan, C. M., A. M. Apergis-Schoute, S. Morein-Zamir, et al. 2015. Functional Neuroimaging of Avoidance Habits in Obsessive-Compulsive Disorder. *Am. J. Psychiatry* **172**:284–293. [10]

Gillan, C. M., M. Kosinski, R. Whelan, E. A. Phelps, and N. D. Daw. 2016. Characterizing a Psychiatric Symptom Dimension Related to Deficits in Goal-Directed Control. *eLife* **5**:e11305. [17]

Gillan, C. M., S. Morein-Zamir, G. P. Urcelay, et al. 2014. Enhanced Avoidance Habits in Obsessive-Compulsive Disorder. *Biol. Psychiatry* **75**:631–638. [10]

Gillan, C. M., M. Papmeyer, S. Morein-Zamir, et al. 2011. Disruption in the Balance between Goal-Directed Behavior and Habit Learning in Obsessive-Compulsive Disorder. *Am. J. Psychiatry* **168**:718–726. [10]

Gilman, S. R., I. Iossifov, D. Levy, et al. 2011. Rare de Novo Variants Associated with Autism Implicate a Large Functional Network of Genes Involved in Formation and Function of Synapses. *Neuron* **70**:898–907. [3]

Gilpin, N. W., M. A. Herman, and M. Roberto. 2014. The Central Amygdala as an Integrative Hub for Anxiety and Alcohol Use Disorders. *Biol. Psychiatry* **77**:859–869. [1]

Giuliano, A. J., H. Li, R. I. Mesholam-Gately, et al. 2012. Neurocognition in the Psychosis Risk Syndrome: A Quantitative and Qualitative Review. *Curr. Pharm. Des.* **18**:399–415. [3]

Glahn, D. C., A. R. Laird, I. Ellison-Wright, et al. 2008. Meta-Analysis of Gray Matter Anomalies in Schizophrenia: Application of Anatomic Likelihood Estimation and Network Analysis. *Biol. Psychiatry* **64**:774–781. [4]

Gläscher, J., N. Daw, P. Dayan, and J. P. O'Doherty. 2010. States versus Rewards: Dissociable Neural Prediction Error Signals Underlying Model-Based and Model-Free Reinforcement Learning. *Neuron* **66**:585–595. [4, 12]

Glassman, R. B. 1987. An Hypothesis About Redundancy and Reliability in the Brains of Higher Species : Analogies with Genes, Internal Organs, and Engineering Systems. *Neurosci. Biobehav. Rev.* **11**:275–285. [9]

Glausier, J. R., K. N. Fish, and D. A. Lewis. 2014. Altered Parvalbumin Basket Cell Inputs in the Dorsolateral Prefrontal Cortex of Schizophrenia Subjects. *Mol. Psychiatry* **19**:30–36. [16]

Glausier, J. R., and D. A. Lewis. 2013. Dendritic Spine Pathology in Schizophrenia. *Neuroscience* **251**:90–107. [16]

Glimcher, P. W. 2011. Understanding Dopamine and Reinforcement Learning: The Dopamine Reward Prediction Error Hypothesis. *PNAS* **108**:15647–15654. [12, 13]

Glue, P., M. R. Donovan, S. Kolluri, and B. Emir. 2010. Meta-Analysis of Relapse Prevention Antidepressant Trials in Depressive Disorders. *Aust. NZ J. Psychiatry* **44**:697–705. [15]

Gold, J. I., and M. N. Shadlen. 2007. The Neural Basis of Decision Making. *Annu. Rev. Neurosci.* **30**:535–574. [5, 6]

Gold, J. M., R. L. Fuller, B. M. Robinson, et al. 2006. Intact Attentional Control of Working Memory Encoding in Schizophrenia. *J. Abnorm. Psychol.* **115**:658–673. [4]

Gold, J. M., B. Hahn, W. W. Zhang, et al. 2010. Reduced Capacity but Spared Precision and Maintenance of Working Memory Representations in Schizophrenia. *Arch. Gen. Psychiatry* **67**:570–577. [4]

Gold, J. M., W. Kool, M. M. Botvinick, et al. 2015. Cognitive Effort Avoidance and Detection in People with Schizophrenia. *Cogn Affect Behav Neurosci* **15**:145–154. [4]

Gold, J. M., G. P. Strauss, J. A. Waltz, et al. 2013. Negative Symptoms of Schizophrenia Are Associated with Abnormal Effort-Cost Computations. *Biol. Psychiatry* **74**:130–136. [4]

Gold, J. M., J. A. Waltz, T. M. Matveeva, et al. 2012. Negative Symptoms and the Failure to Represent the Expected Reward Value of Actions: Behavioral and Computational Modeling Evidence. *Arch. Gen. Psychiatry* **69**:129–138. [4, 5]

Gold, J. M., C. M. Wilk, R. P. McMahon, R. W. Buchanan, and S. J. Luck. 2003. Working Memory for Visual Features and Conjunctions in Schizophrenia. *J. Abnorm. Psychol.* **112**:61–71. [4]

Goldberg, D. 1996. A Dimensional Model for Common Mental Disorders. *Br. J. Psychiatry* **168**:44–49. [8]

Goldman-Rakic, P. S. 1995. Cellular Basis of Working Memory. *Neuron* **14**:477–485. [3]

Goldman-Rakic, P. S. 1996. Regional and Cellular Fractionation of Working Memory. *PNAS* **93**:13473–13480. [3]

Goldman-Rakic, P. S., S. A. Castner, T. H. Svensson, L. J. Siever, and G. V. Williams. 2004. Targeting the Dopamine D1 Receptor in Schizophrenia: Insights for Cognitive Dysfunction. *Psychopharmacology (Berl.)* **174**:3–16. [5, 13]

Goldstein, D. S., and B. S. McEwen. 2002. Allostasis, Homeostats, and the Nature of Stress. *Stress* **5**:55–58. [2]

Gonzalez-Burgos, G., R. Y. Cho, and D. A. Lewis. 2015. Alterations in Cortical Network Oscillations and Parvalbumin Neurons in Schizophrenia. *Biol. Psychiatry* **77**:1031–1040. [16]

Gonzalez-Burgos, G., and D. A. Lewis. 2012. NMDA Receptor Hypofunction, Parvalbumin-Positive Neurons, and Cortical Gamma Oscillations in Schizophrenia. *Schizophr. Bull.* **38**:950–957. [12]

Gonzalez-Pinto, A., M. Gutierrez, F. Mosquera, et al. 1998. First Episode in Bipolar Disorder: Misdiagnosis and Psychotic Symptoms. *J. Affect. Disord.* **50**:41–44. [3]

Goodman, W. K., K. D. Foote, B. D. Greenberg, et al. 2010. Deep Brain Stimulation for Intractable Obsessive Compulsive Disorder: Pilot Study Using a Blinded, Staggered-Onset Design. *Biol. Psychiatry* **67**:535–542. [10]

Gosseries, O., C. Schnakers, D. Ledoux, et al. 2011. Automated EEG Entropy Measurements in Coma, Vegetative State/Unresponsive Wakefulness Syndrome and Minimally Conscious State. *Funct. Neurol.* **26**:25–30. [3]

Gotlib, I. H., and J. Joormann. 2010. Cognition and Depression: Current Status and Future Directions. *Annu Rev Clin Psychol* **6**:285–312. [15]

Gottesman, I. I., and T. D. Gould. 2003. The Endophenotype Concept in Psychiatry: Etymology and Strategic Intentions. *Am. J. Psychiatry* **160**:636–645. [5, 16]

Grace, A. A. 2006. Disruption of Cortical-Limbic Interaction as a Substrate for Comorbidity. *Neurotox. Res.* **10**:93–101. [3]

———. 2010. Dopamine System Dysregulation by the Ventral Subiculum as the Common Pathophysiological Basis for Schizophrenia Psychosis, Psychostimulant Abuse, and Stress. *Neurotox. Res.* **18**:367–376. [3]

Gradin, V. B., P. Kumar, G. Waiter, et al. 2011. Expected Value and Prediction Error Abnormalities in Depression and Schizophrenia. *Brain* **134**:1751–1764. [4]

Grassian, S. 1983. Psychopathological Effects of Solitary Confinement. *Am. J. Psychiatry* **140**:1450–1454. [2]

Green, E. K., D. Grozeva, I. Jones, et al. 2010. The Bipolar Disorder Risk Allele at CACNA1C Also Confers Risk of Recurrent Major Depression and of Schizophrenia. *Mol. Psychiatry* **15**:1016–1022. [1, 3]

Green, M. F. 2006. Cognitive Impairment and Functional Outcome in Schizophrenia and Bipolar Disorder. *J. Clin. Psychiatry* **67(Suppl 9)**:3–8; discussion 36–42. [16]

Green, N., R. Bogacz, J. Huebl, et al. 2013. Reduction of Influence of Task Difficulty on Perceptual Decision Making by STN Deep Brain Stimulation. *Curr. Biol.* **23**:1681–1684. [5]

Greenberg, B. D., L. A. Gabriels, D. A. Malone, Jr., et al. 2010. Deep Brain Stimulation of the Ventral Internal Capsule/Ventral Striatum for Obsessive-Compulsive Disorder: Worldwide Experience. *Mol. Psychiatry* **15**:64–79. [10]

Greenberg, B. D., D. A. Malone, G. M. Friehs, et al. 2006. Three-Year Outcomes in Deep Brain Stimulation for Highly Resistant Obsessive-Compulsive Disorder. *Neuropsychopharmacology* **31**:2384–2393. [10]

Greenzang, C., D. S. Manoach, D. C. Goff, and J. J. Barton. 2007. Task-Switching in Schizophrenia: Active Switching Costs and Passive Carry-over Effects in an Antisaccade Paradigm. *Exp. Brain Res.* **181**:493–502. [4]

Gregory, S. G., J. J. Connelly, A. J. Towers, et al. 2009. Genomic and Epigenetic Evidence for Oxytocin Receptor Deficiency in Autism. *BMC Med.* **7**:62. [3]

Grimm, O., S. Vollstadt-Klein, L. Krebs, M. Zink, and M. N. Smolka. 2012. Reduced Striatal Activation During Reward Anticipation Due to Appetite-Provoking Cues in Chronic Schizophrenia: A fMRI Study. *Schizophr. Res.* **134**:151–157. [4]

Gross, J. J. 1998. Antecedent- and Response-Focused Emotion Regulation: Divergent Consequences for Experience, Expression, and Physiology. *J. Pers. Soc. Psychol.* **74**:224–237. [15]

Grunze, H. C., D. G. Rainnie, M. E. Hasselmo, et al. 1996. NMDA-Dependent Modulation of CA1 Local Circuit Inhibition. *J. Neurosci.* **16**:2034–2043. [16]

Gu, X., P. R. Hof, K. J. Friston, and J. Fan. 2013. Anterior Insular Cortex and Emotional Awareness. *J. Comp. Neurol.* **521**:3371–3388. [12]

Guilmatre, A., C. Dubourg, A.-L. Mosca, et al. 2009. Recurrent Rearrangements in Synaptic and Neurodevelopmental Genes and Shared Biologic Pathways in Schizophrenia, Autism, and Mental Retardation. *Arch. Gen. Psychiatry* **66**:947–956. [3]

Guitart-Masip, M., Q. J. M. Huys, L. Fuentemilla, et al. 2012. Go and No-Go Learning in Reward and Punishment: Interactions between Affect and Effect. *NeuroImage* **62**:154–166. [15]

Gulsuner, S., T. Walsh, A. C. Watts, et al. 2013. Spatial and Temporal Mapping of de Novo Mutations in Schizophrenia to a Fetal Prefrontal Cortical Network. *Cell* **154**:518–529. [16]

Gutman, D. A., P. E. Holtzheimer, T. E. Behrens, H. Johansen-Berg, and H. S. Mayberg. 2009. A Tractography Analysis of Two Deep Brain Stimulation White Matter Targets for Depression. *Biol. Psychiatry* **65**:276–282. [5]

Haase, L., B. Cerf-Ducastel, and C. Murphy. 2009. Cortical Activation in Response to Pure Taste Stimuli During the Physiological States of Hunger and Satiety. *NeuroImage* **44**:1008–1021. [14]

Haddon, J. E., and S. Killcross. 2007. Contextual Control of Choice Performance: Behavioral, Neurobiological, and Neurochemical Influences. *Ann. NY Acad. Sci.* **1104**:250–269. [4]

Haeffel, G. J., L. Y. Abramson, P. C. Brazy, et al. 2007. Explicit and Implicit Cognition: A Preliminary Test of a Dual-Process Theory of Cognitive Vulnerability to Depression. *Behav. Res. Ther.* **45**:1155–1167. [15]

Haeffel, G. J., L. Y. Abramson, Z. R. Voelz, et al. 2005. Negative Cognitive Styles, Dysfunctional Attitudes, and the Remitted Depression Paradigm: A Search for the Elusive Cognitive Vulnerability to Depression Factor among Remitted Depressives. *Emotion* **5**:343–348. [15]

Haeffel, G. J., B. E. Gibb, G. I. Metalsky, et al. 2008. Measuring Cognitive Vulnerability to Depression: Development and Validation of the Cognitive Style Questionnaire. *Clin. Psychol. Rev.* **28**:824–836. [15]

Haeusler, S., and W. Maass. 2007. A Statistical Analysis of Information-Processing Properties of Lamina-Specific Cortical Microcircuit Models. *Cereb. Cortex* **17**:149–162. [7]

Hagmann, P., L. Cammoun, X. Gigandet, et al. 2008. Mapping the Structural Core of Human Cerebral Cortex. *PLoS Biol.* **6**:e159. [16]

Hahn, B., B. M. Robinson, S. T. Kaiser, et al. 2010. Failure of Schizophrenia Patients to Overcome Salient Distractors During Working Memory Encoding. *Biol. Psychiatry* **68**:603–609. [4]

Hajos, M., W. E. Hoffmann, and B. Kocsis. 2008. Activation of Cannabinoid-1 Receptors Disrupts Sensory Gating and Neuronal Oscillation: Relevance to Schizophrenia. *Biol. Psychiatry* **63**:1075–1083. [16]

Haken, H. 1983. Synergetics: An Introduction. Non-Equilibrium Phase Transition and Self-Selforganisation in Physics, Chemistry and Biology. Berlin: Springer-Verlag. [11]

Hammen, C. 1991. Generation of Stress in the Course of Unipolar Depression. *J. Abnorm. Psychol.* **100**:555–561. [15]

Hammer, M. 1993. An Identified Neuron Mediates the Unconditioned Stimulus in Associative Olfactory Learning in Honeybees. *Nature* **366**:59–63. [13]

Hammer, M., and R. Menzel. 1995. Learning and Memory in the Honeybee. *J. Neurosci.* **15**:1617–1630. [13]

Hariri, A. R., V. S. Mattay, A. Tessitore, et al. 2002. Serotonin Transporter Genetic Variation and the Response of the Human Amygdala. *Science* **297**:400–403. [3]

Harkness, K. L., S. M. Monroe, A. D. Simons, and M. Thase. 1999. The Generation of Life Events in Recurrent and Non-Recurrent Depression. *Psychol. Med.* **29**:135–144. [15]

Harlé, K. M., P. Shenoy, J. L. Stewart, et al. 2014. Altered Neural Processing of the Need to Stop in Young Adults at Risk for Stimulant Dependence. *J. Neurosci.* **34**:4567–4580. [14]

Harris, A., T. Hare, and A. Rangel. 2013. Temporally Dissociable Mechanisms of Self-Control: Early Attentional Filtering versus Late Value Modulation. *J. Neurosci.* **33**:18917–18931. [6]

Harrison, P. J., and D. R. Weinberger. 2005. Schizophrenia Genes, Gene Expression, and Neuropathology: On the Matter of Their Convergence. *Mol. Psychiatry* **10**:40–68– image 45. [3]

Härter, M., M. Berger, F. Schneider, and G. Ollenschläger, eds. 2009. Nationale Versorgungs Leitlinie Unipolare Depression. Praxisleitlinien in Psychiatrie und Psychotherapie. Berlin: Springer. [15]

Hartmann, M. N., O. M. Hager, A. V. Reimann, et al. 2015. Apathy but Not Diminished Expression in Schizophrenia Is Associated with Discounting of Monetary Rewards by Physical Effort. *Schizophr. Bull.* **41**:503–512. [4]

Harvey, P. D. 2014. What Is the Evidence for Changes in Cognition and Functioning over the Lifespan in Patients with Schizophrenia? *J. Clin. Psychiatry* **75 Suppl** **2**:34–38. [16]

Harvey, P. D., J. Lombardi, M. Leibman, et al. 1997. Age-Related Differences in Formal Thought Disorder in Chronically Hospitalized Schizophrenic Patients: A Cross-Sectional Study across Nine Decades. *Am. J. Psychiatry* **154**:205–210. [16]

Hastie, T., R. Tibshirani, and J. H. Friedman. 2001. The Elements of Statistical Learning: Data Mining, Inference, and Prediction. New York: Springer. [14]

Hauser, T. U., R. Iannaccone, J. Ball, et al. 2014. Role of the Medial Prefrontal Cortex in Impaired Decision Making in Juvenile Attention-Deficit/Hyperactivity Disorder. *JAMA Psychiatry* **71**:1165–1173. [7]

Hay, N., and S. J. Russell. 2011. Metareasoning for Monte Carlo Tree Search. http://www.eecs.berkeley.edu/Pubs/TechRpts/2011/EECS-2011-119.pdf (accessed July 12, 2016). [15]

Hayes, P. E., and P. Ettigi. 1983. Dexamethasone Suppression Test in Diagnosis of Depressive Illness. *Clin. Pharm.* **2**:538–545. [1]

Hazy, T. E., M. J. Frank, and R. C. O'Reilly. 2006. Banishing the Homunculus: Making Working Memory Work. *Neuroscience* **139**:105–118. [4]

Hebb, D. O. 1949. The Organization of Behavior. New York: Wiley. [13]

Heerey, E. A., K. R. Bell-Warren, and J. M. Gold. 2008. Decision-Making Impairments in the Context of Intact Reward Sensitivity in Schizophrenia. *Biol. Psychiatry* **64**:62–69. [4]

Helmholtz, H. 1860/1962. Handbuch Der Physiologischen Optik (English Translation), vol. 3. Dover, NY: Southall JPC. [7]

Herman, J. P., and W. E. Cullinan. 1997. Neurocircuitry of Stress: Central Control of the Hypothalamo–Pituitary–Adrenocortical Axis. *Trends Neurosci.* **20**:78–84. [12]

Hershberger, W. A. 1986. An Approach through the Looking-Glass. *Anim. Learn. Behav.* **14**:443–451. [15]

Hertäg, L., J. Hass, T. Golovko, and D. Durstewitz. 2012. An Approximation to the Adaptive Exponential Integrate-and-Fire Neuron Model Allows Fast and Predictive Fitting to Physiological Data. *Front. Comput. Neurosci.* **6**:62. [5]

Herzallah, M. M., A. A. Moustafa, J. Y. Natsheh, et al. 2013. Depression Impairs Learning, Whereas the Selective Serotonin Reuptake Inhibitor, Paroxetine, Impairs Generalization in Patients with Major Depressive Disorder. *J. Affect. Disord.* **151**: 484–492. [3]

Hester, R., C. Simoes-Franklin, and H. Garavan. 2007. Post-Error Behavior in Active Cocaine Users: Poor Awareness of Errors in the Presence of Intact Performance Adjustments. *Neuropsychopharmacology* **32**:1974–1984. [14]

Heyman, G. 2009. Addiction: A Disorder of Choice. Cambridge, MA: Harvard Univ. Press. [2]

Higley, M. J. 2014. Localized Gabaergic Inhibition of Dendritic Ca^{2+} Signalling. *Nat. Rev. Neurosci.* **15**:567–572. [16]

Hikida, T., K. Kimura, N. Wada, K. Funabiki, and S. Nakanishi. 2010. Distinct Roles of Synaptic Transmission in Direct and Indirect Striatal Pathways to Reward and Aversive Behavior. *Neuron* **66**:896–907. [6]

Hikosaka, O., and M. Isoda. 2010. Switching from Automatic to Controlled Behavior: Cortico-Basal Ganglia Mechanisms. *Trends Cogn. Sci.* **14**:154–161. [6]

Hill, A., M. Rettenberger, N. Habermann, et al. 2012. The Utility of Risk Assessment Instruments for the Prediction of Recidivism in Sexual Homicide Perpetrators. *J. Interpers. Violence* **27**:3553–3578. [5]

Hill, A. B. 1965. The Environment and Disease: Association or Causation? *Proc. R. Soc. Med.* **58**:295–300. [14]

Hill, J. J., T. Hashimoto, and D. A. Lewis. 2006. Molecular Mechanisms Contributing to Dendritic Spine Alterations in the Prefrontal Cortex of Subjects with Schizophrenia. *Mol. Psychiatry* **11**:557–566. [3]

Hille, B. 2007. Ionic Channels in Excitable Membranes. New York: Sinauer. [13]

Hioki, H., S. Okamoto, M. Konno, et al. 2013. Cell Type-Specific Inhibitory Inputs to Dendritic and Somatic Compartments of Parvalbumin-Expressing Neocortical Interneuron. *J. Neurosci.* **33**:544–555. [16]

Hirano, Y., N. Oribe, S. Kanba, et al. 2015. Spontaneous Gamma Activity in Schizophrenia. *JAMA Psychiatry* **72**:813–821. [16]

Hirsch, S., and R. Bolles. 1980. On the Ability of Prey to Recognize Predators. *Z. Tierpsychol* **54**:71–84. [15]

Hodgkin, A. L., and A. F. Huxley. 1952. A Quantitative Description of Membrane Current and Its Application to Conduction and Excitation in Nerve. *J. Physiol.* **117**:500–544. [9, 12]

Hoel, E. P., L. Albantakis, and G. Tononi. 2013. Quantifying Causal Emergence Shows That Macro Can Beat Micro. *PNAS* **110**:19790–19795. [14]

Hoffman, P., and M. A. L. Ralph. 2011. Reverse Concreteness Effects Are Not a Typical Feature of Semantic Dementia: Evidence for the Hub-and-Spoke Model of Conceptual Representation. *Cereb. Cortex* **21**:2103–2112. [12]

Hoffman, R. E., and I. Cavus. 2002. Slow Transcranial Magnetic Stimulation, Long-Term Depotentiation, and Brain Hyperexcitability Disorders. *Am. J. Psychiatry* **159**:1093–1102. [5]

Hoffman, R. E., R. Gueorguieva, K. A. Hawkins, et al. 2005. Temporoparietal Transcranial Magnetic Stimulation for Auditory Hallucinations: Safety, Efficacy and Moderators in a Fifty Patient Sample. *Biol. Psychiatry* **58**:97–104. [16]

Hoffman, R. E., M. Hampson, K. Wu, et al. 2007. Probing the Pathophysiology of Auditory/Verbal Hallucinations by Combining Functional Magnetic Resonance Imaging and Transcranial Magnetic Stimulation. *Cereb. Cortex* **17**:2733–2743. [5]

Hoftman, G. D., D. W. Volk, H. H. Bazmi, et al. 2015. Altered Cortical Expression of GABA-Related Genes in Schizophrenia: Illness Progression Vs Developmental Disturbance. *Schizophr. Bull.* **41**:180–191. [16]

Hollon, S. D., R. J. DeRubeis, J. Fawcett, et al. 2014. Effect of Cognitive Therapy with Antidepressant Medications Vs Antidepressants Alone on the Rate of Recovery in Major Depressive Disorder: A Randomized Clinical Trial. *JAMA Psychiatry* **71**:1157–1164. [15]

Hollon, S. D., R. J. DeRubeis, R. C. Shelton, et al. 2005. Prevention of Relapse Following Cognitive Therapy Vs Medications in Moderate to Severe Depression. *Arch. Gen. Psychiatry* **62**:417–422. [15]

Hollon, S. D., R. C. Shelton, S. Wisniewski, et al. 2006. Presenting Characteristics of Depressed Outpatients as a Function of Recurrence: Preliminary Findings from the Star*D Clinical Trial. *J. Psychiatr. Res.* **40**:59–69. [15]

Homayoun, H., and B. Moghaddam. 2008. Orbitofrontal Cortex Neurons as a Common Target for Classic and Glutamatergic Antipsychotic Drugs. *PNAS* **105**:18041–18046. [3]

Hong, L. E., M. T. Avila, and G. K. Thaker. 2005. Response to Unexpected Target Changes During Sustained Visual Tracking in Schizophrenic Patients. *Exp. Brain Res.* **165**:125–131. [7]

Horan, W. P., D. L. Braff, K. H. Nuechterlein, et al. 2008. Verbal Working Memory Impairments in Individuals with Schizophrenia and Their First-Degree Relatives: Findings from the Consortium on the Genetics of Schizophrenia. *Schizophr. Res.* **103**:218–228. [4]

Horga, G., T. V. Maia, R. Marsh, et al. 2015. Changes in Corticostriatal Connectivity During Reinforcement Learning in Humans. *Hum. Brain Mapp.* **36**:793–803. [12]

Howes, O. D., S. K. Bose, F. Turkheimer, et al. 2011a. Progressive Increase in Striatal Dopamine Synthesis Capacity as Patients Develop Psychosis: A PET Study. *Mol. Psychiatry* **16**:885–886. [4]

Howes, O. D., S. K. Bose, F. Turkheimer, et al. 2011b. Dopamine Synthesis Capacity before Onset of Psychosis: A Prospective [18f]-DOPA PET Imaging Study. *Am. J. Psychiatry* **168**:1311–1317. [4]

Howes, O. D., J. Kambeitz, E. Kim, et al. 2012. The Nature of Dopamine Dysfunction in Schizophrenia and What This Means for Treatment. *Arch. Gen. Psychiatry* **69**: 776–786. [4]

Howes, O. D., and S. Kapur. 2009. The Dopamine Hypothesis of Schizophrenia: Version III: The Final Common Pathway. *Schizophr. Bull.* **35**:549–562. [4]

Howes, O. D., R. McCutcheon, and J. Stone. 2015. Glutamate and Dopamine in Schizophrenia: An Update for the 21st Century. *J Psychopharmacol* **29**:97–115. [4]

Howes, O. D., A. J. Montgomery, M.-C. Asselin, et al. 2009. Elevated Striatal Dopamine Function Linked to Prodromal Signs of Schizophrenia. *Arch. Gen. Psychiatry* **66**: 13–20. [3, 4]

Howes, O. D., and R. M. Murray. 2014. Schizophrenia: An Integrated Socio-developmental-Cognitive Model. *Lancet* **383**:1677–1687. [4]

Hoy, K. E., N. W. Bailey, S. L. Arnold, and P. B. Fitzgerald. 2015. The Effect of Transcranial Direct Current Stimulation on Gamma Activity and Working Memory in Schizophrenia. *Psychiatry Res.* **228**:191–196. [16]

Hradetzky, E., T. M. Sanderson, T. M. Tsang, et al. 2012. The Methylazoxymethanol Acetate (MAM-E17) Rat Model: Molecular and Functional Effects in the Hippo-campus. *Neuropsychopharmacology* **37**:364–377. [3]

Hrdlicka, M., and I. Dudova. 2015. Atypical Antipsychotics in the Treatment of Early-Onset Schizophrenia. *Neuropsychiatr. Dis. Treat.* **11**:907–913. [11]

Hunt, L., R. J. Dolan, and T. E. Behrens. 2014. Hierarchical Competitions Subserving Multi-Attribute Choice. *Nat. Neurosci.* **17**:1613–1622. [6]

Huttenlocher, P. R. 1979. Synaptic Density in Human Frontal Cortex - Developmental Changes and Effects of Aging. *Brain Res.* **163**:195–205. [16]

Hutton, S. B., B. K. Puri, L. J. Duncan, et al. 1998. Executive Function in First-Episode Schizophrenia. *Psychol. Med.* **28**:463–473. [4]

Huttunen, J., M. Heinimaa, T. Svirskis, et al. 2008. Striatal Dopamine Synthesis in First-Degree Relatives of Patients with Schizophrenia. *Biol. Psychiatry* **63**:114–117. [4]

Huys, Q. J. M. 2007. Reinforcers and Control: Towards a Computational Aetiology of Depression. Ph.D. Thesis, Gatsby Computational Neuroscience Unit, University College London. [2, 17]

Huys, Q. J. M., R. Cools, M. Gölzer, et al. 2011. Disentangling the Roles of Approach, Activation and Valence in Instrumental and Pavlovian Responding. *PLoS Comput. Biol.* **7**:e1002028. [6, 14, 15]

Huys, Q. J. M., N. D. Daw, and P. Dayan. 2015a. Depression: A Decision-Theoretic Analysis. *Annu. Rev. Neurosci.* **38**:1–23. [3, 12, 15, 17]

Huys, Q. J. M., and P. Dayan. 2009. A Bayesian Formulation of Behavioral Control. *Cognition* **113**:314–328. [15]

Huys, Q. J. M., N. Eshel, E. O'Nions, et al. 2012. Bonsai Trees in Your Head: How the Pavlovian System Sculpts Goal-Directed Choices by Pruning Decision Trees. *PLoS Comput. Biol.* **8**:e1002410. [3, 6, 15]

Huys, Q. J. M., M. Guitart-Masip, R. J. Dolan, and P. Dayan. 2015b. Decision-Theoretic Psychiatry. *Clin. Psychol. Sci.* **3**:400–421. [5, 6, 15]

Huys, Q. J. M., N. Lally, P. Faulkner, et al. 2015c. Interplay of Approximate Planning Strategies. *PNAS* **112**:3098–3103. [15]

Huys, Q. J. M., D. A. Pizzagalli, R. Bogdan, and P. Dayan. 2013. Mapping Anhedonia onto Reinforcement Learning: A Behavioural Meta-Analysis. *Biol. Mood Anxiety Disord.* **3**:12. [3]

Huys, Q. J. M., P. N. Tobler, G. Hasler, and S. B. Flagel. 2014. The Role of Learning-Related Dopamine Signals in Addiction Vulnerability. *Prog. Brain Res.* **211**:31–77. [15]

Huys, Q. J. M., J. Vogelstein, and P. Dayan. 2009. Psychiatry: Insights into Depression through Normative Decision-Making Models. In: Advances in Neural Information Processing Systems 21, ed. D. Koller et al., pp. 729–736. MIT Press. [15]

Hyman, S. E. 2003. Forward. In: Advancing DSM: Dilemmas in Psychiatric Diagnosis, ed. K. Phillips et al., pp. xi–xix. Washington, D.C.: American Psychiatric Association. [8]

———. 2007. Can Neuroscience Be Integrated into the DSM-V? *Nat. Rev. Neurosci.* **8**:725–732. [8]

———. 2010. The Diagnosis of Mental Disorders: The Problem of Reification. *Annu. Rev. Clin. Psych.* **6**:155–179. [8, 9, 10]

———. 2012. Revolution Stalled. *Sci. Transl. Med.* **4**:155cm111. [7]

———. 2014. Revitalizing Psychiatric Therapeutics. *Neuropsychopharmacology* **39**:220–229. [1]

Hyman, S. M., P. Paliwal, T. M. Chaplin, et al. 2008. Severity of Childhood Trauma Is Predictive of Cocaine Relapse Outcomes in Women but Not Men. *Drug Alcohol Depend.* **92**:208–216. [3]

Iacoviello, B. M., L. B. Alloy, L. Y. Abramson, and J. Y. Choi. 2010. The Early Course of Depression: A Longitudinal Investigation of Prodromal Symptoms and Their Relation to the Symptomatic Course of Depressive Episodes. *J. Abnorm. Psychol.* **119**:459–467. [15]

ICD-10. 1992. International Statistical Classification of Diseases and Related Health Problems: Tenth Revision, vol. 2, World Health Organization. http://apps.who.int/classifications/icd10/browse/2016/en (accessed May 5, 2016). [2, 16]

Ide, J. S., P. Shenoy, A. J. Yu, and C. S. Li. 2013. Bayesian Prediction and Evaluation in the Anterior Cingulate Cortex. *J. Neurosci.* **33**:2039–2047. [14]

Iglesias, S., C. Mathys, K. H. Brodersen, et al. 2013. Hierarchical Prediction Errors in Midbrain and Basal Forebrain During Sensory Learning. *Neuron* **80**:519–530. [7, 12]

Insel, C., J. Reinen, J. Weber, et al. 2014. Antipsychotic Dose Modulates Behavioral and Neural Responses to Feedback During Reinforcement Learning in Schizophrenia. *Cogn Affect Behav Neurosci* **14**:189–201. [4]

Insel, T. R. 2010. Rethinking Schizophrenia. *Nature* **468**:187–193. [16]

———. 2011. Treatment Development: The Past 50 Years. NIMH Director's Blog. http://www.nimh.nih.gov/about/director/2011/treatment-development-the-past-50-years.shtml (accessed June 10, 2016). [1]

———. 2012. Next-Generation Treatments for Mental Disorders. *Sci. Transl. Med.* **4**:155ps119. [7, 14]

Insel, T. R., and B. N. Cuthbert. 2009. Endophenotypes: Bridging Genomic Complexity and Disorder Heterogeneity. *Biol. Psychiatry* **66**:988–989. [8, 9, 10]

———. 2015. Medicine. Brain Disorders? Precisely. *Science* **348**:499–500. [14, 16]

Insel, T. R., B. N. Cuthbert, M. Garvey, et al. 2010. Research Domain Criteria (RDoC): Toward a New Classification Framework for Research on Mental Disorders. *Am. J. Psychiatry* **167**:748–751. [2, 4, 8, 14, 17]

Intl. Schizophrenia Consortium. 2009. Common Polygenic Variation Contributes to Risk of Schizophrenia and Bipolar Disorder. *Nature* **460**:748–752. [1, 3]

Iordanova, M. D. 2009. Dopaminergic Modulation of Appetitive and Aversive Predictive Learning. *Rev. Neurosci.* **20**:383–404. [14]

Iossifov, I., B. J. O'Roak, S. J. Sanders, et al. 2014. The Contribution of *de Novo* Coding Mutations to Autism Spectrum Disorder. *Nature* **515**:216–221. [1]

Irish, M., J. R. Hodges, and O. Piguet. 2014. Right Anterior Temporal Lobe Dysfunction Underlies Theory of Mind Impairments in Semantic Dementia. *Brain* **137**:1241–1253. [12]

Irwin, D. E., B. Stucky, M. M. Langer, et al. 2010. An Item Response Analysis of the Pediatric PROMIS Anxiety and Depressive Symptoms Scale. *Qual. Life Res.* **21**:195–207. [8]

Iwata, Y., S. Nakajima, T. Suzuki, et al. 2015. Effects of Glutamate Positive Modulators on Cognitive Deficits in Schizophrenia: A Systematic Review and Meta-Analysis of Double-Blind Randomized Controlled Trials. *Mol. Psychiatry* **20**:1151–1160. [16]

Jaaro-Peled, H., A. Hayashi-Takagi, S. Seshadri, et al. 2009. Neurodevelopmental Mechanisms of Schizophrenia: Understanding Disturbed Postnatal Brain Maturation through Neuregulin-1-Erbb4 and DISC1. *Trends Neurosci.* **32**:485–495. [3]

Jacobs, W. J., and L. Nadel. 1998. Neurobiology of Reconstructed Memory. *Psychol. Public Policy Law* **4**:1110–1134. [2, 10]

Jaeger, D., and J. M. Bower. 1999. Synaptic Control of Spiking in Cerebellar Purkinje Cells: Dynamic Current Clamp Based on Model Conductances. *J. Neurosci.* **19**:6090–6101. [12]

Jamadar, S., P. Michie, and F. Karayanidis. 2010. Compensatory Mechanisms Underlie Intact Task-Switching Performance in Schizophrenia. *Neuropsychologia* **48**:1305–1323. [4]

James, G., J. G. Witten, T. Hastie, and R. Tibshirani. 2013. An Introduction to Statistical Learning. Springer Texts in Statistics. New York: Springer. [14]

Jansen, B. H., and V. G. Rit. 1995. Electroencephalogram and Visual Evoked Potential Generation in a Mathematical Model of Coupled Cortical Columns. *Biol. Cybern.* **73**:357–366. [12]

Jardri, R., and S. Denève. 2013. Circular Inferences in Schizophrenia. *Brain* **136**:3227–3241. [5]

Jarrett, R. B., D. Kraft, J. Doyle, et al. 2001. Preventing Recurrent Depression Using Cognitive Therapy with and without a Continuation Phase: A Randomized Clinical Trial. *Arch. Gen. Psychiatry* **58**:381–388. [15]

Jarrett, R. B., A. Minhajuddin, H. Gershenfeld, E. S. Friedman, and M. E. Thase. 2013. Preventing Depressive Relapse and Recurrence in Higher-Risk Cognitive Therapy Responders: A Randomized Trial of Continuation Phase Cognitive Therapy, Fluoxetine, or Matched Pill Placebo. *JAMA Psychiatry* [15]

Javitt, D. C., M. Steinschneider, C. E. Schroeder, and J. C. Arezzo. 1996. Role of Cortical N-Methyl-D-Aspartate Receptors in Auditory Sensory Memory and Mismatch Negativity Generation: Implications for Schizophrenia. *PNAS* **93**:11962–11967. [12]

Jaynes, E. T. 2003. Probability Theory: The Logic of Science: Cambridge Univ. Press. [7, 17]

Jazbec, S., C. Pantelis, T. Robbins, et al. 2007. Intra-Dimensional/Extra-Dimensional Set-Shifting Performance in Schizophrenia: Impact of Distractors. *Schizophr. Res.* **89**:339–349. [4]

Jirsa, V. K., W. C. Stacey, P. P. Quilichini, A. I. Ivanov, and C. Bernard. 2014. On the Nature of Seizure Dynamics. *Brain* **137**:2210–2230. [12]

Jocham, G., T. A. Klein, and M. Ullsperger. 2011. Dopamine-Mediated Reinforcement Learning Signals in the Striatum and Ventromedial Prefrontal Cortex Underlie Value-Based Choices. *J. Neurosci.* **31**:1606–1613. [6]

Johnson, A., and A. D. Redish. 2007. Neural Ensembles in Ca3 Transiently Encode Paths Forward of the Animal at a Decision Point. *J. Neurosci.* **27**:12176–12189. [15]

Johnson, S. A., N. M. Fournier, and L. E. Kalynchuk. 2006. Effect of Different Doses of Corticosterone on Depression-Like Behavior and HPA Axis Responses to a Novel Stressor. *Behav. Brain Res.* **168**:280–288. [12]

Joormann, J. 2004. Attentional Bias in Dysphoria: The Role of Inhibitory Processes. *Cogn. Emot.* **18**:125–147. [15]

———. 2006. Differential Effects of Rumination and Dysphoria on the Inhibition of Irrelevant Emotional Material: Evidence from a Negative Priming Task. *Cogn. Ther. Res.* **30**:149–160. [15]

Jovanovic, T., and K. J. Ressler. 2010. How the Neurocircuitry and Genetics of Fear Inhibition May Inform Our Understanding of PTSD. *Am. J. Psychiatry* **167**:648–662. [10]

Joyce, E., S. Hutton, S. Mutsatsa, et al. 2002. Executive Dysfunction in First-Episode Schizophrenia and Relationship to Duration of Untreated Psychosis: The West London Study. *Br. J. Psychiatry Suppl.* **43**:s38–44. [4]

Juckel, G., F. Schlagenhauf, M. Koslowski, et al. 2006a. Dysfunction of Ventral Striatal Reward Prediction in Schizophrenic Patients Treated with Typical, Not Atypical, Neuroleptics. *Psychopharmacology (Berl.)* **187**:222–228. [4]

———. 2006b. Dysfunction of Ventral Striatal Reward Prediction in Schizophrenia. *NeuroImage* **29**:409–416. [4]

Judd, L. L., and H. S. Akiskal. 2000. Delineating the Longitudinal Structure of Depressive Illness: Beyond Clinical Subtypes and Duration Thresholds. *Pharmacopsychiatry* **33**:3–7. [15]

Judd, L. L., H. S. Akiskal, J. D. Maser, et al. 1998. Major Depressive Disorder: A Prospective Study of Residual Subthreshold Depressive Symptoms as Predictor of Rapid Relapse. *J. Affect. Disord.* **50**:97–108. [15]

Judd, L. L., M. J. Paulus, P. J. Schettler, et al. 2000. Does Incomplete Recovery from First Lifetime Major Depressive Episode Herald a Chronic Course of Illness? *Am. J. Psychiatry* **157**:1501–1504. [15]

Just, M., V. Cherkassky, T. Keller, and N. Minshew. 2004. Cortical Activation and Synchronization During Sentence Comprehension in High-Functioning Autism: Evidence of Underconnectivity. *Brain* **127**:1811–1821. [3]

Kaabi, B., J. Gelernter, S. W. Woods, et al. 2006. Genome Scan for Loci Predisposing to Anxiety Disorders Using a Novel Multivariate Approach: Strong Evidence for a Chromosome 4 Risk Locus. *Am. J. Hum. Genet.* **78**:543–553. [3]

Kaddurah-Daouk, R., and R. Weinshilboum. 2015. Pharmacometabolomics Research Network: Metabolomic Signatures for Drug Response Phenotypes: Pharmaco-metabolomics Enables Precision Medicine. *Clin. Pharmacol. Ther.* **98**:71–75. [5]

Kalin, N. H., S. C. Risch, D. S. Janowsky, and D. L. Murphy. 1981. Use of Dexamethasone Suppression Test in Clinical Psychiatry. *J. Clin. Psychopharmacol.* **1**:64–69. [1]

Kalivas, P. W. 2004. Glutamate Systems in Cocaine Addiction. *Curr. Opin. Pharm.* **4**:23–29. [12]

———. 2009. The Glutamate Homeostasis Hypothesis of Addiction. *Nat. Rev. Neurosci.* **10**:561–572. [12]

Kalivas, P. W., and N. D. Volkow. 2014. The Neural Basis of Addiction: A Pathology of Motivation and Choice. *Am. J. Psychiatry* **162**:1403–1413. [12]

Kalivas, P. W., N. D. Volkow, and J. Seamans. 2005. Unmanageable Motivation in Addiction: A Pathology in Prefrontal-Accumbens Glutamate Transmission. *Neuron* **45**:647–650. [12]

Kalman, R. E. 1964. When Is a Linear Control System Optimal? *J. Basic Engineer.* **4**:51–60. [14]

Kambeitz, J., A. Abi-Dargham, S. Kapur, and O. D. Howes. 2014. Alterations in Cortical and Extrastriatal Subcortical Dopamine Function in Schizophrenia: Systematic Review and Meta-Analysis of Imaging Studies. *Br. J. Psychiatry* **204**:420–429. [4, 16]

Kapur, S., A. G. Phillips, and T. R. Insel. 2012. Why Has It Taken So Long for Biological Psychiatry to Develop Clinical Tests and What to Do About It? *Mol. Psychiatry* **17**:1174–1179. [7, 12]

Karayanidis, F., R. Nicholson, U. Schall, et al. 2006. Switching between Univalent Task-Sets in Schizophrenia: Erp Evidence of an Anticipatory Task-Set Reconfiguration Deficit. *Clin. Neurophysiol.* **117**:2172–2190. [4]

Karayiorgou, M., M. A. Morris, B. Morrow, et al. 1995. Schizophrenia Susceptibility Associated with Interstitial Deletions of Chromosome 22q11. *PNAS* **92**:7612–7616. [1]

Karayiorgou, M., T. J. Simon, and J. A. Gogos. 2010. 22q11.2 Microdeletions: Linking DNA Structural Variation to Brain Dysfunction and Schizophrenia. *Nat. Rev. Neurosci.* **11**:402–416. [5]

Kass, R. E., and D. Steffey. 1989. Approximate Bayesian Inference in Conditionally Independent Hierarchical Models (Parametric Empirical Bayes Models). *J. Am. Stat. Assoc.* **407**:717–726. [11]

Kaufman, J., J. Gelernter, J. J. Hudziak, A. R. Tyrka, and J. D. Coplan. 2015. The Research Domain Criteria (RDoC) Project and Studies of Risk and Resilience in Maltreated Children. *J. Am. Acad. Child Adolesc. Psychiatry* **54**:617–625. [11]

Kay, S. R. 1990. Positive-Negative Symptom Assessment in Schizophrenia: Psychometric Issues and Scale Comparison. *Psych. Quart.* **61**:163–178. [11]

Kaymaz, N., J. van Os, A. J. M. Loonen, and W. A. Nolen. 2008. Evidence That Patients with Single versus Recurrent Depressive Episodes Are Differentially Sensitive to Treatment Discontinuation: A Meta-Analysis of Placebo-Controlled Randomized Trials. *J. Clin. Psychiatry* **69**:1423–1436. [15]

Kayser, A., J. M. Mitchell, D. Weinstein, and M. J. Frank. 2015. Dopamine, Locus of Control, and the Exploration-Exploitation Tradeoff. *Neuropsychopharmacology* **40**:454–462. [6]

Keck, P. E., Jr., and H. K. Manji. 2002. Current and Emerging Treatments for Acute Mania and Long-Term Prophylaxis in Bipolar Disorder. In: Neuropsychopharmacology: The Fifth Generation of Progress, ed. K. L. Davis et al., pp. 1109–1118. Philadelphia: Lippincott Williams and Wilkins. [1]

Kegeles, L. S., A. Abi-Dargham, W. G. Frankle, et al. 2010. Increased Synaptic Dopamine Function in Associative Regions of the Striatum in Schizophrenia. *Arch. Gen. Psychiatry* **67**:231–239. [4, 16]

Keller, M. B., P. W. Lavori, C. E. Lewis, and G. L. Klerman. 1983. Predictors of Relapse in Major Depressive Disorder. *JAMA* **250**:3299–3304. [15]

Keller, M. B., M. H. Trivedi, M. E. Thase, et al. 2007. The Prevention of Recurrent Episodes of Depression with Venlafaxine for Two Years (PREVENT) Study: Outcomes from the 2-Year and Combined Maintenance Phases. *J. Clin. Psychiatry* **68**:1246–1256. [15]

Kendler, K. S. 1996. Major Depression and Generalized Anxiety Disorder: Same Genes, (Partly) Different Environments—Revisited. *Br. J. Psychiatry* **168**:68–75. [8]

———. 2001. Twin Studies of Psychiatric Illness. *Arch. Gen. Psych.* **58**:1005–1014. [8]

———. 2006. Reflections on the Relationship between Psychiatric Genetics and Psychiatric Nosology. *Am. J. Psychiatry* **163**:1138–1146. [1]

Kendler, K. S. 2012. The Dappled Nature of Causes of Psychiatric Illness: Replacing the Organic-Functional/Hardware-Software Dichotomy with Empirically Based Pluralism. *Mol. Psychiatry* **17**:377–388. [14]

Kendler, K. S., C. G. Davis, and R. C. Kessler. 1997. The Familial Aggregation of Common Psychiatric and Substance Use Disorders in the National Comorbidity Survey: A Family History Study. *Br. J. Psychiatry* **170**:541–548. [8]

Kendler, K. S., and C. O. Gardner. 2010. Dependent Stressful Life Events and Prior Depressive Episodes in the Prediction of Major Depression: The Problem of Causal Inference in Psychiatric Epidemiology. *Arch. Gen. Psychiatry* **67**:1120–1127. [14]

———. 2014. Sex Differences in the Pathways to Major Depression: A Study of Opposite-Sex Twin Pairs. *Am. J. Psychiatry* **171**:426–435. [14]

Kendler, K. S., and C. O. Gardner, Jr. 1998. Boundaries of Major Depression: An Evaluation of DSM-IV Criteria. *Am. J. Psychiatry* **155**:172–177. [10]

Kendler, K. S., L. M. Karkowski, and C. A. Prescott. 1999. Causal Relationship between Stressful Life Events and the Onset of Major Depression. *Am. J. Psychiatry* **156**:837–841. [15]

Kendler, K. S., R. C. Kessler, E. E. Walters, et al. 1995. Stressful Life Events, Genetic Liability, and Onset of an Episode of Major Depression in Women. *Am. J. Psychiatry* **152**:833–842. [15]

Kendler, K. S., and M. C. Neale. 2010. Endophenotype: A Conceptual Analysis. *Mol. Psychiatry* **15**:789–797. [5]

Kendler, K. S., L. M. Thornton, and C. O. Gardner. 2000. Stressful Life Events and Previous Episodes in the Etiology of Major Depression in Women: An Evaluation of the "Kindling" Hypothesis. *Am. J. Psychiatry* **157**:1243–1251. [15]

———. 2001. Genetic Risk, Number of Previous Depressive Episodes, and Stressful Life Events in Predicting Onset of Major Depression. *Am. J. Psychiatry* **158**:582–586. [15]

Kendler, K. S., and D. Walsh. 1995. Gender and Schizophrenia. Results of an Epidemiologically-Based Family Study. *Br. J. Psychiatry* **167**:184–192. [16]

Kennedy, N., R. Abbott, and E. S. Paykel. 2003. Remission and Recurrence of Depression in the Maintenance Era: Long-Term Outcome in a Cambridge Cohort. *Psychol. Med.* **33**:827–838. [15]

Kenny, E. M., P. Cormican, S. Furlong, et al. 2014. Excess of Rare Novel Loss-of-Function Variants in Synaptic Genes in Schizophrenia and Autism Spectrum Disorders. *Mol. Psychiatry* **19**:872–879. [3]

Kent, L., J. Emerton, V. Bhadravathi, et al. 2008. X-Linked Ichthyosis (Steroid Sulfatase Deficiency) Is Associated with Increased Risk of Attention Deficit Hyperactivity Disorder, Autism and Social Communication Deficits. *J. Med. Genet.* **45**:519–524. [3]

Kéri, S., A. Juhasz, A. Rimanoczy, et al. 2005a. Habit Learning and the Genetics of the Dopamine D3 Receptor: Evidence from Patients with Schizophrenia and Healthy Controls. *Behav. Neurosci.* **119**:687–693. [4]

Kéri, S., O. Kelemen, G. Szekeres, et al. 2000. Schizophrenics Know More Than They Can Tell: Probabilistic Classification Learning in Schizophrenia. *Psychol. Med.* **30**:149–155. [4]

Kéri, S., A. A. Moustafa, C. E. Myers, G. Benedek, and M. A. Gluck. 2010. A-Synuclein Gene Duplication Impairs Reward Learning. *PNAS* **107**:15992–15994. [2]

Kéri, S., O. Nagy, O. Kelemen, C. E. Myers, and M. A. Gluck. 2005b. Dissociation between Medial Temporal Lobe and Basal Ganglia Memory Systems in Schizophrenia. *Schizophr. Res.* **77**:321–328. [4]

Keshavan, M. S., G. Berger, R. B. Zipursky, S. J. Wood, and C. Pantelis. 2005. Neurobiology of Early Psychosis. *Br. J. Psychiatry Suppl.* **48**:S8–18. [16]

Kessler, R. C., A. Sonnega, E. Bromet, M. Hughes, and C. B. Nelson. 1995. Posttraumatic Stress Disorder in the National Comorbidity Survey. *Arch. Gen. Psychiatry* **52**:1048–1060. [10]

Kety, S. S., D. Rosenthal, P. H. Wender, and F. Schulsinger. 1971. Mental Illness in the Biological and Adoptive Families of Adopted Schizophrenics. *Am. J. Psychiatry* **128**:302–306. [8]

Kheirbek, M. A., K. C. Klemenhagen, A. Sahay, and R. Hen. 2012. Neurogenesis and Generalization: A New Approach to Stratify and Treat Anxiety Disorders. *Nat. Neurosci.* **15**:1613–1620. [12]

Khin, N. A., Y. F. Chen, Y. Yang, P. Yang, and T. P. Laughren. 2011. Exploratory Analyses of Efficacy Data from Major Depressive Disorder Trials Submitted to the Us Food and Drug Administration in Support of New Drug Applications. *J. Clin. Psychiatry* **72**:464–472. [10]

Kiebel, S. J., O. David, and K. J. Friston. 2006. Dynamic Causal Modelling of Evoked Responses in EEG/MEG with Lead Field Parameterization. *NeuroImage* **30**:1273–1284. [12]

Kieffaber, P. D., E. S. Kappenman, M. Bodkins, et al. 2006. Switch and Maintenance of Task Set in Schizophrenia. *Schizophr. Res.* **84**:345–358. [4]

Kilb, W. 2012. Development of the Gabaergic System from Birth to Adolescence. *Neuroscientist* **18**:613–630. [16]

Killcross, S., and E. Coutureau. 2003. Coordination of Actions and Habits in the Medial Prefrontal Cortex of Rats. *Cereb. Cortex* **13**:400–408. [15]

Kim, I. H., M. A. Rossi, D. K. Aryal, et al. 2015. Spine Pruning Drives Antipsychotic-Sensitive Locomotion via Circuit Control of Striatal Dopamine. *Nat. Neurosci.* **18**:883–891. [16]

Kim, J., D. C. Glahn, K. H. Nuechterlein, and T. D. Cannon. 2004. Maintenance and Manipulation of Information in Schizophrenia: Further Evidence for Impairment in the Central Executive Component of Working Memory. *Schizophr. Res.* **68**:173–187. [4]

Kim-Cohen, J., A. Caspi, T. E. Moffitt, et al. 2003. Prior Juvenile Diagnoses in Adults with Mental Disorder: Developmental Follow-Back of a Prospective-Longitudinal Cohort. *Arch. Gen. Psychiatry* **60**:709–717. [10]

Kimoto, S., M. M. Zaki, H. H. Bazmi, and D. A. Lewis. 2015. Altered Markers of Cortical Gamma-Aminobutyric Acid Neuronal Activity in Schizophrenia: Role of the NARP Gene. *JAMA Psychiatry* **72**:747–756. [16]

Kinon, B. J., L. Chen, H. Ascher-Svanum, et al. 2010. Early Response to Antipsychotic Drug Therapy as a Clinical Marker of Subsequent Response in the Treatment of Schizophrenia. *Neuropsychopharmacology* **35**:581. [12]

Kinon, B. J., B. A. Millen, L. Zhang, and D. L. McKinzie. 2015. Exploratory Analysis for a Targeted Patient Population Responsive to the Metabotropic Glutamate 2/3 Receptor Agonist Pomaglumetad Methionil in Schizophrenia. *Biol. Psychiatry* **78**:754–762. [16]

Kirby, K. N., N. M. Petry, and W. K. Bickel. 1999. Heroin Addicts Have Higher Discount Rates for Delayed Rewards Than Non-Drug-Using Controls. *J. Exp. Psychol.* **128**:78–87. [2]

Kirk, U., X. Gu, A. H. Harvey, P. Fonagy, and P. R. Montague. 2014. Mindfulness Training Modulates Value Signals in Ventromedial Prefrontal Cortex through Input from Insular Cortex. *NeuroImage* **100**:254–262. [12]

Kirsch, P., S. Ronshausen, D. Mier, and B. Gallhofer. 2007. The Influence of Anti-psychotic Treatment on Brain Reward System Reactivity in Schizophrenia Patients. *Pharmacopsychiatry* **40**:196–198. [4]

Kishida, K. T., B. King-Casas, and P. R. Montague. 2010. Neuroeconomic Approaches to Mental Disorders. *Neuron* **67**:543–554. [2]

Klein-Flugge, M. C., L. T. Hunt, D. R. Bach, R. J. Dolan, and T. E. Behrens. 2011. Dissociable Reward and Timing Signals in Human Midbrain and Ventral Striatum. *Neuron* **72**:654–664. [14]

Klingner, C. M., K. Langbein, M. Dietzek, et al. 2014. Thalamocortical Connectivity During Resting State in Schizophrenia. *Eur. Arch. Psychiatry Clin. Neurosci.* **264**:111–119. [16]

Knight, B. W., A. Omurtag, and L. Sirovich. 2000. The Approach of a Neuron Population Firing Rate to a New Equilibrium: An Exact Theoretical Result. *Neural Comput.* **12**: 1045–1055. [12]

Koch, C., and G. Laurent. 1999. Complexity and the Nervous System. *Science* **284**:96–98. [5]

Koch, K., C. Schachtzabel, G. Wagner, et al. 2009. Altered Activation in Association with Reward-Related Trial-and-Error Learning in Patients with Schizophrenia. *NeuroImage* **50**:223–232. [4]

Kocsis, J. H., M. E. Thase, M. H. Trivedi, et al. 2007. Prevention of Recurrent Episodes of Depression with Venlafaxine ER in a 1-Year Maintenance Phase from the PREVENT Study. *J. Clin. Psychiatry* **68**:1014–1023. [15]

Kohannim, O., X. Hua, D. P. Hibar, et al. 2010. Boosting Power for Clinical Trials Using Classifiers Based on Multiple Biomarkers. *Neurobiol. Aging* **31**:1429–1442. [14]

Koller, D., and N. Friedman. 2009. Probabilistic Graphical Models. Cambridge, MA: MIT Press. [3]

Konopaske, G. T., K. A. Dorph-Petersen, R. A. Sweet, et al. 2008. Effect of Chronic Antipsychotic Exposure on Astrocyte and Oligodendrocyte Numbers in Macaque Monkeys. *Biol. Psychiatry* **63**:759–765. [16]

Konorski, J. 1949. Conditioned Reflexes and Neuron Organization. *Calif. Med.* **70**:311. [13]

Korb, A. S., A. M. Hunter, I. A. Cook, and A. F. Leuchter. 2009. Rostral Anterior Cingulate Cortex Theta Current Density and Response to Antidepressants and Placebo in Major Depression. *Clin. Neurophysiol.* **120**:1313–1319. [15]

Koshino, H., R. K. Kana, T. A. Keller, et al. 2008. fMRI Investigation of Working Memory for Faces in Autism: Visual Coding and Underconnectivity with Frontal Areas. *Cereb. Cortex* **18**:289–300. [3]

Kraemer, H., A. Noda, and R. O'Hara. 2004. Categorical versus Dimensional Approaches to Diagnosis: Methodological Challenges. *J. Psychiatr. Res.* **38**:17–25. [8]

Kraepelin, E. 1919. Dementia Praecox and Paraphrenia (trans. R. M. Barclay). Edinburgh: Livingston. [9]

Kraepelin, E., and A. R. Diefendorf. 1907. Clinical Psychiatry: A Textbook for Students and Physicians (7th edition). London: MacMillan. [9]

Kravitz, A. V., L. D. Tye, and A. C. Kreitzer. 2012. Distinct Roles for Direct and Indirect Pathway Striatal Neurons in Reinforcement. *Nat. Neurosci.* **15**:816–818. [2, 6]

Kremer, B., P. Goldberg, S. E. Andrew, et al. 1994. A Worldwide Study of the Huntington's Disease Mutation: The Sensitivity and Specificity of Measuring Cag Repeats. *N. Engl. J. Med.* **330**:1401–1406. [2]

Krueger, R. F. 1999. The Structure of Common Mental Disorders. *JAMA Psychiatry* **56**:921–926. [2]

Krueger, R. F., and A. W. I. MacDonald. 2005. Dimensional Approaches to Understanding and Treating Psychosis. *Psychiatr. Ann.* **35**:31–34. [9]

Krueger, R. F., and K. E. Markon. 2006. Reinterpreting Comorbidity: A Model-Based Approach to Understanding and Classifying Psychopathology. *Annu. Rev. Clin. Psych.* **2**:111–133. [5]

Krueger, R. F., K. E. Markon, C. J. Patrick, and W. G. Iacono. 2005. Externalizing Psychopathology in Adulthood: A Dimensional-Spectrum Conceptualization and Its Implications for DSM-V. *J. Abnorm. Psychol.* **114**:537–550. [8]

Krystal, J. H. 2015. Deconstructing N-Methyl-D-Aspartate Glutamate Receptor Contributions to Cortical Circuit Functions to Construct Better Hypotheses About the Pathophysiology of Schizophrenia. *Biol. Psychiatry* **77**:508–510. [16]

Krystal, J. H., W. Abi-Saab, E. Perry, et al. 2005. Preliminary Evidence of Attenuation of the Disruptive Effects of the NMDA Glutamate Receptor Antagonist, Ketamine, on Working Memory by Pretreatment with the Group II Metabotropic Glutamate Receptor Agonist, LY354740, in Healthy Human Subjects. *Psychopharmacology (Berl.)* **179**:303–309. [16]

Krystal, J. H., D. C. D'Souza, D. Mathalon, et al. 2003. NMDA Receptor Antagonist Effects, Cortical Glutamatergic Function, and Schizophrenia: Toward a Paradigm Shift in Medication Development. *Psychopharmacology (Berl.)* **169**:215–233. [16]

Krystal, J. H., G. Sanacora, and R. S. Duman. 2013. Rapid-Acting Glutamatergic Antidepressants: The Path to Ketamine and Beyond. *Biol. Psychiatry* **73**:1133–1141. [5]

Krystal, J. H., and M. W. State. 2014. Psychiatric Disorders: Diagnosis to Therapy. *Cell* **157**:201–214. [11, 16]

Kumar, P., G. Waiter, T. Ahearn, et al. 2008. Abnormal Temporal Difference Reward-Learning Signals in Major Depression. *Brain* **131**:2084–2093. [3]

Kupfer, D. J., M. B. First, and D. A. Regier. 2002. Introduction. In: A Research Agenda for DSM-V, ed. D. J. Kupfer et al., pp. xv–xxiii. Washington, D.C.: American Psychiatric Association. [8]

Kurth-Nelson, Z., G. Barnes, D. Sejdinovic, R. Dolan, and P. Dayan. 2015. Temporal Structure in Associative Retrieval. *eLife* **4**: [15]

Kurth-Nelson, Z., and A. D. Redish. 2012. Modeling Decision-Making Systems in Addiction. In: Computational Neuroscience of Drug Addiction, ed. B. Gutkin and S. H. Ahmed, pp. 163–188. New York: Springer. [2]

Kuyken, W., E. Watkins, E. Holden, et al. 2010. How Does Mindfulness-Based Cognitive Therapy Work? *Behav. Res. Ther.* **48**:1105–1112. [15]

Kwan, A. C., and Y. Dan. 2012. Dissection of Cortical Microcircuits by Single-Neuron Stimulation *in Vivo*. *Curr. Biol.* **22**:1459–1467. [3]

LaBuda, M. C., I. I. Gottesman, and D. L. Pauls. 1993. Usefulness of Twin Studies for Exploring the Etiology of Childhood and Adolescent Psychiatric Disorders. *Am. J. Med. Genet.* **48**:47–59. [3]

Ladd, C. O., K. V. Thrivikraman, R. L. Huot, and P. M. Plotsky. 2005. Differential Neuroendocrine Responses to Chronic Variable Stress in Adult Long Evans Rats Exposed to Handling-Maternal Separation as Neonates. *Psychoneuroendocrinology* **30**:520–533. [3]

Lambo, M. E., and G. G. Turrigiano. 2013. Synaptic and Intrinsic Homeostatic Mechanisms Cooperate to Increase L2/3 Pyramidal Neuron Excitability During a Late Phase of Critical Period Plasticity. *J. Neurosci.* **33**:8810–8819. [16]

Lamers, F., P. van Oppen, H. C. Comijs, et al. 2011. Comorbidity Patterns of Anxiety and Depressive Disorders in a Large Cohort Study: The Netherlands Study of Depression and Anxiety (NESDA). *J. Clin. Psychiatry* **72**:341–348. [1]

Lane, H. Y., Y. C. Liu, C. L. Huang, et al. 2008. Sarcosine (N-Methylglycine) Treatment for Acute Schizophrenia: A Randomized, Double-Blind Study. *Biol. Psychiatry* **63**:9–12. [16]

Langston, J. W., P. Ballard, J. W. Tetrud, and I. Irwin. 1983. Chronic Parkinsonism in Humans Due to a Product of Meperidine-Analog Synthesis. *Science* **219**:979–980. [2]

Langston, J. W., L. S. Forno, C. S. Rebert, and I. Irwin. 1984. Selective Nigral Toxicity after Systemic Administration of 1-Methyl-4-Phenyl-1,2,5,6-Tetrahydropyrine (MPTP) in the Squirrel Monkey. *Brain Res.* **292**:390–394. [2]

Langston, J. W., and J. Palfreman. 2013. The Case of the Frozen Addicts. Amsterdam: IOS Press. [2]

Lapish, C. C., E. Balaguer-Ballester, J. K. Seamans, J. K. A. G. Phillips, and D. Durstewitz. 2015. Amphetamine Exerts Dose-Dependent Changes in Prefrontal Cortex Attractor Dynamics During Working Memory. *J. Neurosci.* **35**:10172–10187. [5]

Lapish, C. C., S. Kroener, D. Durstewitz, A. Lavin, and J. K. Seamans. 2007. The Ability of the Mesocortical Dopamine System to Operate in Distinct Temporal Modes. *Psychopharmacology (Berl.)* **191**:609–625. [4]

Laplace, P.-S. 1774. Mémoire Sur la Probabilité Des Causes Par Les Évènemens. *Mém. Acad. Roy. Sci.* **6**:621–656. [7]

Laruelle, M., A. Abi-Dargham, R. Gil, L. Kegeles, and R. Innis. 1999. Increased Dopamine Transmission in Schizophrenia: Relationship to Illness Phases. *Biol. Psychiatry* **46**:56–72. [4]

Laruelle, M., A. Abi-Dargham, C. H. van Dyck, et al. 1996. Single Photon Emission Computerized Tomography Imaging of Amphetamine-Induced Dopamine Release in Drug-Free Schizophrenic Subjects. *PNAS* **93**:9235–9240. [4, 16]

Lawrie, S. M., C. Buechel, H. C. Whalley, et al. 2002. Reduced Frontotemporal Functional Connectivity in Schizophrenia Associated with Auditory Hallucinations. *Biol. Psychiatry* **51**:1008–1011. [3]

Lawson, R. P., G. Rees, and K. J. Friston. 2014. An Aberrant Precision Account of Autism. *Front. Hum. Neurosci.* **8**:302. [7]

Lazarus, M. S., K. Krishnan, and Z. J. Huang. 2015. Gad67 Deficiency in Parvalbumin Interneurons Produces Deficits in Inhibitory Transmission and Network Disinhibition in Mouse Prefrontal Cortex. *Cereb. Cortex* **25**:1290–1296. [16]

Lazarus, R. S. 2006. Stress and Emotion: A New Synthesis. New York: Springer. [15]

Ledford, H. 2014. Medical Research: If Depression Were Cancer. *Nature* **515**:182–184. [3]

LeDoux, J. E. 1996. The Emotional Brain. New York: Simon and Schuster. [10]

———. 2000. Emotion Circuits in the Brain. *Annu. Rev. Neurosci.* **23**:155–184. [3]

Lee, J., and S. Park. 2005. Working Memory Impairments in Schizophrenia: A Meta-Analysis. *J. Abnorm. Psychol.* **114**:599–611. [4]

Lee, S. W., S. Shimojo, and J. P. O'Doherty. 2014. Neural Computations Underlying Arbitration between Model-Based and Model-Free Learning. *Neuron* **81**:687–699. [4, 10]

Lee, Y. M., and K.-U. Lee. 2011. Time to Discontinuation among the Three Second-Generation Antidepressants in a Naturalistic Outpatient Setting of Depression. *Psychiatry Clin. Neurosci.* **65**:630–637. [15]

Lempert, K. M., P. W. Glimcher, and E. A. Phelps. 2015. Emotional Arousal and Discount Rate in Intertemporal Choice Are Reference Dependent. *J. Exp. Psychol. Gen.* **144**:366. [12]

Lencz, T., B. Cornblatt, and R. M. Bilder. 2001. Neurodevelopmental Models of Schizophrenia: Pathophysiologic Synthesis and Directions for Intervention Research. *Psychopharmacol. Bull.* **35**:95–125. [16]

Lepage, M., M. Bodnar, and C. R. Bowie. 2014. Neurocognition: Clinical and Functional Outcomes in Schizophrenia. *Can. J. Psychiatry* **59**:5–12. [4]

Lesch, K. P., D. Bengel, A. Heils, et al. 1996. Association of Anxiety-Related Traits with a Polymorphism in the Serotonin Transporter Gene Regulatory Region. *Science* **274**:1527–1531. [3]

Lesch, K. P., S. Selch, T. J. Renner, et al. 2011. Genome-Wide Copy Number Variation Analysis in Attention-Deficit/Hyperactivity Disorder: Association with Neuropeptide Y Gene Dosage in an Extended Pedigree. *Mol. Psychiatry* **16**:491–503. [3]

Lesh, T. A., T. A. Niendam, M. J. Minzenberg, and C. S. Carter. 2011. Cognitive Control Deficits in Schizophrenia: Mechanisms and Meaning. *Neuropsychopharmacology* **36**:316–338. [4]

Letzkus, J. J., S. B. E. Wolff, E. M. M. Meyer, et al. 2011. A Disinhibitory Microcircuit for Associative Fear Learning in the Auditory Cortex. *Nature* **480**:331–335. [3]

Leuchter, A. F., I. A. Cook, L. B. Marangell, et al. 2009. Comparative Effectiveness of Biomarkers and Clinical Indicators for Predicting Outcomes of SSRI Treatment in Major Depressive Disorder: Results of the BRITE-MD Study. *Psychiatry Res.* **169**:124–131. [15]

Levine, S. Z., I. Lurie, R. Kohn, and I. Levav. 2011a. Trajectories of the Course of Schizophrenia: From Progressive Deterioration to Amelioration over Three Decades. *Schizophr. Res.* **126**:184–191. [3]

Levine, S. Z., J. Rabinowitz, H. Ascher-Svanum, D. E. Faries, and A. H. Lawson. 2011b. Extent of Attaining and Maintaining Symptom Remission by Antipsychotic Medication in the Treatment of Chronic Schizophrenia: Evidence from the CATIE Study. *Schizophr. Res.* **133**:42–46. [1]

Levy, D., M. Ronemus, B. Yamrom, et al. 2011. Rare *de Novo* and Transmitted Copy-Number Variation in Autistic Spectrum Disorders. *Neuron* **70**:886–897. [1, 3]

Lewinsohn, P. M., T. E. Joiner, Jr., and P. Rohde. 2001. Evaluation of Cognitive Diathesis-Stress Models in Predicting Major Depressive Disorder in Adolescents. *J. Abnorm. Psychol.* **110**:203–215. [15]

Lewis, D. A. 2014. Inhibitory Neurons in Human Cortical Circuits: Substrate for Cognitive Dysfunction in Schizophrenia. *Curr. Opin. Neurobiol.* **26**:22–26. [3]

Lewis, D. A., R. Y. Cho, C. S. Carter, et al. 2008a. Subunit-Selective Modulation of GABA Type a Receptor Neurotransmission and Cognition in Schizophrenia. *Am. J. Psychiatry* **165**:1585–1593. [4]

Lewis, D. A., A. A. Curley, J. R. Glausier, and D. W. Volk. 2012. Cortical Parvalbumin Interneurons and Cognitive Dysfunction in Schizophrenia. *Trends Neurosci.* **35**:57–67. [3]

Lewis, D. A., T. Hashimoto, and H. M. Morris. 2008b. Cell and Receptor Type-Specific Alterations in Markers of GABA Neurotransmission in the Prefrontal Cortex of Subjects with Schizophrenia. *Neurotox. Res.* **14**:237–248. [16]

Lewis, D. A., and P. Levitt. 2002. Schizophrenia as a Disorder of Neurodevelopment. *Annu. Rev. Neurosci.* **25**:409–432. [3]

Leyton, M., and P. Vezina. 2014. Dopamine Ups and Downs in Vulnerability to Addictions: A Neurodevelopmental Model. *Trends Pharmacol. Sci.* **35**:268–276. [11]

Li, C.-S. R., T. R. Kosten, and R. Sinha. 2005. Sex Differences in Brain Activation During Stress Imagery in Abstinent Cocaine Users: A Functional Magnetic Resonance Imaging Study. *Biol. Psychiatry* **57**:487–494. [3]

Li, C.-Y. T., M.-M. Poo, and Y. Dan. 2009. Burst Spiking of a Single Cortical Neuron Modifies Global Brain State. *Science* **324**:643–646. [3]

Li, L. M., K. Uehara, and T. Hanakawa. 2015. The Contribution of Interindividual Factors to Variability of Response in Transcranial Direct Current Stimulation Studies. *Front. Cell. Neurosci.* **9**:181. [5]

Lichtenstein, P., B. H. Yip, C. Björk, et al. 2009. Common Genetic Determinants of Schizophrenia and Bipolar Disorder in Swedish Families: A Population-Based Study. *Lancet* **373**:234–239. [3]

Lieberman, J. A., D. Perkins, A. Belger, et al. 2001. The Early Stages of Schizophrenia: Speculations on Pathogenesis, Pathophysiology, and Therapeutic Approaches. *Biol. Psychiatry* **50**:884–897. [3, 16]

Lieberman, J. A., T. S. Stroup, J. P. McEvoy, et al. 2005. Effectiveness of Antipsychotic Drugs in Patients with Chronic Schizophrenia. *New Engl. J. Med.* **353**:1209–1223. [1]

Liebowitz, M. R., F. M. Quitkin, J. W. Stewart, et al. 1988. Antidepressant Specificity in Atypical Depression. *Arch. Gen. Psychiatry* **45**:129–137. [1]

Lieder, F., N. Goodman, and Q. J. M. Huys. 2013. Learned Helplessness and Generalization. Cognitive Science Conference. [15]

Liljeholm, M., and J. P. O'Doherty. 2012. Contributions of the Striatum to Learning, Motivation, and Performance: An Associative Account. *Trends Cogn. Sci.* **16**:467–475. [6]

Lim, S. L., J. P. O'Doherty, and A. Rangel. 2011. The Decision Value Computations in the vmPFC and Striatum Use a Relative Value Code That Is Guided by Visual Attention. *J. Neurosci.* **31**:13214–13223. [6]

Lin, A. C., A. M. Bygrave, A. de Calignon, T. Lee, and G. Miesenbock. 2014. Sparse, Decorrelated Odor Coding in the Mushroom Body Enhances Learned Odor Discrimination. *Nat. Neurosci.* **17**:559–568. [16]

Lindamer, L. A., J. B. Lohr, M. J. Harris, and D. V. Jeste. 1997. Gender, Estrogen, and Schizophrenia. *Psychopharmacol. Bull.* **33**:221–228. [16]

Lionel, A. C., J. Crosbie, N. Barbosa, et al. 2011. Rare Copy Number Variation Discovery and Cross-Disorder Comparisons Identify Risk Genes for ADHD. *Sci. Transl. Med.* **3**:95ra75–95ra75. [3]

Lisman, J. 2012. Excitation, Inhibition, Local Oscillations, or Large-Scale Loops: What Causes the Symptoms of Schizophrenia? *Curr. Opin. Neurobiol.* **22**:537–544. [16]

Lisman, J. E., J. T. Coyle, R. W. Green, et al. 2008. Circuit-Based Framework for Understanding Neurotransmitter and Risk Gene Interactions in Schizophrenia. *Trends Neurosci.* **31**:234–242. [12]

Liu, R. T., and L. B. Alloy. 2010. Stress Generation in Depression: A Systematic Review of the Empirical Literature and Recommendations for Future Study. *Clin. Psychol. Rev.* **30**:582–593. [15]

Llinas, R. R. 2001. I of the Vortex. Cambridge, MA: MIT Press. [2]

Lo, C. C., and X. J. Wang. 2006. Cortico-Basal Ganglia Circuit Mechanism for a Decision Threshold in 630 Reaction Time Tasks. *Nat. Neurosci.* **9**:956–963. [6]

Lodge, D. J., and A. A. Grace. 2006. The Hippocampus Modulates Dopamine Neuron Responsivity by Regulating the Intensity of Phasic Neuron Activation. *Neuropsychopharmacology* **31**:1356–1361. [4]

———. 2007. Aberrant Hippocampal Activity Underlies the Dopamine Dysregulation in an Animal Model of Schizophrenia. *J. Neurosci.* **27**:11424–11430. [4]

———. 2008. Hippocampal Dysfunction and Disruption of Dopamine System Regulation in an Animal Model of Schizophrenia. *Neurotox. Res.* **14**:97–104. [3]

———. 2011a. Developmental Pathology, Dopamine, Stress and Schizophrenia. *Int. J. Dev. Neurosci.* **29**:207–213. [4]

———. 2011b. Hippocampal Dysregulation of Dopamine System Function and the Pathophysiology of Schizophrenia. *Trends Pharmacol. Sci.* **32**:507–513. [16]

Logothetis, N. K., O. Eschenko, Y. Murayama, et al. 2012. Hippocampal-Cortical Interaction During Periods of Subcortical Silence. *Nature* **491**:547–553. [3]

Lorenz, E. N. 1963. Deterministic Nonperiodic Flow. *J. Atmos. Sci.* **20**:130–141. [11]

Loza, W., and G. K. Dhaliwal. 2005. Predicting Violence among Forensic-Correctional Populations: The Past 2 Decades of Advancements and Future Endeavors. *J. Interpers. Violence* **20**:188–194. [5]

Lui, S., L. Yao, Y. Xiao, et al. 2015. Resting-State Brain Function in Schizophrenia and Psychotic Bipolar Probands and Their First-Degree Relatives. *Psychol. Med.* **45**:97–108. [16]

Luksys, G., W. Gerstner, and C. Sandi. 2009. Stress, Genotype and Norepinephrine in the Prediction of Mouse Behavior Using Reinforcement Learning. *Nat. Neurosci.* **12**:1180–1186. [3]

Lupien, S. J., S. Parent, A. C. Evans, et al. 2011. Larger Amygdala but No Change in Hippocampal Volume in 10-Year-Old Children Exposed to Maternal Depressive Symptomatology since Birth. *PNAS* **108**:14324–14329. [5]

Luscher, C., R. A. Nicoll, R. C. Malenka, and D. Muller. 2000. Synaptic Plasticity and Dynamic Modulation of the Postsynaptic Membrane. *Nat. Neurosci.* **3**:545–550. [12]

Lusser, R. 1958. Reliability through Safety Margins. Cameron Station Alexandria, Virginia: Defense Documentation Center for Scientific and Technical Information, http://www.dtic.mil/dtic/tr/fulltext/u2/212476.pdf (accessed December 7, 2015). [9]

Lyon, G. J., A. Abi-Dargham, H. Moore, et al. 2011. Presynaptic Regulation of Dopamine Transmission in Schizophrenia. *Schizophr. Bull.* **37**:108–117. [4]

Ma, N., and A. J. Yu. 2015. Statistical Learning and Adaptive Decision-Making Underlie Human Response Time Variability in Inhibitory Control. *Front. Psychol.* **6**:1046. [3]

Ma, S. H., and J. D. Teasdale. 2004. Mindfulness-Based Cognitive Therapy for Depression: Replication and Exploration of Differential Relapse Prevention Effects. *J. Consult. Clin. Psychol.* **72**:31–40. [15]

Ma, W. J., J. M. Beck, P. E. Latham, and A. Pouget. 2006. Bayesian Inference with Probabilistic Population Codes. *Nat. Neurosci.* **9**:1432–1438. [5]

MacDonald, A. W., III. 2013. What Kind of a Thing Is Schizophrenia? Specific Causation and General Failure Modes. In: Schizophrenia: Evolution and Synthesis, ed. S. M. Silverstein et al., pp. 25–48. Cambridge, MA: MIT Press. [9]

MacDonald, M. L., Y. Ding, J. Newman, et al. 2015. Altered Glutamate Protein Co-Expression Network Topology Linked to Spine Loss in the Auditory Cortex of Schizophrenia. *Biol. Psychiatry* **77**:959–968. [16]

Mack, A. H., L. Forman, R. Brown, and A. Frances. 1994. A Brief History of Psychiatric Classification. From the Ancients to DSM-IV. *Psychiatr. Clin. North Am.* **17**:515–523. [8]

MacKay, D. J. 1995. Free-Energy Minimisation Algorithm for Decoding and Crypto-analysis. *Electronics Lett.* **31**:445–447. [11]

Maia, T. V. 2015. Introduction to the Series on Computational Psychiatry. *Clin. Psychol. Sci.* **3**:374–377. [3]

Maia, T. V., and M. Cano-Colino. 2015. The Role of Serotonin in Orbitofrontal Function and Obsessive-Compulsive Disorder. *Clin. Psychol. Sci.* **3**:460–482. [6, 15]

Maia, T. V., and M. J. Frank. 2011. From Reinforcement Learning Models to Psychiatric and Neurological Disorders. *Nat. Neurosci.* **14**:154–162. [2, 6, 12, 15]

Maia, T. V., and J. L. McClelland. 2012. A Neurocomputational Approach to Obsessive-Compulsive Disorder. *Trends Cogn. Sci.* **16**:14–15. [2]

Maier, R., G. Moser, G. B. Chen, et al. 2015. Joint Analysis of Psychiatric Disorders Increases Accuracy of Risk Prediction for Schizophrenia, Bipolar Disorder, and Major Depressive Disorder. *Am. J. Hum. Genet.* **96**:283–294. [16]

Maier, S. F., and L. R. Watkins. 2005. Stressor Controllability and Learned Helplessness: The Roles of the Dorsal Raphe Nucleus, Serotonin, and Corticotropin-Releasing Factor. *Neurosci. Biobehav. Rev.* **29**:829–841. [15]

Malaspina, D., S. Harlap, S. Fennig, et al. 2001. Advancing Paternal Age and the Risk of Schizophrenia. *Arch. Gen. Psychiatry* **58**:361–367. [4]

Maldonado-Aviles, J. G., A. A. Curley, T. Hashimoto, et al. 2009. Altered Markers of Tonic Inhibition in the Dorsolateral Prefrontal Cortex of Subjects with Schizophrenia. *Am. J. Psychiatry* **166**:450–459. [16]

Malhotra, D., S. McCarthy, J. J. Michaelson, et al. 2011. High Frequencies of *de Novo* CNVs in Bipolar Disorder and Schizophrenia. *Neuron* **72**:951–963. [16]

Mann, J. J., D. Currier, B. Stanley, et al. 2006. Can Biological Tests Assist Prediction of Suicide in Mood Disorders? *Int. J. Neuropsychopharmacology* **9**:465–474. [3]

Manoach, D. S., A. K. Lee, M. S. Hamalainen, et al. 2013. Anomalous Use of Context During Task Preparation in Schizophrenia: A Magnetoencephalography Study. *Biol. Psychiatry* **73**:967–975. [4]

Manoach, D. S., K. A. Lindgren, M. V. Cherkasova, et al. 2002. Schizophrenic Subjects Show Deficient Inhibition but Intact Task Switching on Saccadic Tasks. *Biol. Psychiatry* **51**:816–826. [4]

Mantione, M., D. Nieman, M. Figee, et al. 2015. Cognitive Effects of Deep Brain Stimulation in Patients with Obsessive-Compulsive Disorder. *J. Psychiatry Neurosci.* **40**:140210. [10]

Markham, J. A., and J. I. Koenig. 2011. Prenatal Stress: Role in Psychotic and Depressive Diseases. *Psychopharmacology* **214**:89–106. [3]

Markon, K. E., M. Chmielewski, and C. J. Miller. 2011. The Reliability and Validity of Discrete and Continuous Measures of Psychopathology: A Quantitative Review. *Psychol. Bull.* **137**:856–879. [9]

Markon, K. E., R. F. Krueger, and D. Watson. 2005. Delineating the Structure of Normal and Abnormal Personality: An Integrative Hierarchical Approach. *J. Pers. Soc. Psychol.* **88**:139–157. [9]

Markram, H. 2012. The Human Brain Project. *Sci. Am.* **306**:50–55. [5]

Maroco, J., D. Silva, A. Rodrigues, et al. 2011. Data Mining Methods in the Prediction of Dementia: A Real-Data Comparison of the Accuracy, Sensitivity and Specificity of Linear Discriminant Analysis, Logistic Regression, Neural Networks, Support Vector Machines, Classification Trees and Random Forests. *BMC Res. Notes* **4**:299. [14]

Marr, D. 1982. Vision. A Computational Investigation into the Human Representation and Processing of Visual Information. Cambridge, MA: MIT Press. [5, 12]

Marreiros, A. C., S. J. Kiebel, J. Daunizeau, L. M. Harrison, and K. J. Friston. 2009. Population Dynamics under the Laplace Assumption. *NeuroImage* **44**:701–714. [12]

Marsman, A., M. P. van den Heuvel, D. W. Klomp, et al. 2013. Glutamate in Schizophrenia: A Focused Review and Meta-Analysis of ^{1}H-MRS Studies. *Schizophr. Bull.* **39**:120–129. [4, 16]

Mataix-Cols, D. 2014. Clinical Practice. Hoarding Disorder. *New Engl. J. Med.* **370**:2023–2030. [10]

Mataix-Cols, D., S. Wooderson, N. Lawrence, et al. 2004. Distinct Neural Correlates of Washing, Checking, and Hoarding Symptom Dimensions in Obsessive-Compulsive Disorder. *Arch. Gen. Psychiatry* **61**:564–576. [10]

Mathys, C., J. Daunizeau, K. J. Friston, and K. E. Stephan. 2011. A Bayesian Foundation for Individual Learning under Uncertainty. *Front. Hum. Neurosci.* **5**:39. [7]

Mattheisen, M., J. F. Samuels, Y. Wang, et al. 2015. Genome-Wide Association Study in Obsessive-Compulsive Disorder: Results from the OCGAS. *Mol. Psychiatry* **20**:337–344. [10]

Mayberg, H. S. 2009. Targeted Electrode-Based Modulation of Neural Circuits for Depression. *J. Clin. Invest.* **119**:717–725. [15]

Mayberg, H. S., M. Liotti, S. K. Brannan, et al. 1999. Reciprocal Limbic-Cortical Function and Negative Mood: Converging PET Findings in Depression and Normal Sadness. *Am. J. Psychiatry* **156**:675–682. [1]

Mayberg, H. S., A. M. Lozano, V. Voon, et al. 2005. Deep Brain Stimulation for Treatment-Resistant Depression. *Neuron* **45**:651–660. [2, 3]

Mayer-Schönberger, V., and K. Cukier. 2013. Big Data: A Revolution That Will Transform How We Live, Work, and Think. New York: Houghton Mifflin Harcourt. [2]

Mayr, E., and W. B. Provine, eds. 1980/1998. The Evolutionary Synthesis: Perspectives on the Unification of Biology. Cambridge, MA: Harvard Univ. Press. [13]

McClelland, J. L., M. M. Botvinick, D. C. Noelle, et al. 2010. Letting Structure Emerge: Connectionist and Dynamical Systems Approaches to Cognition. *Trends Cogn. Sci.* **14**:348–356. [5, 12]

McClelland, J. L., B. L. McNaughton, and R. C. O'Reilly. 1995. Why There Are Complementary Learning Systems in the Hippocampus and Neocortex: Insights from the Successes and Failures of Connectionist Models of Learning and Memory. *Psychol. Rev.* **102**:419–457. [6]

McClelland, J. L., and T. T. Rogers. 2003. The Parallel Distributed Processing Approach to Semantic Cognition. *Nat. Rev. Neurosci.* **4**:310–322. [5, 12, 17]

McClelland, J. L., and D. E. Rumelhart. 1986. Parallel Distributed Processing Volume 1. Explorations in the Microstructure of Cognition: Foundations. Cambridge, MA: MIT Press. [9]

McClintock, S. M., J. Choi, Z. D. Deng, et al. 2014. Multifactorial Determinants of the Neurocognitive Effects of Electroconvulsive Therapy. *J. ECT* **30**:165–176. [5]

McClure, S. M., G. S. Berns, and P. R. Montague. 2003a. Temporal Prediction Errors in a Passive Learning Task Activate Human Striatum. *Neuron* **38**:339–346. [13]

McClure, S. M., N. D. Daw, and P. R. Montague. 2003b. A Computational Substrate for Incentive Salience. *Trends Neurosci.* **26**:423–428. [13]

McClure, S. M., D. I. Laibson, G. Loewenstein, and J. D. Cohen. 2004. Separate Neural Systems Value Immediate and Delayed Monetary Rewards. *Science* **306**:503–507. [6]

McCrone, P. 2008. Paying the Price: The Cost of Mental Health Care in England to 2026. London: King's Fund Publishing. [3]

McCulloch, W. S., and W. Pitts. 1943. A Logical Calculus of the Ideas Immanent in Nervous Activity. *Bull. Math. Biophysics* **5**:115–133. [12]

McDonald, W. M., W. V. McCall, and C. Epstein. 2002. Electroconvulsive Therapy: Sixty Years of Progress and a Comparison with Transcranial Magnetic Stimulation an Vagal Nerve Stimulation. In: Neuropsychopharmacology: The Fifth Generation of Progress, ed. K. L. Davis et al., pp. 1079–1108. Philadelphia: Lippincott Williams and Wilkins. [1]

McEwen, B. S. 2003. Mood Disorders and Allostatic Load. *Biol. Psychiatry* **54**: [5]

———. 2006. Protective and Damaging Effects of Stress Mediators: Central Role of the Brain. *Dialogues Clin. Neurosci.* **8**:367–381. [3]

———. 2015. Biomarkers for Assessing Population and Individual Health and Disease Related to Stress and Adaptation. *Metabolism* **64**:S2–S10. [3]

McEwen, B. S., and E. Stellar. 1993. Stress and the Individual. Mechanisms Leading to Disease. *Arch. Intern. Med.* **153**:2093–2101. [16]

McGlashan, T. H. 1988. A Selective Review of Recent North American Long-Term Followup Studies of Schizophrenia. *Schizophr. Bull.* **14**:515–539. [3]

McGlinchey, J. B., M. Zimmerman, D. Young, and I. Chelminski. 2006. Diagnosing Major Depressive Disorder VIII: Are Some Symptoms Better Than Others? *J. Nerv. Ment. Dis.* **194**:785–790. [15]

McGrath, C. L., M. E. Kelley, P. E. Holtzheimer, et al. 2013. Toward a Neuroimaging Treatment Selection Biomarker for Major Depressive Disorder. *JAMA Psychiatry* **70**:821–829. [15]

McGrath, P. J., J. W. Stewart, F. M. Quitkin, et al. 2006. Predictors of Relapse in a Prospective Study of Fluoxetine Treatment of Major Depression. *Am. J. Psychiatry* **163**:1542–1548. [15]

McGuffin, P. 1979. Is Schizophrenia an Hla-Associated Disease? *Psychol. Med.* **9**:721–728. [3]

McHugh, P. R. 2005. Striving for Coherence: Psychiatry's Efforts over Classification *JAMA* **293**:2526–2528. [8]

McHugh, P. R., and P. R. Slavney. 1998. The Perspectives of Psychiatry. Baltimore: Johns Hopkins Univ. Press. [2]

McNally, R. J. 2007. Mechanisms of Exposure Therapy: How Neuroscience Can Improve Psychological Treatments for Anxiety Disorders. *Clin. Psychol. Rev.* **27**: 750–759. [5]

Meiran, N., J. Levine, N. Meiran, and A. Henik. 2000. Task Set Switching in Schizophrenia. *Neuropsychology* **14**:471–482. [4]

Mellsop, G., D. Menkes, and S. El-Badri. 2007. Releasing Psychiatry from the Constraints of Categorical Diagnosis. *Australas. Psychiatry* **15**:3–5. [8]

Meltzer, H. Y. 1992. Treatment of the Neuroleptic-Nonresponsive Schizophrenic Patient. *Schizophr. Bull.* **18**:515–542. [16]

Menzies, L., S. R. Chamberlain, A. R. Laird, et al. 2008. Integrating Evidence from Neuroimaging and Neuropsychological Studies of Obsessive-Compulsive Disorder: The Orbitofronto-Striatal Model Revisited. *Neurosci. Biobehav. Rev.* **32**:525–549. [10]

Menzies, P. 2012. The Causal Structure of Mechanisms. *Stud. Hist. Philos. Biol. Biomed. Sci.* **43**:796–805. [14]

Mesholam-Gately, R. I., A. J. Giuliano, K. P. Goff, S. V. Faraone, and L. J. Seidman. 2009. Neurocognition in First-Episode Schizophrenia: A Meta-Analytic Review. *Neuropsychology* **23**:315–336. [4]

Meyer-Lindenberg, A., R. K. Olsen, P. D. Kohn, et al. 2005. Regionally Specific Disturbance of Dorsolateral Prefrontal-Hippocampal Functional Connectivity in Schizophrenia. *Arch. Gen. Psychiatry* **62**:379–386. [3]

Meyer-Lindenberg, A., and D. R. Weinberger. 2006. Intermediate Phenotypes and Genetic Mechanisms of Psychiatric Disorders. *Nat. Rev. Neurosci.* **7**:818–827. [5]

Michino, M., T. Beuming, P. Donthamsetti, et al. 2015. What Can Crystal Structures of Aminergic Receptors Tell Us About Designing Subtype-Selective Ligands? *Pharmacol. Rev.* **67**:198–213. [5]

Milad, M. R., S. C. Furtak, J. L. Greenberg, et al. 2013. Deficits in Conditioned Fear Extinction in Obsessive-Compulsive Disorder and Neurobiological Changes in the Fear Circuit. *JAMA Psychiatry* **70**:608–618. [10]

Millan, M. J., Y. Agid, M. Brüne, et al. 2012. Cognitive Dysfunction in Psychiatric Disorders: Characteristics, Causes and the Quest for Improved Therapy. *Nat. Rev. Drug Discov.* **11**:141–168. [12]

Millar, J. K., J. C. Wilson-Annan, S. Anderson, et al. 2000. Disruption of Two Novel Genes by a Translocation Co-Segregating with Schizophrenia. *Hum. Mol. Genet.* **9**:1415–1423. [3]

Miller, B., E. Messias, J. Miettunen, et al. 2011. Meta-Analysis of Paternal Age and Schizophrenia Risk in Male versus Female Offspring. *Schizophr. Bull.* **37**:1039–1047. [4]

Mineka, S., D. W. Watson, and L. A. Clark. 1998. Psychopathology: Comorbidity of Anxiety and Unipolar Mood Disorders. *Annu. Rev. Psychol.* **49**:377–412. [8]

Miner, M. A. 1945. Cumulative Damage in Fatigue. *J. Appl. Mech.* **12**:159–164. [9]

Minzenberg, M. J., A. J. Firl, J. H. Yoon, et al. 2010. Gamma Oscillatory Power Is Impaired During Cognitive Control Independent of Medication Status in First-Episode Schizophrenia. *Neuropsychopharmacology* **35**:2590–2599. [4]

Minzenberg, M. J., A. R. Laird, S. Thelen, C. S. Carter, and D. C. Glahn. 2009. Meta-Analysis of 41 Functional Neuroimaging Studies of Executive Function in Schizophrenia. *Arch. Gen. Psychiatry* **66**:811–822. [4]

Misiak, B., D. Frydecka, M. Zawadzki, M. Krefft, and A. Kiejna. 2014. Refining and Integrating Schizophrenia Pathophysiology - Relevance of the Allostatic Load Concept. *Neurosci. Biobehav. Rev.* **45**:183–201. [11]

Mitchell, A. J., J. B. McGlinchey, D. Young, I. Chelminski, and M. Zimmerman. 2009. Accuracy of Specific Symptoms in the Diagnosis of Major Depressive Disorder in Psychiatric out-Patients: Data from the MIDAS Project. *Psychol. Med.* **39**:1107–1116. [15]

Mitchell, K., Z. J. Huang, and B. Moghaddam. 2011. Following the Genes: A Framework for Animal Modeling of Psychiatric Disorders. *BMC Biol.* **9**:76. [3]

Miyamoto, S., G. E. Duncan, D. C. Goff, and J. A. Lieberman. 2002. Therapeutics of Schizophrenia. In: Neuropsychopharmacology: The Fifth Generation of Progress, ed. K. L. Davis et al., pp. 775–807. Philadelphia: Lippincott Williams and Wilkins. [1]

Moghaddam, B., and B. W. Adams. 1998. Reversal of Phencyclidine Effects by a Group II Metabotropic Glutamate Receptor Agonist in Rats. *Science* **281**:1349–1352. [16]

Moghaddam, B., B. W. Adams, A. Verma, and D. Daly. 1997. Activation of Glutamatergic Neurotransmission by Ketamine: A Novel Step in the Pathway from NMDA Receptor Blockade to Dopaminergic and Cognitive Disruptions Associated with the Prefrontal Cortex. *J. Neurosci.* **17**:2921–2927. [16]

Monroe, S. M., E. J. Bromet, M. M. Connell, and S. C. Steiner. 1986. Social Support, Life Events, and Depressive Symptoms: A 1-Year Prospective Study. *J. Consult. Clin. Psychol.* **54**:424–431. [15]

Monroe, S. M., and K. L. Harkness. 2011. Recurrence in Major Depression: A Conceptual Analysis. *Psychol. Rev.* **118**:655–674. [15]

Montague, P. R., P. Dayan, S. J. Nowlan, A. Pouget, and T. J. Sejnowski. 1993. Using Aperiodic Reinforcement for Directed Self-Organization During Development. In: Neural Information Processing Systems 5, ed. S. J. Hanson et al., pp. 969–976. San Mateo: Morgan Kaufmann. [13]

Montague, P. R., P. Dayan, C. Person, and T. J. Sejnowski. 1995. Bee Foraging in Uncertain Environments Using Predictive Hebbian Learning. *Nature* **377**:725–728. [5, 13]

Montague, P. R., P. Dayan, and T. J. Sejnowski. 1996. A Framework for Mesencephalic Dopamine Systems Based on Predictive Hebbian Learning. *J. Neurosci.* **16**:1936–1947. [5, 6, 12, 13, 15]

Montague, P. R., R. J. Dolan, K. J. Friston, and P. Dayan. 2012. Computational Psychiatry. *Trends Cogn. Sci.* **16**:72–80. [2, 4, 6, 11, 12, 15, 16]

Montague, P. R., S. E. Hyman, and J. D. Cohen. 2004. Computational Roles for Dopamine in Behavioural Control. *Nature* **431**:760–767. [13]

Monterosso, J. R., A. R. Aron, X. Cordova, J. Xu, and E. D. London. 2005. Deficits in Response Inhibition Associated with Chronic Methamphetamine Abuse. *Drug Alcohol Depend.* **79**:273–277. [14]

Moons, K. G., A. P. Kengne, D. E. Grobbee, et al. 2012a. Risk Prediction Models: II. External Validation, Model Updating, and Impact Assessment. *Heart* **98**:691–698. [14]

Moons, K. G., A. P. Kengne, M. Woodward, et al. 2012b. Risk Prediction Models: I. Development, Internal Validation, and Assessing the Incremental Value of a New (Bio)Marker. *Heart* **98**:683–690. [14]

Moore, T. H. M., S. Zammit, A. Lingford-Hughes, et al. 2007. Cannabis Use and Risk of Psychotic or Affective Mental Health Outcomes: A Systematic Review. *Lancet* **370**:319–328. [3]

Moradi, E., A. Pepe, C. Gaser, H. Huttunen, and J. Tohka. 2015. Machine Learning Framework for Early MRI-Based Alzheimer's Conversion Prediction in MCI Subjects. *NeuroImage* **104**:398–412. [14]

Moran, R. J., P. Campo, M. Symmonds, et al. 2013. Free Energy, Precision and Learning: The Role of Cholinergic Neuromodulation. *J. Neurosci.* **33**:8227–8236. [12]

Moran, R. J., M. W. Jones, A. J. Blockeel, et al. 2015. Losing Control under Ketamine: Suppressed Cortico-Hippocampal Drive Following Acute Ketamine in Rats. *Neuropsychopharmacology* **40**:268–277. [12]

Moran, R. J., K. E. Stephan, R. J. Dolan, and K. J. Friston. 2011. Consistent Spectral Predictors for Dynamic Causal Models of Steady-State Responses. *NeuroImage* **55**:1694–1708. [12]

Morgan, C. A., S. Wang, S. M. Southwick, et al. 2000. Plasma Neuropeptide-Y Concentrations in Humans Exposed to Military Survival Training. *Biol. Psychiatry* **47**:902–909. [3]

Mori, T., T. Ohnishi, R. Hashimoto, et al. 2007. Progressive Changes of White Matter Integrity in Schizophrenia Revealed by Diffusion Tensor Imaging. *Psychiatry Res.* **154**:133–145. [16]

Morris, C., and H. Lecar. 1981. Voltage Oscillations in the Barnacle Giant Muscle Fiber. *Biophys. J.* **35**:193. [12]

Morris, H. M., T. Hashimoto, and D. A. Lewis. 2008a. Alterations in Somatostatin mRNA Expression in the Dorsolateral Prefrontal Cortex of Subjects with Schizophrenia or Schizoaffective Disorder. *Cereb. Cortex* **18**:1575–1587. [16]

Morris, R. G. M. 1981. Spatial Localization Does Not Require the Presence of Local Cues. *Learn. Motiv.* **12**:239–260. [2]

Morris, R. W., A. Vercammen, R. Lenroot, et al. 2012. Disambiguating Ventral Striatum fMRI-Related BOLD Signal During Reward Prediction in Schizophrenia. *Mol. Psychiatry* **17**:235, 280–239. [4, 12]

Morris, S. E., and B. N. Cuthbert. 2012. Research Domain Criteria: Cognitive Systems, Neural Circuits, and Dimensions of Behavior. *Dialogues Clin. Neurosci.* **14**:29–37. [4, 10]

Morris, S. E., E. A. Heerey, J. M. Gold, and C. B. Holroyd. 2008b. Learning-Related Changes in Brain Activity Following Errors and Performance Feedback in Schizophrenia. *Schizophr. Res.* **99**:274–285. [4]

Morrison, A. P., S. L. K. Stewart, P. French, et al. 2011. Early Detection and Intervention Evaluation for People at High-Risk of Psychosis-2 (EDIE-2): Trial Rationale, Design and Baseline Characteristics. *Early Interv. Psychiatry* **5**:24–32. [3]

Mosteller, F. 1974. Robert R. Bush, Early Career. *J. Math. Psychol.* **11**:163–178. [13]

Moustafa, A. A., and M. A. Gluck. 2011. Computational Cognitive Models of Prefrontal-Striatal-Hippocampal Interactions in Parkinson's Disease and Schizophrenia. *Neural Netw.* **24**:575–591. [2]

Mowrer, O. H. 1947. On the Dual Nature of Learning: A Reinterpretation of Conditioning and Problem-Solving. *Harv. Educ. Rev.* **17**:102–150. [15]

Mukai, J., M. Tamura, K. Fénelon, et al. 2015. Molecular Substrates of Altered Axonal Growth and Brain Connectivity in a Mouse Model of Schizophrenia. *Neuron* **86**:680–695. [5]

Mukerjee, S. 2010. The Emperor of All Maladies: A Biography of Cancer. New York: Scribner. [16]

Mulert, C., G. Juckel, M. Brunnmeier, et al. 2007. Prediction of Treatment Response in Major Depression: Integration of Concepts. *J. Affect. Disord.* **98**:215–225. [15]

Murphy, D. L., P. R. Moya, M. A. Fox, et al. 2013. Anxiety and Affective Disorder Comorbidity Related to Serotonin and Other Neurotransmitter Systems: Obsessive-Compulsive Disorder as an Example of Overlapping Clinical and Genetic Heterogeneity. *Phil. Trans. R. Soc. B* **368**:20120435. [10]

Murray, G. K., P. R. Corlett, L. Clark, et al. 2008. Substantia Nigra/Ventral Tegmental Reward Prediction Error Disruption in Psychosis. *Mol. Psychiatry* **13**:267–276. [4]

Murray, J. D., A. Anticevic, M. Gancsos, et al. 2014. Linking Microcircuit Dysfunction to Cognitive Impairment: Effects of Disinhibition Associated with Schizophrenia in a Cortical Working Memory Model. *Cereb. Cortex* **24**:859–872. [16]

Must, A., Z. Szabo, N. Bodi, et al. 2006. Sensitivity to Reward and Punishment and the Prefrontal Cortex in Major Depression. *J. Affect. Disord.* **90**:209–215. [14]

Muzerelle, A., S. Scotto-Lomassese, J. F. Bernard, M. Soiza-Reilly, and P. Gaspar. 2016. Conditional Anterograde Tracing Reveals Distinct Targeting of Individual Serotonin Cell Groups (B5–B9) to the Forebrain and Brainstem. *Brain Struct. Funct.* **221**:535–561. [11]

Nadel, L., and W. J. Jacobs. 1996. The Role of the Hippocampus in PTSD, Panic, and Phobia. In: The Hippocampus: Functions and Clinical Relevance, ed. N. Kato, pp. 455–463. Amsterdam: Elsevier. [2, 10]

Nadel, L., and M. Moscovitch. 1997. Memory Consolidation, Retrograde Amnesia and the Hippocampal Complex. *Curr. Opin. Neurobiol.* **7**:217–227. [10]

Narrow, W. E., D. E. Clarke, S. J. Kuramoto, et al. 2013. DSM-5 Field Trials in the United States and Canada, Part III: Development and Reliability Testing of a Cross-Cutting Symptom Assessment for DSM-5. *Am. J. Psychiatry* **170**:71–82. [7]

National Institute of Mental Health. 2008. The National Institute of Mental Health Strategic Plan. NIH Publication 8-6368. Bethesda, MD: National Institute of Mental Health [8].

Nelson, M. D., A. J. Saykin, L. A. Flashman, and H. J. Riordan. 1998. Hippocampal Volume Reduction in Schizophrenia as Assessed by Magnetic Resonance Imaging: A Meta-Analytic Study. *Arch. Gen. Psychiatry* **55**:433–440. [3]

Nelson, S. B., and V. Valakh. 2015. Excitatory/Inhibitory Balance and Circuit Homeostasis in Autism Spectrum Disorders. *Neuron* **87**:684–698. [3]

Neves-Pereira, M., J. K. Cheung, A. Pasdar, et al. 2005. BDNF Gene Is a Risk Factor for Schizophrenia in a Scottish Population. *Mol. Psychiatry* **10**:208–212. [3]

Neymotin, S. A., M. T. Lazarewicz, M. Sherif, et al. 2011. Ketamine Disrupts Theta Modulation of Gamma in a Computer Model of Hippocampus. *J. Neurosci.* **31**:11733–11743. [9]

Nicodemus, K. K., A. J. Law, E. Radulescu, et al. 2010. Biological Validation of Increased Schizophrenia Risk with NRG1, ERBB4, and AKT1 Epistasis via Functional Neuroimaging in Healthy Controls. *Arch. Gen. Psychiatry* **67**:991–1001. [3]

Nielsen, M. O., E. Rostrup, S. Wulff, et al. 2012a. Improvement of Brain Reward Abnormalities by Antipsychotic Monotherapy in Schizophrenia. *Arch. Gen. Psychiatry* **69**:1195–1204. [4]

Nielsen, M. O., E. Rostrup, S. Wulff, et al. 2012b. Alterations of the Brain Reward System in Antipsychotic Naive Schizophrenia Patients. *Biol. Psychiatry* **71**:898–905. [4]

Niv, Y., N. D. Daw, D. Joel, and P. Dayan. 2007. Tonic Dopamine: Opportunity Costs and the Control of Response Vigor. *Psychopharmacology* **191**:507–520. [4, 6, 15]

Niv, Y., and P. R. Montague. 2008. Theoretical and Empirical Studies of Learning. In: Neuroeconomics: Decision Making and the Brain, ed. P. W. Glimcher et al., pp. 329–249. New York: Academic Press. [13]

Niv, Y., and G. Schoenbaum. 2008. Dialogs on Prediction Errors. *Trends Cogn. Sci.* **12**:265–272. [13]

Nock, M. K., and M. R. Banaji. 2007. Prediction of Suicide Ideation and Attempts among Adolescents Using a Brief Performance-Based Test. *J. Consult. Clin. Psychol.* **75**:707–715. [3]

Nock, M. K., I. Hwang, N. Sampson, et al. 2009. Cross-National Analysis of the Associations among Mental Disorders and Suicidal Behavior: Findings from the WHO World Mental Health Surveys. *PLoS Med.* **6**:e1000123. [3]

Nock, M. K., J. M. Park, C. T. Finn, et al. 2010. Measuring the Suicidal Mind: Implicit Cognition Predicts Suicidal Behavior. *Psychol. Sci.* **21**:511–517. [3]

Nolen-Hoeksema, S. 1991. Responses to Depression and Their Effects on the Duration of Depressive Episodes. *J. Abnorm. Psychol.* **100**:569–582. [15]

Nolen-Hoeksema, S., J. Larson, and C. Grayson. 1999. Explaining the Gender Difference in Depressive Symptoms. *J. Pers. Soc. Psychol.* **77**:1061–1072. [15]

Noorani, I., and R. H. S. Carpenter. 2012. Antisaccades as Decisions: LATER Model Predicts Latency Distributions and Error Responses. *Eur. J. Neurosci.* **37**:330–338. [6]

Nuechterlein, K. H., K. L. Subotnik, M. F. Green, et al. 2011. Neurocognitive Predictors of Work Outcome in Recent-Onset Schizophrenia. *Schizophr. Bull.* **37 Suppl 2**: S33–40. [4]

Nugent, T. F., 3rd, D. H. Herman, A. Ordonez, et al. 2007. Dynamic Mapping of Hippocampal Development in Childhood Onset Schizophrenia. *Schizophr. Res.* **90**: 62–70. [16]

Oades, R. D. 1997. Stimulus Dimension Shifts in Patients with Schizophrenia, with and without Paranoid Hallucinatory Symptoms, or Obsessive Compulsive Disorder: Strategies, Blocking and Monoamine Status. *Behav. Brain Res.* **88**:115–131. [4]

Obeso, J., and J. Guridi. 2001. Deep-Brain Stimulation of the Subthalamic Nucleus or the Pars Interna of the Globus Pallidus in Parkinson's Disease. *New Engl. J. Med.* **345**:956–962. [2]

Odeh, M. S., R. A. Zeiss, and M. T. Huss. 2006. Cues They Use: Clinicians' Endorsement of Risk Cues in Predictions of Dangerousness. *Behav. Sci. Law* **24**:147–156. [5]

Odgers, C. L., M. M. Moretti, and N. D. Reppucci. 2005. Examining the Science and Practice of Violence Risk Assessment with Female Adolescents. *Law Hum. Behav.* **29**:7–27. [5]

O'Doherty, J. P., P. Dayan, K. Friston, H. Critchley, and R. J. Dolan. 2003. Temporal Difference Models and Reward-Related Learning in the Human Brain. *Neuron* **38**: 329–337. [13]

O'Doherty, J. P., P. Dayan, J. Schultz, et al. 2004. Dissociable Roles of Ventral and Dorsal Striatum in Instrumental Conditioning. *Science* **304**:452–454. [5, 13]

O'Doherty, J. P., A. Hampton, and H. Kim. 2007. Model-Based fMRI and Its Application to Reward Learning and Decision Making. *Ann. NY Acad. Sci.* **1104**:35–53. [12]

Odum, A. L., and A. A. L. Baumann. 2010. Delay Discounting: State and Trait Variable. In: Impulsivity: The Behavioral and Neurological Science of Discounting, ed. G. Madden and W. Bickel, pp. 39–65. Washington, D.C.: American Psychological Association. [2]

Odum, A. L., G. J. Madden, and W. K. Bickel. 2002. Discounting of Delayed Health Gains and Losses by Current, Never- and Ex-Smokers of Cigarettes. *Nicotine Tob. Res.* **4**:295–303. [2]

Oglodek, E., A. Szota, M. Just, D. Mos, and A. Araszkiewicz. 2014. The Role of the Neuroendocrine and Immune Systems in the Pathogenesis of Depression. *Pharmacol. Rep.* **66**:776–781. [11]

O'Keefe, J. 2015. Spatial Cells in the Hippocampal Formation. Nobel Lecture, Dec. 7, 2014. https://www.nobelprize.org/nobel_prizes/medicine/laureates/2014/okeefe-lecture.pdf (accessed May 5, 2016). [2]

O'Keefe, J., and J. Dostrovsky. 1971. The Hippocampus as a Spatial Map. Preliminary Evidence from Unit Activity in the Freely Moving Rat. *Brain Res.* **34**:171–175. [2]

O'Keefe, J., and L. Nadel. 1978. The Hippocampus as a Cognitive Map. Oxford: Clarendon Press. [2]

Olabi, B., I. Ellison-Wright, A. M. McIntosh, et al. 2011. Are There Progressive Brain Changes in Schizophrenia? A Meta-Analysis of Structural Magnetic Resonance Imaging Studies. *Biol. Psychiatry* **70**:88–96. [4, 16]

Olfson, M., S. C. Marcus, M. Tedeschi, and G. J. Wan. 2006. Continuity of Antidepressant Treatment for Adults with Depression in the United States. *Am. J. Psychiatry* **163**:101–108. [15]

Olney, J. W., and N. B. Farber. 1995. Glutamate Receptor Dysfunction and Schizophrenia. *Arch. Gen. Psychiatry* **52**:998–1007. [16]

Olsen, S. R., D. S. Bortone, H. Adesnik, and M. Scanziani. 2012. Gain Control by Layer Six in Cortical Circuits of Vision. *Nature* **483**:47–52. [3]

Oquendo, M. A., A. Barrera, S. P. Ellis, et al. 2004. Instability of Symptoms in Recurrent Major Depression: A Prospective Study. *Am. J. Psychiatry* **161**:255–261. [5]

O'Reilly, R. C., T. S. Braver, and J. D. Cohen. 1999. A Biologically-Based Computational Model of Working Memory. In: Models of Working Memory: Mechanisms of Active Maintenance and Executive Control, ed. A. Miyake and P. Shah, pp. 375–411. New York: Cambridge Univ. Press. [4]

O'Reilly, R. C., and M. J. Frank. 2006. Making Working Memory Work: A Computational Model of Learning in the Prefrontal Cortex and Basal Ganglia. *Neural Comput.* **18**:283–328. [4]

O'Reilly, R. C., and J. L. McClelland. 1994. Hippocampal Conjunctive Encoding, Storage, and Recall: Avoiding a Trade-Off. *Hippocampus* **4**:661–682. [2]

Otto, A. R., A. Skatova, S. Madlon-Kay, and N. D. Daw. 2015. Cognitive Control Predicts Use of Model-Based Reinforcement Learning. *J. Cogn. Neurosci.* **27**:319–333. [4]

Owen, M. J. 2014. New Approaches to Psychiatric Diagnostic Classification. *Neuron* **84**:564–571. [7]

Pace, T. W. W., T. C. Mletzko, O. Alagbe, et al. 2006. Increased Stress-Induced Inflammatory Responses in Male Patients with Major Depression and Increased Early Life Stress. *Am. J. Psychiatry* **163**:1630–1633. [12]

Padman, R., X. Bai, and E. M. Airoldi. 2007. A New Machine Learning Classifier for High Dimensional Healthcare Data. *Stud. Health Technol. Inform.* **129**:664–668. [14]

Palminteri, S. M., M. Khamassi, M. Joffily, and G. Coricelli. 2015. Contextual Modulation of Value Signals in Reward and Punishment Learning. *Nat. Commun.* **6**: 8096. [5]

Pantelis, C., F. Z. Barber, T. R. Barnes, et al. 1999. Comparison of Set-Shifting Ability in Patients with Chronic Schizophrenia and Frontal Lobe Damage. *Schizophr. Res.* **37**:251–270. [4]

Pantelis, C., D. Velakoulis, P. D. McGorry, et al. 2003. Neuroanatomical Abnormalities before and after Onset of Psychosis: A Cross-Sectional and Longitudinal MRI Comparison. *Lancet* **361**:281–288. [3]

Papageorgiou, C., and A. Wells. 2002. Positive Beliefs About Depressive Rumination: Development and Preliminary Validation of a Self-Report Scale. *Behav. Ther.* **32**: 13–26. [15]

———. 2003. An Empirical Test of a Clinical Metacognitive Model of Rumination and Depression. *Cogn. Ther. Res.* **27**:261–273. [15]

Parker, G. 2005. Beyond Major Depression. *Psychol. Med.* **35**:467–474. [8]

Parsons, R. G., and K. J. Ressler. 2013. Implications of Memory Modulation for Post-Traumatic Stress and Fear Disorders. *Nature* **16**:146–153. [3]

Parsons, T. D., and A. A. Rizzo. 2008. Affective Outcomes of Virtual Reality Exposure Therapy for Anxiety and Specific Phobias: A Meta-Analysis. *J. Behav. Ther. Exp. Psychiatry* **39**:250–261. [12]

Patil, S. T., L. Zhang, F. Martenyi, et al. 2007. Activation of mGlu2/3 Receptors as a New Approach to Treat Schizophrenia: A Randomized Phase 2 Clinical Trial. *Nat. Med.* **13**:1102–1107. [16]

Patten, S. B., J. V. A. Williams, D. H. Lavorato, A. G. M. Bulloch, and G. MacQueen. 2012. Depressive Episode Characteristics and Subsequent Recurrence Risk. *J. Affect. Disord.* **140**:277–284. [15]

Paulus, M. P. 2007. Decision-Making Dysfunctions in Psychiatry: Altered Homeostatic Processing? *Science* **318**:602–606. [14]

Paulus, M. P., and J. Y. Angela. 2012. Emotion and Decision-Making: Affect-Driven Belief Systems in Anxiety and Depression. *Trends Cogn. Sci.* **16**:476–483. [12]

Paulus, M. P., and M. B. Stein. 2006. An Insular View of Anxiety. *Biol. Psychiatry* **60**: 383–387. [12]

———. 2010. Interoception in Anxiety and Depression. *Brain Struct. Funct.* **214**:451–463. [12]

Pavlov, I. P. 1927. Conditioned Reflexes: An Investigation of the Physiological Activity of the Cerebral Cortex. New York: Dover Publications. [13]

Payne, J. D., E. D. Jackson, S. Hoscheidt, et al. 2007. Stress Administered Prior to Encoding Impairs Neutral but Enhances Emotional Long-Term Episodic Memories. *Learn. Mem.* **14**:861–868. [2]

Pearl, J. 1988. Probabilistic Reasoning in Intelligent Systems: Networks of Plausible Inference. Burlington, MA: Morgan Kaufmann. [17]

———. 2009a. Causal Inference in Statistics: An Overview. *Stat. Surv.* **3**:96–146. [14]

———. 2009b. Causality: Models, Reasoning and Inference, 2nd edition. Cambridge: Cambridge Univ. Press. [3, 14, 17]

———. 2010. An Introduction to Causal Inference. *Int. J. Biostat.* **6**:Article 7. [14]

Pearson, R. M., J. Heron, K. Button, et al. 2015. Cognitive Styles and Future Depressed Mood in Early Adulthood: The Importance of Global Attributions. *J. Affect. Disord.* **171**:60–67. [15]

Peeters, F., J. Berkhof, J. Rottenberg, and N. A. Nicolson. 2010. Ambulatory Emotional Reactivity to Negative Daily Life Events Predicts Remission from Major Depressive Disorder. *Behav. Res. Ther.* **48**:754–760. [15]

Pekkonen, E., J. Hirvonen, I. P. Jääskeläinen, S. Kaakkola, and J. Huttunen. 2001. Auditory Sensory Memory and the Cholinergic System: Implications for Alzheimer's Disease. *NeuroImage* **14**:376–382. [12]

Peled, A. 2009. Neuroscientific Psychiatric Diagnosis. *Med. Hypotheses* **73**:220–229. [11]

Pencina, M. J., and R. B. D'Agostino, Sr. 2012. Thoroughly Modern Risk Prediction? *Sci. Transl. Med.* **4**:131fs110. [14]

Pendyam, S., A. Mohan, P. W. Kalivas, and S. S. Nair. 2009. Computational Model of Extracellular Glutamate in the Nucleus Accumbens Incorporates Neuroadaptations by Chronic Cocaine. *Neuroscience* **158**:1266–1276. [12]

Penny, W. D., K. E. Stephan, A. Mechelli, and K. J. Friston. 2004. Comparing Dynamic Causal Models. *NeuroImage* **22**:1157–1172. [10]

Peralta, V., and M. J. Cuesta. 2000. Clinical Models of Schizophrenia: A Critical Approach to Competing Conceptions. *Psychopathology* **33**:252–258. [8]

Peralta, V., M. J. Cuesta, C. Giraldo, A. Cardenas, and F. Gonzales. 2002. Classifying Psychotic Disorder: Issues Regarding Categorical vs. Dimensional Approaches and Time Frame to Assess Symptoms. *Eur. Arch. Psychiatry Clin. Neurosci.* **252**:12–18. [8]

Perlis, R. H. 2013. A Clinical Risk Stratification Tool for Predicting Treatment Resistance in Major Depressive Disorder. *Biol. Psychiatry* **74**:7–14. [14]

Persons, J. B. 1986. The Advantages of Studying Psychological Phenomena Rather Than Psychiatric Diagnoses. *Am. Psychol.* **41**:1252–1260. [9]

Pessiglione, M., B. Seymour, G. Flandin, R. J. Dolan, and C. D. Frith. 2006. Dopamine-Dependent Prediction Errors Underpin Reward-Seeking Behaviour in Humans. *Nature* **442**:1042–1045. [6]

Petanjek, Z., M. Judas, G. Simic, et al. 2011. Extraordinary Neoteny of Synaptic Spines in the Human Prefrontal Cortex. *PNAS* **108**:13281–13286. [16]

Peters, J., and C. Büchel. 2010. Episodic Future Thinking Reduces Reward Delay Discounting through an Enhancement of Prefrontal-Mediotemporal Interactions. *Neuron* **66**:138–148. [2]

Petrides, G., M. Fink, M. M. Husain, et al. 2001. ECT Remission Rates in Psychotic versus Nonpsychotic Depressed Patients: A Report from Core. *J. ECT* **17**:244–253. [1]

Petry, N. M. 2012. Contingency Management for Substance Abuse Treatment: A Guide for Implementing This Evidence-Based Practice. New York: Routledge. [2]

Pettersson-Yeo, W., S. Benetti, A. F. Marquand, et al. 2013. Using Genetic, Cognitive and Multi-Modal Neuroimaging Data to Identify Ultra-High-Risk and First-Episode Psychosis at the Individual Level. *Psychol. Med.* **43**:2547–2562. [3]

Pettorruso, M., L. De Risio, M. Di Nicola, et al. 2014. Allostasis as a Conceptual Framework Linking Bipolar Disorder and Addiction. *Front. Psychiatry* **5**:173. [11]

Pevsner, J. 2005. Bioinformatics and Functional Genomics. Chichester: John Wiley & Sons. [2]

Pfeffer, C. K., M. Xue, M. He, Z. J. Huang, and M. Scanziani. 2013. Inhibition of Inhibition in Visual Cortex: The Logic of Connections between Molecularly Distinct Interneurons. *Nat. Neurosci.* **16**:1068–1076. [16]

Pfeiffer, B. E., and D. J. Foster. 2013. Hippocampal Place-Cell Sequences Depict Future Paths to Remembered Goals. *Nature* **497**:74–79. [15]

Phillips, J. 2013. Review of "the Conceptual Evolution of DSM-5," by D. A. Regier Et Al, Eds. (2011). *J. Nerv. Ment. Disease* **201**:828–829. [8]

Phillips, L. J., D. Velakoulis, C. Pantelis, et al. 2002. Non-Reduction in Hippocampal Volume Is Associated with Higher Risk of Psychosis. *Schizophr. Res.* **58**:145–158. [3]

Pies, R. 2007. How "Objective" Are Psychiatric Diagnoses? (Guess Again). *Psychiatry (Edgmont)* **4**:18–22. [1]

Piet, J., and E. Hougaard. 2011. The Effect of Mindfulness-Based Cognitive Therapy for Prevention of Relapse in Recurrent Major Depressive Disorder: A Systematic Review and Meta-Analysis. *Clin. Psychol. Rev.* **31**:1032–1040. [15]

Pilkonis, P. A., S. W. Choi, S. P. Reise, et al. 2011. Item Banks for Measuring Emotional Distress from the Patient-Reported Outcomes Measurement Information System (PROMIS): Depression, Anxiety, and Anger. *Assessment* **18**:263–283. [8]

Pine, D. S., and N. A. Fox. 2015. Childhood Antecedents and Risk for Adult Mental Disorders. *Annu. Rev. Psychol.* **66**:459–485. [10]

Pisani, A. R., P. A. Wyman, M. Petrova, et al. 2012. Emotion Regulation Difficulties, Youth–Adult Relationships, and Suicide Attempts among High School Students in Underserved Communities. *J. Youth Adolesc.* **42**:807–820. [3]

Pitman, R. K. 1987. A Cybernetic Model of Obsessive-Compulsive Psychopathology. *Compr. Psychiatry* **28**:334–343. [2]

Pizzagalli, D., R. D. Pascual-Marqui, J. B. Nitschke, et al. 2001. Anterior Cingulate Activity as a Predictor of Degree of Treatment Response in Major Depression: Evidence from Brain Electrical Tomography Analysis. *Am. J. Psychiatry* **158**:405–415. [15]

Pizzagalli, D. A., A. L. Jahn, and J. P. O'Shea. 2005. Toward an Objective Characterization of an Anhedonic Phenotype: A Signal-Detection Approach. *Biol. Psychiatry* **57**:319–327. [3]

Plaut, D. C., and T. Shallice. 1993. Deep Dyslexia: A Case Study of Connectionist Neuropsychology. *Cogn. Neuropsychol.* **10**:377–500. [5]

Plichta, M. M., and A. Scheres. 2014. Ventral–Striatal Responsiveness During Reward Anticipation in ADHD and Its Relation to Trait Impulsivity in the Healthy Population: A Meta-Analytic Review of the fMRI Literature. *Neurosci. Biobehav. Rev.* **38**:125–134. [12]

Plotsky, P. M., M. J. Owens, and C. B. Nemeroff. 1998. Psychoneuroendocrinology of Depression. Hypothalamic-Pituitary-Adrenal Axis. *Psychiatr. Clin. North Am.* **21**:293–307. [1]

Pocklington, A. J., E. Rees, J. T. Walters, et al. 2015. Novel Findings from CNVs Implicate Inhibitory and Excitatory Signaling Complexes in Schizophrenia. *Neuron* **86**:1203–1214. [16]

Poggio, T., and E. Bizzi. 2004. Generalization in Vision and Motor Control. *Nature* **431**:768–774. [2]

Pokorny, A. D. 1983. Prediction of Suicide in Psychiatric Patients. Report of a Prospective Study. *Arch. Gen. Psychiatry* **40**:249–257. [3]

Pokos, V., and D. J. Castle. 2006. Prevalence of Comorbid Anxiety Disorders in Schizophrenia Spectrum Disorders: A Literature Review. *Curr. Psychiatry Rev.* **2**:285–307. [9]

Poland, D. 1993. Cooperative Catalysis and Chemical Chaos: A Chemical Model for the Lorenz Equations. *Physica D* **65** 86–99. [11]

Polikov, V. S., P. A. Tresco, and W. M. Reichert. 2005. Response of Brain Tissue to Chronically Implanted Neural Electrodes. *J. Neurosci. Methods* **148**:1–18. [9]

Posternak, M. A., D. A. Solomon, A. C. Leon, et al. 2006. The Naturalistic Course of Unipolar Major Depression in the Absence of Somatic Therapy. *J. Nerv. Ment. Dis.* **194**:324–329. [15]

Potenza, M. N., K.-i. A. Hong, C. M. Lacadie, et al. 2012. Neural Correlates of Stress-Induced and Cue-Induced Drug Craving: Influences of Sex and Cocaine Dependence. *Am. J. Psychiatry* **169**:406–414. [3]

Prata, D., A. Mechelli, and S. Kapur. 2014. Clinically Meaningful Biomarkers for Psychosis: A Systematic and Quantitative Review. *Neurosci. Biobehav. Rev.* **45**:134–141. [14]

Praveen, P., and H. Fröhlich. 2013. Boosting Probabilistic Graphical Model Inference by Incorporating Prior Knowledge from Multiple Sources. *PLoS One* **8**:e67410. [9]

Preston, G. A., and D. R. Weinberger. 2005. Intermediate Phenotypes in Schizophrenia: A Selective Review. *Dialogues Clin. Neurosci.* **7**:165–179. [5]

Preti, A., and M. Cella. 2010. Randomized-Controlled Trials in People at Ultra High Risk of Psychosis: A Review of Treatment Effectiveness. *Schizophr. Res.* **123**:30–36. [10]

Qiu, A., M. Vaillant, P. Barta, J. T. Ratnanather, and M. I. Miller. 2007. Region-of-Interest-Based Analysis with Application of Cortical Thickness Variation of Left Planum Temporale in Schizophrenia and Psychotic Bipolar Disorder. *Hum. Brain Mapp.* **29**:973–985. [16]

Quattrocki, E., and K. Friston. 2014. Autism, Oxytocin and Interoception. *Neurosci. Biobehav. Rev.* **47**:410–430. [7]

Quian Quiroga, R., and S. Panzeri. 2009. Extracting Information from Neuronal Populations: Information Theory and Decoding Approaches. *Nat. Rev. Neurosci.* **10**:173–185. [3]

Quinn, J. J., P. K. Hitchcott, E. A. Umeda, A. P. Arnold, and J. R. Taylor. 2007. Sex Chromosome Complement Regulates Habit Formation. *Nat. Neurosci.* **10**:1398–1400. [3]

Rae, A., and P. Lindsay. 2004. A Behaviour-Based Method for Fault Tree Generation. In: Proceedings of the 22nd International System Safety Conference, Unionville, VA: System Safety Society Publications. http://staff.itee.uq.edu.au/pal/papers/ISSS04.pdf (accessed March 2, 2016). [9]

Ragland, J. D., A. R. Laird, C. Ranganath, et al. 2009. Prefrontal Activation Deficits During Episodic Memory in Schizophrenia. *Am. J. Psychiatry* **166**:863–874. [4]

Rajkowska, G., J. J. Miguel-Hidalgo, J. Wei, et al. 1999. Morphometric Evidence for Neuronal and Glial Prefrontal Cell Pathology in Major Depression. *Biol. Psychiatry* **45**:1085–1098. [3]

Ramon, S., B. Healy, and N. Renouf. 2007. Recovery from Mental Illness as an Emergent Concept and Practice in Australia and the UK. *Int. J. Soc. Psychiatry* **53**:108–122. [12]

Rangel, A., C. Camerer, and P. R. Montague. 2008. A Framework for Studying the Neurobiology of Value-Based Decision Making. *Nat. Rev. Neurosci.* **9**:545–556. [2, 17]

Rao, R. P. N., and D. H. Ballard. 1999. Predictive Coding in the Visual Cortex: A Functional Interpretation of Some Extra-Classical Receptive-Field Effects. *Nat. Neurosci.* **2**:79–87. [7]

Rao, S. G., G. V. Williams, and P. S. Goldman-Rakic. 2000. Destruction and Creation of Spatial Tuning by Disinhibition: GABA-a Blockade of Prefrontal Cortical Neurons Engaged by Working Memory. *J. Neurosci.* **20**:485–494. [16]

Rapkin, A. J., and S. A. Winer. 2008. The Pharmacologic Management of Premenstrual Dysphoric Disorder. *Expert Opin. Pharmacother.* **9**:429–445. [10]

Ratcliff, R., and M. J. Frank. 2012. Reinforcement-Based Decision Making in Cortico-striatal Circuits: Mutual Constraints by Neurocomputational and Diffusion Models. *Neural Comput.* **24**:1186–1229. [5, 6]

Ratcliff, R., and G. McKoon. 2008. The Diffusion Decision Model: Theory and Data for Two-Choice Decision Tasks. *Neural Comput.* **20**:873–922. [6]

Rauch, S. L., M. A. Jenike, N. M. Alpert, et al. 1994. Regional Cerebral Blood Flow Measured During Symptom Provocation in Obsessive-Compulsive Disorder Using Oxygen 15-Labeled Carbon Dioxide and Positron Emission Tomography. *Arch. Gen. Psychiatry* **51**:62–70. [10]

Ravizza, S. M., K. C. Moua, D. Long, and C. S. Carter. 2010. The Impact of Context Processing Deficits on Task-Switching Performance in Schizophrenia. *Schizophr. Res.* **116**:274–279. [4]

Raychaudhuri, S., R. M. Plenge, E. J. Rossin, et al. 2009. Identifying Relationships among Genomic Disease Regions: Predicting Genes at Pathogenic Snp Associations and Rare Deletions. *PLoS Genet.* **5**:e1000534. [3]

Real, L. A. 1992. Animal Choice Behavior and the Evolution of Cognitive Architecture. *Science* **253**:980–986. [13]

Rector, T. S., B. C. Taylor, and T. J. Wilt. 2012. Chapter 12: Systematic Review of Prognostic Tests. *J. Gen. Intern. Med.* **27(Suppl 1)**:94–101. [14]

Redish, A. D. 1999. Beyond the Cognitive Map: From Place Cells to Episodic Memory. Cambridge, MA: MIT Press. [2]

———. 2004. Addiction as a Computational Process Gone Awry. *Science* **306**:1944–1947. [12, 13, 17]

———. 2013. The Mind within the Brain: How We Make Decisions and How Those Decisions Go Wrong. Oxford: Oxford Univ. Press. [2, 9, 10]

Redish, A. D., S. Jensen, and A. Johnson. 2008. A Unified Framework for Addiction: Vulnerabilities in the Decision Process. *Behav. Brain Sci.* **31**:415–487. [2, 10, 12, 17]

Redish, A. D., S. Jensen, A. Johnson, and Z. Kurth-Nelson. 2007. Reconciling Reinforcement Learning Models with Behavioral Extinction and Renewal: Implications for Addiction, Relapse, and Problem Gambling. *Psychol. Rev.* **114**:784–805. [2]

Redish, A. D., and A. Johnson. 2007. A Computational Model of Craving and Obsession. *Ann. NY Acad. Sci.* **1104**:324–339. [11]

Redish, A. D., and D. S. Touretzky. 1997. The Role of the Hippocampus in Solving the Morris Water Maze. *Neural Comput.* **10**:73–112. [2]

Reggia, J. A., and D. Montgomery. 1996. A Computational Model of Visual Hallucinations in Migraine. *Comp. Biol. Med.* **26**:133–141. [2]

Regier, D. A., W. E. Narrow, D. E. Clarke, et al. 2013. DSM-5 Field Trials in the United States and Canada, Part II: Test-Retest Reliability of Selected Categorical Diagnoses. *Am. J. Psychiatry* **170**:59–70. [7]

Regier, D. A., W. E. Narrow, E. A. Kuhl, and D. J. Kupfer, eds. 2011. The Conceptual Evolution of DSM-5. Arlington: American Psychiatric Publishing. [8]

Regier, P. S., and A. D. Redish. 2015. Contingency Management and Deliberative Decision-Making Processes. *Front. Psychiatry* **6**:76. [2]

Reinen, J., E. E. Smith, C. Insel, et al. 2014. Patients with Schizophrenia Are Impaired When Learning in the Context of Pursuing Rewards. *Schizophr. Res.* **152**:309–310. [4]

Rescorla, R. A., and A. R. Wagner. 1972. A Theory of Pavlovian Conditioning: Variations in the Effectiveness of Reinforcement. New York: Appleton-Century-Crofts. [13]

Reske, M., D. C. Delis, and M. P. Paulus. 2011. Evidence for Subtle Verbal Fluency Deficits in Occasional Stimulant Users: Quick to Play Loose with Verbal Rules. *J. Psychiatr. Res.* **45**:361–368. [14]

Ressler, K. J., and H. S. Mayberg. 2007. Targeting Abnormal Neural Circuits in Mood and Anxiety Disorders: From the Laboratory to the Clinic. *Nat. Neurosci.* **10**:1116–1124. [1]

Rieke, F., D. Warland, R. de Ruyter van Steveninck, and W. Bialek. 1997. Spikes: Exploring the Neural Code. Cambridge, MA: MIT Press. [3]

Rigotti, M., O. Barak, M. R. Warden, et al. 2013. The Importance of Mixed Selectivity in Complex Cognitive Tasks. *Nature* **497**:585–590. [6]

Ripke, S., B. M. Neale, A. Corvin, et al. 2014. Biological Insights from 108 Schizophrenia-Associated Genetic Loci. *Nature* **511**:421–427. [16]

Robbins, T. 2002. The 5-Choice Serial Reaction Time Task: Behavioural Pharmacology and Functional Neurochemistry. *Psychopharmacology* **163**:362–380. [3]

Robert, C. 2007. The Bayesian Choice: From Decision-Theoretic Foundations to Computational Implementation. New York: Springer. [7]

Roberts, B. M., D. E. Holden, C. L. Shaffer, et al. 2010. Prevention of Ketamine-Induced Working Memory Impairments by AMPA Potentiators in a Nonhuman Primate Model of Cognitive Dysfunction. *Behav. Brain Res.* **212**:41–48. [16]

Robins, E., and S. B. Guze. 1970. Establishment of Diagnostic Validity in Psychiatric Illness: Its Application to Schizophrenia. *Am. J. Psychiatry* **126**:983–987. [7, 8]

Robinson, O. J., R. Cools, C. O. Carlisi, B. J. Sahakian, and W. C. Drevets. 2012. Ventral Striatum Response During Reward and Punishment Reversal Learning in Unmedicated Major Depressive Disorder. *Am. J. Psychiatry* **169**:152–159. [12]

Rodrigues, S., J. Gonçalves, and J. R. Terry. 2007. Existence and Stability of Limit Cycles in a Macroscopic Neuronal Population Model. *Physica D* **233**:39–65. [12]

Roesch, M. R., D. J. Calu, G. R. Esber, and G. Schoenbaum. 2010. All That Glitters ... Dissociating Attention and Outcome Expectancy from Prediction Errors Signals. *J. Neurophysiol.* **104**:587–595. [3]

Rogasch, N. C., Z. J. Daskalakis, and P. B. Fitzgerald. 2014. Cortical Inhibition, Excitation, and Connectivity in Schizophrenia: A Review of Insights from Transcranial Magnetic Stimulation. *Schizophr. Bull.* **40**:685–696. [16]

Rogers, T. T., and J. L. McClelland. 2008. Précis of Semantic Cognition: A Parallel Distributed Processing Approach. *Behav. Brain Sci.* **31**:689–714. [12]

Rogers, T. T., J. L. McClelland, K. Patterson, M. A. Lambon-Ralph, and J. R. Hodges. 1999. A Recurrent Connectionist Model of Semantic Dementia. Poster, Cogn. Neurosci. Soc. Annual Meeting. http://psych.wisc.edu/Rogers/posters/CNS99_sd-model.pdf (accessed July 10, 2016). [12]

Roiser, J. P., R. Elliott, and B. J. Sahakian. 2012. Cognitive Mechanisms of Treatment in Depression. *Neuropsychopharmacology* **37**:117–136. [15]

Roiser, J. P., O. D. Howes, C. A. Chaddock, E. M. Joyce, and P. McGuire. 2013. Neural and Behavioral Correlates of Aberrant Salience in Individuals at Risk for Psychosis. *Schizophr. Bull.* **39**:1328–1336. [4]

Rolls, E. T., and G. Deco. 2015. Stochastic Cortical Neurodynamics Underlying the Memory and Cognitive Changes in Aging. *Neurobiol. Learn. Mem.* **118**:150–161. [4]

Rose, N. R. 2014. Learning from Myocarditis: Mimicry, Chaos and Black Holes. *F1000Prime Rep.* **6**:25. [11]

Rosell, D. R., L. C. Zaluda, M. M. McClure, et al. 2015. Effects of the D1 Dopamine Receptor Agonist Dihydrexidine (DAR-0100A) on Working Memory in Schizotypal Personality Disorder. *Neuropsychopharmacology* **40**:446–453. [5]

Rosen, A. M., T. Spellman, and J. A. Gordon. 2015. Electrophysiological Endophenotypes in Rodent Models of Schizophrenia and Psychosis. *Biol. Psychiatry* **77**:1041–1049. [1]

Rosenthal, R. W. 1993. Suicide Attempts and Signalling Games. *Math. Social Sci.* **26**:25–33. [3]

Rotem, A., A. Neef, N. E. Neef, et al. 2014. Solving the Orientation Specific Constraints in Transcranial Magnetic Stimulation by Rotating Fields. *PLoS One* **9**:e86794. [5]

Rothbaum, B. O., and M. Davis. 2003. Applying Learning Principles to the Treatment of Post-Trauma Reactions. *Ann. NY Acad. Sci.* **8**:112–121. [5]

Rothman, K. J., and S. Greenland. 2005. Causation and Causal Inference in Epidemiology. *Am. J. Public Health* **95(Suppl 1)**:144–150. [14]

Rothschild, A. J., B. W. Dunlop, D. L. Dunner, et al. 2009. Assessing Rates and Predictors of Tachyphylaxis During the Prevention of Recurrent Episodes of Depression with Venlafaxine ER for Two Years (PREVENT) Study. *Psychopharmacol. Bull.* **42**:5–20. [15]

Rothwell, P. E., M. V. Fuccillo, S. Maxeiner, et al. 2014. Autism-Associated Neuroligin-3 Mutations Commonly Impair Striatal Circuits to Boost Repetitive Behaviors. *Cell* **158**:198–212. [3]

Rubinov, M., S. A. Knock, C. J. Stam, et al. 2009. Small-World Properties of Nonlinear Brain Activity in Schizophrenia. *Hum. Brain Mapp.* **30**:403–416. [12]

Rudolph, M., and A. Destexhe. 2006. Analytical Integrate-and-Fire Neuron Models with Conductance-Based Dynamics for Event-Driven Simulation Strategies. *Neural Comput.* **18**:2146–2210. [12]

Rujescu, D., A. Ingason, S. Cichon, et al. 2009. Disruption of the Neurexin 1 Gene Is Associated with Schizophrenia. *Hum. Mol. Genet.* **18**:988–996. [3]

Rush, A. J., M. H. Trivedi, S. R. Wisniewski, et al. 2006a. Acute and Longer-Term Outcomes in Depressed Outpatients Requiring One or Several Treatment Steps: A Star*D Report. *Am. J. Psychiatry* **163**:1905–1917. [1, 15]

Rush, A. J., M. H. Trivedi, S. R. Wisniewski, et al. 2006b. Bupropion-SR, Sertraline, or Venlafaxine-XR after Failure of SSRIs for Depression. *N. Engl. J. Med.* **354**:1231–1242. [10]

Rutter, M., J. Kim-Cohen, and B. Maughan. 2006. Continuities and Discontinuities in Psychopathology between Childhood and Adult Life. *J. Child Psychol. Psychiatry* **47**:276–295. [10, 17]

Salamone, J. D., and M. Correa. 2012. The Mysterious Motivational Functions of Mesolimbic Dopamine. *Neuron* **76**:470–485. [4]

Salamone, J. D., M. Correa, A. M. Farrar, E. J. Nunes, and M. Pardo. 2009. Dopamine, Behavioral Economics, and Effort. *Front. Behav. Neurosci.* **3**:13. [4]

Salamone, J. D., M. Correa, E. J. Nunes, P. A. Randall, and M. Pardo. 2012. The Behavioral Pharmacology of Effort-Related Choice Behavior: Dopamine, Adenosine and Beyond. *J. Exp. Anal. Behav.* **97**:125–146. [4]

Salisbury, D. F., N. Kuroki, K. Kasai, M. E. Shenton, and R. W. McCarley. 2007. Progressive and Interrelated Functional and Structural Evidence of Post-Onset Brain Reduction in Schizophrenia. *Arch. Gen. Psychiatry* **64**:521–529. [16]

Salo, R., T. E. Nordahl, K. Possin, et al. 2002. Preliminary Evidence of Reduced Cognitive Inhibition in Methamphetamine-Dependent Individuals. *Psychiatry Res.* **111**:65–74. [14]

Samson, R. D., M. J. Frank, and J. M. Fellous. 2010. Computational Models of Reinforcement Learning: The Role of Dopamine as a Reward Signal. *Cogn. Neurodyn.* **4**:91–105. [4]

Samuel, A. L. 1959. Some Studies in Machine Learning Using the Game of Checkers. *IBM J. Res. Dev.* **3**:210–229. [13]

Sanchez-Morla, E. M., A. Barabash, V. Martinez-Vizcaino, et al. 2009. Comparative Study of Neurocognitive Function in Euthymic Bipolar Patients and Stabilized Schizophrenic Patients. *Psychiatry Res.* **169**:220–228. [16]

Sanislow, C. A., D. S. Pine, K. J. Quinn, et al. 2010. Developing Constructs for Psychopathology Research: Research Domain Criteria. *J. Abnorm. Psychol.* **119**:631–639. [8]

Sapolsky, R. M. 2000. Glucocorticoids and Hippocampal Atrophy in Neuropsychiatric Disorders. *Arch. Gen. Psychiatry* **57**:925–935. [3]

Saxena, S., A. L. Brody, K. M. Maidment, and L. R. Baxter, Jr. 2007. Paroxetine Treatment of Compulsive Hoarding. *J. Psychiatr. Res.* **41**:481–487. [10]

Schadt, E. E., M. D. Linderman, J. Sorenson, L. Lee, and G. P. Nolan. 2010. Computational Solutions to Large-Scale Data Management and Analysis. *Nat. Rev. Genet.* **11**:647–657. [2]

Schatzberg, A. F., and C. B. Nemeroff. 1995. The American Psychiatric Press Textbook of Psychopharmacology. Washington, D.C.: American Psychiatric Association. [2]

Schizophrenia Working Group of the Psychiatric Genomics Consortium. 2014. Biological Insights from 108 Schizophrenia-Associated Genetic Loci. *Nature* **511**: 421–427. [1, 10]

Schlagenhauf, F., Q. J. Huys, L. Deserno, et al. 2014. Striatal Dysfunction During Reversal Learning in Unmedicated Schizophrenia Patients. *NeuroImage* **89**:171–180. [4]

Schlagenhauf, F., G. Juckel, M. Koslowski, et al. 2008. Reward System Activation in Schizophrenic Patients Switched from Typical Neuroleptics to Olanzapine. *Psychopharmacology (Berl.)* **196**:673–684. [4]

Schlagenhauf, F., P. Sterzer, K. Schmack, et al. 2009. Reward Feedback Alterations in Unmedicated Schizophrenia Patients: Relevance for Delusions. *Biol. Psychiatry* **65**:1032–1039. [4]

Schmidt, A., R. Bachmann, M. Kometer, et al. 2012. Mismatch Negativity Encoding of Prediction Errors Predicts S-Ketamine-Induced Cognitive Impairments. *Neuropsychopharmacology* **37**:865–875. [12]

Schneider, M., M. Debbane, A. S. Bassett, et al. 2014. Psychiatric Disorders from Childhood to Adulthood in 22q11.2 Deletion Syndrome: Results from the International Consortium on Brain and Behavior in 22q11.2 Deletion Syndrome. *Am. J. Psychiatry* **171**:627–639. [1]

Schobel, S. A., N. H. Chaudhury, U. A. Khan, et al. 2013. Imaging Patients with Psychosis and a Mouse Model Establishes a Spreading Pattern of Hippocampal Dysfunction and Implicates Glutamate as a Driver. *Neuron* **78**:81–93. [16]

Schroll, H., C. Beste, and F. H. Hamker. 2015. Combined Lesions of Direct and Indirect Basal Ganglia Pathways but Not Changes in Dopamine Levels Explain Learning Deficits in Patients with Huntingtons Disease. *Eur. J. Neurosci.* **41**:1227–1244. [2]

Schulkin, J. 2010. Social Allostasis: Anticipatory Regulation of the Internal Milieu. *Front. Evol. Neurosci.* **2**:111. [12]

Schüll, N. D. 2012. Addiction by Design: Machine Gambling in Las Vegas. Princeton: Princeton Univ. Press. [2]

Schultz, W. 2007. Multiple Dopamine Functions at Different Time Courses. *Annu. Rev. Neurosci.* **30**:259–288. [3, 4]

———. 2013. Updating Dopamine Reward Signals. *Curr. Opin. Neurobiol.* **23**:229–238. [6]

Schultz, W., P. Apicella, and T. Ljungberg. 1993. Responses of Monkey Dopamine Neurons to Reward and Conditioned Stimuli During Successive Steps of Learning a Delayed Response Task. *J. Neurosci.* **13**:900–913. [13]

Schultz, W., P. Dayan, and P. R. Montague. 1997. A Neural Substrate of Prediction and Reward. *Science* **275**:1593–1599. [5, 6, 12, 13, 14]

Schultz, W., and A. Dickinson. 2000. Neuronal Coding of Prediction Errors. *Annu. Rev. Neurosci.* **23**:473–500. [14]

Schwarz, E., and S. Bahn. 2008. The Utility of Biomarker Discovery Approaches for the Detection of Disease Mechanisms in Psychiatric Disorders. *Br. J. Pharmacol.* **153(Suppl 1)**:S133–136. [1]

Scott, I. A., and P. B. Greenberg. 2010. Cautionary Tales in the Interpretation of Studies of Tools for Predicting Risk and Prognosis. *Intern. Med. J.* **40**:803–812. [14]

Seamans, J. K., N. Gorelova, D. Durstewitz, and C. R. Yang. 2001. Bidirectional Dopamine Modulation of Gabaergic Inhibition in Prefrontal Cortical Pyramidal Neurons. *J. Neurosci.* **21**:3628–3638. [4]

Seamans, J. K., and C. R. Yang. 2004. The Principal Features and Mechanisms of Dopamine Modulation in the Prefrontal Cortex. *Prog. Neurobiol.* **74**:1–58. [2, 4, 5]

Sebat, J., B. Lakshmi, D. Malhotra, et al. 2007. Strong Association of *de Novo* Copy Number Mutations with Autism. *Science* **316**:445–449. [1, 3]

Sebat, J., D. L. Levy, and S. E. McCarthy. 2009. Rare Structural Variants in Schizophrenia: One Disorder, Multiple Mutations; One Mutation, Multiple Disorders. *Trends Genet.* **25**:528–535. [16]

Segal, Z. V., P. Bieling, T. Young, et al. 2010. Antidepressant Monotherapy Vs Sequential Pharmacotherapy and Mindfulness-Based Cognitive Therapy, or Placebo, for Relapse Prophylaxis in Recurrent Depression. *Arch. Gen. Psychiatry* **67**:1256–1264. [15]

Segal, Z. V., M. Gemar, and S. Williams. 1999. Differential Cognitive Response to a Mood Challenge Following Successful Cognitive Therapy or Pharmacotherapy for Unipolar Depression. *J. Abnorm. Psychol.* **108**:3–10. [15]

Segal, Z. V., S. Kennedy, M. Gemar, et al. 2006. Cognitive Reactivity to Sad Mood Provocation and the Prediction of Depressive Relapse. *Arch. Gen. Psychiatry* **63**:749–755. [15]

Seidman, L. J., A. J. Giuliano, E. C. Meyer, et al. 2010. Neuropsychology of the Prodrome to Psychosis in the NAPLS Consortium: Relationship to Family History and Conversion to Psychosis. *Arch. Gen. Psychiatry* **67**:578–588. [3, 10]

Sejnowski, T. J., C. Koch, and P. S. Churchland. 1988. Computational Neuroscience. *Science* **241**:1299–1306. [16]

Selemon, L. D., and G. Rajkowska. 2003. Cellular Pathology in the Dorsolateral Prefrontal Cortex Distinguishes Schizophrenia from Bipolar Disorder. *Curr. Mol. Med.* **3**:427–436. [16]

Semple, D. M., A. M. McIntosh, and S. M. Lawrie. 2005. Cannabis as a Risk Factor for Psychosis: Systematic Review. *J. Psychopharm.* **19**:187–194. [3]

Senço, N. M., Y. Huang, G. D'Urso, et al. 2015. Transcranial Direct Current Stimulation in Obsessive-Compulsive Disorder: Emerging Clinical Evidence and Considerations for Optimal Montage of Electrodes. *Expert Rev. Med. Devices* **12**:381–391. [5]

Serre, T., A. Oliva, and T. Poggio. 2007. A Feedforward Architecture Accounts for Rapid Categorization. *PNAS* **104**:6424–6429. [2]

Servan-Schreiber, D., C. S. Carter, R. M. Bruno, and J. D. Cohen. 1998. Dopamine and the Mechanisms of Cognition: Part II. D-Amphetamine Effects in Human Subjects Performing a Selective Attention Task. *Biol. Psychiatry* **43**:723–729. [12]

Seth, A. K., K. Suzuki, and H. D. Critchley. 2011. An Interoceptive Predictive Coding Model of Conscious Presence. *Front. Psychol.* **2**:395. [12]

Seu, E., S. M. Groman, A. P. Arnold, and J. D. Jentsch. 2014. Sex Chromosome Complement Influences Operant Responding for a Palatable Food in Mice. *Genes Brain Behav.* **13**:527–534. [3]

Shay, J. 1994. Achilles in Vietnam: Combat Trauma and the Undoing of Character. New York: Scribner. [10]

Sheffield, J. M., J. M. Gold, M. E. Strauss, et al. 2014. Common and Specific Cognitive Deficits in Schizophrenia: Relationships to Function. *Cogn. Affect. Behav. Neurosci.* **14**:161–174. [4]

Sheline, Y. I., D. M. Barch, J. L. Price, et al. 2009. The Default Mode Network and Self-Referential Processes in Depression. *PNAS* **106**:1942–1947. [15]

Shelton, M. A., J. T. Newman, H. Gu, et al. 2015. Loss of Microtubule-Associated Protein 2 Immunoreactivity Linked to Dendritic Spine Loss in Schizophrenia. *Biol. Psychiatry* **78**:359–360. [16]

Shenoy, P., R. P. N. Rao, and A. Yu. 2010. A Rational Decision Making Framework for Inhibitory Control. *Adv. Neural Inf. Process Syst.* **23**:1–9. [14]

Shenoy, P., and A. J. Yu. 2011. Rational Decision-Making in Inhibitory Control. *Front. Hum. Neurosci.* **5**:48. [3, 14]

Shewhart, W. A. 1938. Application of Statistical Methods to Manufacturing Problems. *J. Franklin Inst.* **226**:163–186. [9]

Shi, C., X. Yu, E. F. Cheung, D. H. Shum, and R. C. Chan. 2014. Revisiting the Therapeutic Effect of rTMS on Negative Symptoms in Schizophrenia: A Meta-Analysis. *Psychiatry Res.* **215**:505–513. [16]

Shipp, S., R. A. Adams, and K. J. Friston. 2013. Reflections on Agranular Architecture: Predictive Coding in the Motor Cortex. *Trends Neurosci.* **36**:706–716. [7]

Shohamy, D., C. E. Myers, K. D. Geghman, J. Sage, and M. A. Gluck. 2006. L-dopa Impairs Learning, but Not Generalization, in Parkinson's Disease. *Neuropsychologia* **44**:774–784. [2]

Shors, T. J. 2004. Learning During Stressful Times. *Learn. Mem.* **11**:137–144. [10]

Shotbolt, P., P. R. Stokes, S. F. Owens, et al. 2011. Striatal Dopamine Synthesis Capacity in Twins Discordant for Schizophrenia. *Psychol. Med.* **41**:2331–2338. [4]

Shulman, J. M., P. L. De Jager, and M. B. Feany. 2011. Parkinson's Disease: Genetics and Pathogenesis. *Annu. Rev. Pathol.* **6**:193–222. [2]

Siegle, G. J., W. K. Thompson, A. Collier, et al. 2012. Toward Clinically Useful Neuroimaging in Depression Treatment: Prognostic Utility of Subgenual Cingulate Activity for Determining Depression Outcome in Cognitive Therapy across Studies, Scanners, and Patient Characteristics. *Arch. Gen. Psychiatry* **69**:913–924. [15]

Siekmeier, P. J., M. E. Hasselmo, M. W. Howard, and J. T. Coyle. 2007. Modeling of Context-Dependent Retrieval in Hippocampal Region CA1: Implications for Cognitive Function in Schizophrenia. *Schizophr. Res.* **89**:177–190. [12]

Sigurdsson, T., K. L. Stark, M. Karayiorgou, J. A. Gogos, and J. A. Gordon. 2010. Impaired Hippocampal–Prefrontal Synchrony in a Genetic Mouse Model of Schizophrenia. *Nature* **464**:763–767. [5]

Silverstein, S. M., B. Moghaddam, and T. Wykes, eds. 2013. Schizophrenia: Evolution and Synthesis. Strüngmann Forum Reports 13, vol. J. Lupp, series ed. Cambridge, MA: MIT Press. [2]

Simmons, A., S. C. Matthews, M. B. Stein, and M. P. Paulus. 2004. Anticipation of Emotionally Aversive Visual Stimuli Activates Right Insula. *Neuroreport* **15**:2261–2265. [12]

Simon, J. J., A. Biller, S. Walther, et al. 2009. Neural Correlates of Reward Processing in Schizophrenia: Relationship to Apathy and Depression. *Schizophr. Res.* **118**:154–161. [4]

Singer, T. 2008. Understanding Others: Brain Mechanisms of Theory of Mind and Empathy. In: Neuroeconomics: Decision Making and the Brain, ed. P. W. Glimcher et al., pp. 251–268. London: Academic Press/Elsevier. [2]

Singh, S. P., and V. Singh. 2011. Meta-Analysis of the Efficacy of Adjunctive NMDA Receptor Modulators in Chronic Schizophrenia. *CNS Drugs* **25**:859–885. [16]

Sinyor, M., A. Schaffer, and A. Levitt. 2010. The Sequenced Treatment Alternatives to Relieve Depression (Star*D) Trial: A Review. *Can. J. Psychiatry* **55**:126–135. [1]

Slifstein, M., E. van de Giessen, J. Van Snellenberg, et al. 2015. Deficits in Prefrontal Cortical and Extrastriatal Dopamine Release in Schizophrenia: A Positron Emission Tomographic Functional Magnetic Resonance Imaging Study. *JAMA Psychiatry* **72**:316–324. [16]

Smieskova, R., P. Fusar-Poli, P. Allen, et al. 2010. Neuroimaging Predictors of Transition to Psychosis: A Systematic Review and Meta-Analysis. *Neurosci. Biobehav. Rev.* **34**:1207–1222. [3, 10]

Smith, E. E., T. S. Eich, D. Cebenoyan, and C. Malapani. 2011. Intact and Impaired Cognitive-Control Processes in Schizophrenia. *Schizophr. Res.* **126**:132–137. [4]

Smith, M. A., J. Brandt, and R. Shadmehr. 2000. Motor disorder in Huntington's Disease Begins as a Dysfunction in Error Feedback Control. *Nature* **403**:544–549. [2]

Smith, M. E. 2005. Bilateral Hippocampal Volume Reduction in Adults with Post-Traumatic Stress Disorder: A Meta-Analysis of Structural MRI Studies. *Hippocampus* **15**:798–807. [1]

Smittenaar, P. 2015. Action Control in Uncertain Environments. Ph.D. Thesis, University College London. http://discovery.ucl.ac.uk/id/eprint/1467263 (accessed July 7, 2016). [6]

Smoller, J. W., and M. T. Tsuang. 1998. Panic and Phobic Anxiety: Defining Phenotypes for Genetic Studies. *Am. J. Psychiatry* **155**:1152–1162. [8]

Sobin, C., and H. A. Sackeim. 1997. Psychomotor Symptoms of Depression. *Am. J. Psychiatry* **154**:4–17. [14]

Sohal, V. S., F. Zhang, O. Yizhar, and K. Deisseroth. 2009. Parvalbumin Neurons and Gamma Rhythms Enhance Cortical Circuit Performance. *Nature* **459**:698–702. [4]

Soliman, F., C. E. Glatt, K. G. Bath, et al. 2010. A Genetic Variant BDNF Polymorphism Alters Extinction Learning in Both Mouse and Human. *Science* **327**:863–866. [3]

Soltesz, I., and K. Staley, eds. 2008. Comptuational Neuroscience in Epilepsy. London: Academic Press/Elsevier. [2]

Somlai, Z., A. A. Moustafa, S. Keri, C. E. Myers, and M. A. Gluck. 2011. General Functioning Predicts Reward and Punishment Learning in Schizophrenia. *Schizophr. Res.* **127**:131–136. [4]

Soubrie, P. 1986. Reconciling the Role of Central Serotonin Neurons in Human and Animal Behaviour. *Behav. Brain Sci.* **9**:319–364. [15]

Spellman, T., and J. Gordon. 2015. Synchrony in Schizophrenia: A Window into Circuit-Level Pathophysiology. *Curr. Opin. Neurobiol.* **30**:17–23. [1]

Spencer, K. M. 2011. Baseline Gamma Power During Auditory Steady-State Stimulation in Schizophrenia. *Front. Hum. Neurosci.* **5**:190. [16]

Spiga, F., L. R. Harrison, S. A. Wood, et al. 2008. Effect of the Glucocorticoid Receptor Antagonist Org 34850 on Fast and Delayed Feedback of Corticosterone Release. *J. Endocrinol.* **196**:323–330. [12]

Spitzer, R. L., J. Endicott, and E. Robins. 1975. Research Diagnostic Criteria (RDC), New York: Biometrics Research, New York State Psychiatric Institute. http://garfield.library.upenn.edu/classics1989/A1989U310000002.pdf (accessed Aug. 12, 2016). [9]

Squire, L. R. 2004. Memory Systems of the Brain: A Brief History and Current Perspective. *Neurobiol. Learn. Mem.* **82**:171–177. [10]

Stanislow, C. A., D. S. Pine, K. J. Quinn, et al. 2010. Developing Constructs for Psychopathology Research: Research Domain Criteria. *J. Abnorm. Psychol.* **119**:631–639. [9]

Stark, K. L., B. Xu, A. Bagchi, et al. 2008. Altered Brain Microrna Biogenesis Contributes to Phenotypic Deficits in a 22q11-Deletion Mouse Model. *Nat. Genet.* **40**:751–760. [5]

Stauffer, V. L., B. A. Millen, S. Andersen, et al. 2013. Pomaglumetad Methionil: No Significant Difference as an Adjunctive Treatment for Patients with Prominent Negative Symptoms of Schizophrenia Compared to Placebo. *Schizophr. Res.* **150**:434–441. [16]

St Clair, D., D. Blackwood, W. Muir, et al. 1990. Association within a Family of a Balanced Autosomal Translocation with Major Mental Illness. *Lancet* **336**:13–16. [3]

Stefansson, H., R. A. Ophoff, S. Steinberg, et al. 2009. Common Variants Conferring Risk of Schizophrenia. *Nature* **460**:744–747. [3]

Stephan, K. E., T. Baldeweg, and K. J. Friston. 2006. Synaptic Plasticity and Dysconnection in Schizophrenia. *Biol. Psychiatry* **59**:929–939. [12]

Stephan, K. E., K. J. Friston, and C. D. Frith. 2009a. Dysconnection in Schizophrenia: From Abnormal Synaptic Plasticity to Failures of Self-Monitoring. *Schizophr. Bull.* **35**:509–527. [12]

Stephan, K. E., S. Iglesias, J. Heinzle, and A. O. Diaconescu. 2015. Translational Perspectives for Computational Neuroimaging. *Neuron* **87**:716–732. [12]

Stephan, K. E., and C. Mathys. 2014. Computational Approaches to Psychiatry. *Curr. Opin. Neurobiol.* **25**:85–92. [6, 11, 12]

Stephan, K. E., W. D. Penny, J. Daunizeau, R. J. Moran, and K. J. Friston. 2009b. Bayesian Model Selection for Group Studies. *NeuroImage* **46**:1004–1017. [11]

Sterling, P. 2014. Homeostasis vs. Allostasis: Implications for Brain Function and Mental Disorders. *JAMA Psychiatry* **71**:1192–1193. [12]

Stewart, S. E., D. Yu, J. M. Scharf, et al. 2013. Genome-Wide Association Study of Obsessive-Compulsive Disorder. *Mol. Psychiatry* **18**:788–798. [10]

Stilo, S. A., and R. M. Murray. 2010. The Epidemiology of Schizophrenia: Replacing Dogma with Knowledge. *Dialogues Clin. Neurosci.* **12**:305–315. [4]

Stokes, C. C., C. M. Teeter, and J. S. Isaacson. 2014. Single Dendrite-Targeting Interneurons Generate Branch-Specific Inhibition. *Front. Neural Circuits* **8**:139. [16]

Stone, J. M., F. Day, H. Tsagaraki, et al. 2009. Glutamate Dysfunction in People with Prodromal Symptoms of Psychosis: Relationship to Gray Matter Volume. *Biol. Psychiatry* **66**:533–539. [16]

Stone, J. M., O. D. Howes, A. Egerton, et al. 2010. Altered Relationship between Hippocampal Glutamate Levels and Striatal Dopamine Function in Subjects at Ultra High Risk of Psychosis. *Biol. Psychiatry* **68**:599–602. [4]

Strauss, G. P., M. J. Frank, J. A. Waltz, et al. 2011. Deficits in Positive Reinforcement Learning and Uncertainty-Driven Exploration Are Associated with Distinct Aspects of Negative Symptoms in Schizophrenia. *Biol. Psychiatry* **69**:424–431. [3, 6]

Strobl, C., J. Malley, and G. Tutz. 2009. An Introduction to Recursive Partitioning: Rationale, Application, and Characteristics of Classification and Regression Trees, Bagging, and Random Forests. *Psychol. Methods* **14**:323–348. [14]

Sturgill, J. F., and J. S. Isaacson. 2015. Somatostatin Cells Regulate Sensory Response Fidelity via Subtractive Inhibition in Olfactory Cortex. *Nat. Neurosci.* **18**:531–535. [16]

Stutzmann, G. E., B. S. McEwen, and J. E. LeDoux. 1998. Serotonin Modulation of Sensory Inputs to the Lateral Amygdala: Dependency on Corticosterone. *J. Neurosci.* **18**:9529–9538. [12]

Substance Abuse and Mental Health Services Administration. 2014. Results from the 2013 National Survey on Drug Use and Health: Mental Health Findings. NSDUH Series H-49, HHS Publ. No. (SMA) 14-4887. Rockville, MD: Substance Abuse and Mental Health Services Administration. [3]

Sui, J., T. Adali, Q. Yu, J. Chen, and V. D. Calhoun. 2012. A Review of Multivariate Methods for Multimodal Fusion of Brain Imaging Data. *J. Neurosci. Methods* **204**:68–81. [5]

Sui, J., G. Pearlson, A. Caprihan, et al. 2011. Discriminating Schizophrenia and Bipolar Disorder by Fusing fMRI and DTI in a Multimodal CCA+ Joint ICA Model. *NeuroImage* **57**:839–855. [16]

Sullivan, P. F., K. S. Kendler, and M. C. Neale. 2003. Schizophrenia as a Complex Trait: Evidence from a Meta-Analysis of Twin Studies. *Arch. Gen. Psychiatry* **60**:1187–1192. [3]

Sun, C., M.-C. Cheng, R. Qin, et al. 2011. Identification and Functional Characterization of Rare Mutations of the Neuroligin-2 Gene (NLGN2) Associated with Schizophrenia. *Hum. Mol. Genet.* **20**:3042–3051. [3]

Sussmann, J. E., G. Lymer, J. McKirdy, et al. 2009. White Matter Abnormalities in Bipolar Disorder and Schizophrenia Detected Using Diffusion Tensor Magnetic Resonance Imaging. *Bipolar Disorders* **11**:11–18. [3]

Sutton, R. S. 1988. Learning to Predict by the Methods of Temporal Difference. *Mach. Learn.* **3**:9–44. [13]

Sutton, R. S., and A. G. Barto. 1981. Toward a Modern Theory of Adaptive Networks: Expectation and Prediction. *Psychol. Rev.* **88**:135–170. [13]

———. 1987. A Temporal-Difference Model of Classical Conditioning. Proc. 9th Annual Conf. of the Cognitive Science Society. Hillsdale, NJ: Lawrence Erlbaum Associates. [13]

———. 1998. Reinforcement Learning. Cambridge, MA: MIT Press. [5, 12, 13, 15]

Swedo, S. E., M. B. Schapiro, C. L. Grady, et al. 1989. Cerebral Glucose Metabolism in Childhood-Onset Obsessive-Compulsive Disorder. *Arch. Gen. Psychiatry* **46**:518–523. [10]

Szymanski, S., J. A. Lieberman, J. M. Alvir, et al. 1995. Gender Differences in Onset of Illness, Treatment Response, Course, and Biologic Indexes in First-Episode Schizophrenic Patients. *Am. J. Psychiatry* **152**:698–703. [16]

Tabibnia, G., J. R. Monterosso, K. Baicy, et al. 2011. Different Forms of Self-Control Share a Neurocognitive Substrate. *J. Neurosci.* **31**:4805–4810. [14]

Tanaka, S. 2006. Dopaminergic Control of Working Memory and Its Relevance to Schizophrenia: A Circuit Dynamics Perspective. *Neuroscience* **139**:153–171. [2]

Tandon, R., M. S. Keshavan, and H. A. Nasrallah. 2008. Schizophrenia, "Just the Facts" What We Know in 2008. 2. Epidemiology and Etiology. *Schizophr. Res.* **102**:1–18. [4]

Tang, S. W., and D. Helmeste. 2008. Paroxetine. *Expert Opin. Pharmacother.* **9**:787–794. [8]

Tarur Padinjareveettil, A. M., J. Rogers, C. Loo, and D. Martin. 2015. Transcranial Direct Current Stimulation to Enhance Cognitive Remediation in Schizophrenia. *Brain Stimul.* **8**:307–309. [16]

Tatard-Leitman, V. M., C. R. Jutzeler, J. Suh, et al. 2015. Pyramidal Cell Selective Ablation of N-Methyl-D-Aspartate Receptor 1 Causes Increase in Cellular and Network Excitability. *Biol. Psychiatry* **77**:556–568. [16]

Taylor, S. F., and I. F. Tso. 2015. GABA Abnormalities in Schizophrenia: A Methodological Review of *in Vivo* Studies. *Schizophr. Res.* **167**:84–90. [4]

Teasdale, J. D. 1988. Cognitive Vulnerability to Persistent Depression. *Cogn. Emot.* **2**: 247–274. [15]

Teasdale, J. D., R. G. Moore, H. Hayhurst, et al. 2002. Metacognitive Awareness and Prevention of Relapse in Depression: Empirical Evidence. *J. Consult. Clin. Psychol.* **70**:275–287. [15]

Tebbenkamp, A. T., A. J. Willsey, M. W. State, and N. Sestan. 2014. The Developmental Transcriptome of the Human Brain: Implications for Neurodevelopmental Disorders. *Curr. Opin. Neurol.* **27**:149–156. [16]

Thaker, G. K., D. E. Ross, R. W. Buchanan, H. M. Adami, and D. R. Medoff. 1999. Smooth Pursuit Eye Movements to Extra-Retinal Motion Signals: Deficits in Patients with Schizophrenia. *Psychiatry Res.* **88**:209–219. [7]

Thase, M. E. 2013. The Multifactorial Presentation of Depression in Acute Care. *J. Clin. Psychiatry* **74**:3–8. [5]

Thiagarajan, T. C., M. A. Lebedev, M. A. Nicolelis, and D. Plenz. 2010. Coherence Potentials: Loss-Less, All-or-None Network Events in the Cortex. *PLoS Biol.* **8**: e1000278. [3]

Thomas, A. J., S. Davis, C. Morris, et al. 2014. Increase in Interleukin-1β in Late-Life Depression. *Am. J. Psychiatry* **162**:175–177. [12]

Thompson, A., B. Nelson, and A. Yung. 2011. Predictive Validity of Clinical Variables in the "at Risk" for Psychosis Population: International Comparison with Results from the North American Prodrome Longitudinal Study. *Schizophr. Res.* **126**:51–57. [10]

Thompson, P. M., C. Vidal, J. N. Giedd, et al. 2001. Mapping Adolescent Brain Change Reveals Dynamic Wave of Accelerated Gray Matter Loss in Very Early-Onset Schizophrenia. *PNAS* **98**:11650–11655. [16]

Tien, A. Y., and W. W. Eaton. 1992. Psychopathologic Precursors and Sociodemographic Risk Factors for the Schizophrenia Syndrome. *Arch. Gen. Psychiatry* **49**:37–46. [9]

Timms, A. E., M. O. Dorschner, J. Wechsler, et al. 2013. Support for the N-Methyl-D-Aspartate Receptor Hypofunction Hypothesis of Schizophrenia from Exome Sequencing in Multiplex Families. *JAMA Psychiatry* **70**:582–590. [16]

Tobler, P. N., C. D. Fiorillo, and W. Schultz. 2005. Adaptive Coding of Reward Value by Dopamine Neurons. *Science* **307**:1642–1645. [13]

Tocchetto, A., G. A. Salum, C. Blaya, et al. 2011. Evidence of Association between Val66Met Polymorphism at BDNF Gene and Anxiety Disorders in a Community Sample of Children and Adolescents. *Neurosci. Lett.* **502**:197–200. [3]

Todd, M., Y. Niv, and J. D. Cohen. 2009. Learning to Use Working Memory in Partially Observable Environments through Dopaminergic Reinforcement. In: Advances in Neural Information Processing Systems 21, ed. D. Koller et al., pp. 1689–1696. Cambridge, MA: MIT Press. [6]

Todorov, E., and M. I. Jordan. 2002. Optimal Feedback Control as a Theory of Motor Coordination. *Nat. Neurosci.* **5**:1226–1235. [14]

Tononi, G. 2004. An Information Integration Theory of Consciousness. *BMC Neurosci.* **5**:42. [3]

Tougaard, J. 2002. Signal Detection Theory, Detectability and Stochastic Resonance Effects. *Biol. Cybern.* **87**:79–90. [2]

Tozzi, F., L. M. Thornton, K. L. Klump, et al. 2005. Symptom Fluctuation in Eating Disorders: Correlates of Diagnostic Crossover. *Am. J. Psychiatry* **162**:732–740. [3]

Treadway, M. T., J. W. Buckholtz, R. L. Cowan, et al. 2012. Dopaminergic Mechanisms of Individual Differences in Human Effort-Based Decision-Making. *J. Neurosci.* **32**:6170–6176. [4]

Treadway, M. T., J. W. Buckholtz, A. N. Schwartzman, W. E. Lambert, and D. H. Zald. 2009. Worth the "EEfRT"? The Effort Expenditure for Rewards Task as an Objective Measure of Motivation and Anhedonia. *PLoS One* **4**:e6598. [4]

Treadway, M. T., J. S. Peterman, D. H. Zald, and S. Park. 2015. Impaired Effort Allocation in Patients with Schizophrenia. *Schizophr. Res.* **161**:382–385. [4]

Treadway, M. T., and D. H. Zald. 2011. Reconsidering Anhedonia in Depression: Lessons from Translational Neuroscience. *Neurosci. Biobehav. Rev.* **35**:537–555. [14]

Trifilieff, P., B. Feng, E. Urizar, et al. 2013. Increasing Dopamine D2 Receptor Expression in the Adult Nucleus Accumbens Enhances Motivation. *Mol. Psychiatry* **18**:1025–1033. [4]

Tu, P. C., Y. C. Lee, Y. S. Chen, et al. 2015. Network-Specific Cortico-Thalamic Dysconnection in Schizophrenia Revealed by Intrinsic Functional Connectivity Analyses. *Schizophr. Res.* **166**:137–143. [16]

Tundo, A., J. R. Calabrese, L. Proietti, and R. de Fillippis. 2015. Variation in Response to Short-Term Antidepressant Treatment between Patients with Continuous and Non-Continuous Cycling Bipolar Disorders. *J. Affect. Disord.* **174**:126–130. [10]

Tuominen, H. J., J. Tiihonen, and K. Wahlbeck. 2005. Glutamatergic Drugs for Schizophrenia: A Systematic Review and Meta-Analysis. *Schizophr. Res.* **72**:225–234. [16]

———. 2006. Glutamatergic Drugs for Schizophrenia. *Cochrane Database Syst. Rev.* Cd003730. [16]

Turner, D. C., L. Clark, E. Pomarol-Clotet, et al. 2004. Modafinil Improves Cognition and Attentional Set Shifting in Patients with Chronic Schizophrenia. *Neuropsychopharmacology* **29**:1363–1373. [4]

Turrigiano, G., L. F. Abbott, and E. Marder. 1994. Activity-Dependent Changes in the Intrinsic Properties of Cultured Neurons. *Science* **264**:974–977. [16]

Tye, K. M., and K. Deisseroth. 2012. Optogenetic Investigation of Neural Circuits Underlying Brain Disease in Animal Models. *Nat. Rev. Neurosci.* **13**:251–266. [2]

Tye, K. M., J. J. Mirzabekov, M. R. Warden, et al. 2013. Dopamine Neurons Modulate Neural Encoding and Expression of Depression-Related Behaviour. *Nature* **493**:537–541. [1]

Tye, K. M., R. Prakash, S. Y. Kim, et al. 2011. Amygdala Circuitry Mediating Reversible and Bidirectional Control of Anxiety. *Nature* **471**:358–362. [1]

Tyson, P. J., K. R. Laws, K. H. Roberts, and A. M. Mortimer. 2004. Stability of Set-Shifting and Planning Abilities in Patients with Schizophrenia. *Psychiatry Res.* **129**:229–239. [4]

Ullman, S., and T. Poggio. 2010. Vision: A Computational Investigation into the Human Representation and Processing of Visual Information. Cambridge MA: MIT Press. [12]

Umbricht, D., and S. Krljes. 2005. Mismatch Negativity in Schizophrenia: A Meta-Analysis. *Schizophr. Res.* **76**:1–23. [12]

Umbricht, D., L. Schmid, R. Koller, et al. 2000. Ketamine-Induced Deficits in Auditory and Visual Context-Dependent Processing in Healthy Volunteers: Implications for Models of Cognitive Deficits in Schizophrenia. *Arch. Gen. Psychiatry* **57**:1139–1147. [12]

Urban, N. B. L., L. S. Kegeles, M. Slifstein, et al. 2010. Sex Differences in Striatal Dopamine Release in Young Adults after Oral Alcohol Challenge: A Positron Emission Tomography Imaging Study with [^{11}C]Raclopride. *Biol. Psychiatry* **68**: 689–696. [3]

USA Suicide. 2013. USA Suicide: 2013 Official Final Data. http://www.suicidology.org/Portals/14/docs/Resources/FactSheets/2013datapgsv3.pdf (accessed Feb. 29, 2016). [3]

Van den Bergh, B. R. H., and A. Marcoen. 2004. High Antenatal Maternal Anxiety Is Related to ADHD Symptoms, Externalizing Problems, and Anxiety in 8- and 9-Year-Olds. *Child Dev.* **75**:1085–1097. [3]

van der Gaag, M., D. H. Nieman, J. Rietdijk, et al. 2012. Cognitive Behavioral Therapy for Subjects at Ultrahigh Risk for Developing Psychosis: A Randomized Controlled Clinical Trial. *Schizophr. Bull.* **38**:1180–1188. [3]

van der Gaag, M., F. Smit, A. Bechdolf, et al. 2013. Preventing a First Episode of Psychosis: Meta-Analysis of Randomized Controlled Prevention Trials of 12 Month and Longer-Term Follow-Ups. *Schizophr. Res.* **149**:56–62. [10, 16]

van der Meer, F. J., E. Velthorst, C. J. Meijer, M. W. Machielsen, and L. de Haan. 2012a. Cannabis Use in Patients at Clinical High Risk of Psychosis: Impact on Prodromal Symptoms and Transition to Psychosis. *Curr. Pharm. Des.* **18**:5036–5044. [16]

van der Meer, M. A. A., A. Johnson, N. C. Schmitzer-Torbert, and A. D. Redish. 2010. Triple Dissociation of Information Processing in Dorsal Striatum, Ventral Striatum, and Hippocampus on a Learned Spatial Decision Task. *Neuron* **67**:25–32. [2, 6]

van der Meer, M. A. A., Z. Kurth-Nelson, and A. D. Redish. 2012b. Information Processing in Decision-Making Systems. *Neuroscientist* **18**:342–359. [2]

Van Os, J., C. Gilvarry, R. Bale, et al. 1999. A Comparison of the Utility of Dimensional and Categorical Representations of Psychosis. *Psychol. Med.* **29**:595–606. [9, 10]

Van Os, J., B. P. Rutten, and R. Poulton. 2008. Gene-Environment Interactions in Schizophrenia: Review of Epidemiological Findings and Future Directions. *Schizophr. Bull.* **34**:1066–1082. [4]

Vassos, E., C. B. Pedersen, R. M. Murray, D. A. Collier, and C. M. Lewis. 2012. Meta-Analysis of the Association of Urbanicity with Schizophrenia. *Schizophr. Bull.* **38**:1118–1123. [4]

Vaswani, M., F. K. Linda, and S. Ramesh. 2003. Role of Selective Serotonin Reuptake Inhibitors in Psychiatric Disorders: A Comprehensive Review. *Prog. Neuropsychopharmacol. Biol. Psychiatry* **27**:85–102. [8]

Velakoulis, D., S. J. Wood, M. T. H. Wong, et al. 2006. Hippocampal and Amygdala Volumes According to Psychosis Stage and Diagnosis: A Magnetic Resonance Imaging Study of Chronic Schizophrenia, First-Episode Psychosis, and Ultra-High-Risk Individuals. *Arch. Gen. Psychiatry* **63**:139–149. [3]

Verhagen, M., A. van der Meij, P. A. M. van Deurzen, et al. 2010. Meta-Analysis of the BDNF Val66Met Polymorphism in Major Depressive Disorder: Effects of Gender and Ethnicity. *Mol. Psychiatry* **15**:260–271. [3]

Viceconti, M., P. Hunter, and R. Hose. 2015. Big Data, Big Knowledge: Big Data for Personalized Healthcare. *IEEE J. Biomed. Health Inform.* **19**:1209–1215. [11]

Vidal, C. N., J. L. Rapoport, K. M. Hayashi, et al. 2006. Dynamically Spreading Frontal and Cingulate Deficits Mapped in Adolescents with Schizophrenia. *Arch. Gen. Psychiatry* **63**:25–34. [16]

Videbech, P., and B. Ravnkilde. 2004. Hippocampal Volume and Depression: A Meta-Analysis of MRI Studies. *Am. J. Psychiatry* **161**:1957–1966. [1]

Viguera, A. C., R. J. Baldessarini, and J. Friedberg. 1998. Discontinuing Antidepressant Treatment in Major Depression. *Harv. Rev. Psychiatry* **5**:293–306. [15]

Vijayraghavan, S., M. Wang, S. G. Birnbaum, G. V. Williams, and A. F. Arnsten. 2007. Inverted-U Dopamine D1 Receptor Actions on Prefrontal Neurons Engaged in Working Memory. *Nat. Neurosci.* **10**:376–384. [4]

Vita, A., and L. de Peri. 2007. Hippocampal and Amygdala Volume Reductions in First-Episode Schizophrenia. *Br. J. Psychiatry* **190**:271. [4]

Vita, A., L. de Peri, G. Deste, and E. Sacchetti. 2012. Progressive Loss of Cortical Gray Matter in Schizophrenia: A Meta-Analysis and Meta-Regression of Longitudinal MRI Studies. *Transl. Psychiatry* **2**:e190. [4]

Vittengl, J. R., L. A. Clark, T. W. Dunn, and R. B. Jarrett. 2007. Reducing Relapse and Recurrence in Unipolar Depression: A Comparative Meta-Analysis of Cognitive-Behavioral Therapy's Effects. *J. Consult. Clin. Psychol.* **75**:475. [15]

Volk, D. W., and D. A. Lewis. 2013. Prenatal Ontogeny as a Susceptibility Period for Cortical GABA Neuron Disturbances in Schizophrenia. *Neuroscience* **248**:154–164. [16]

———. 2014. Early Developmental Disturbances of Cortical Inhibitory Neurons: Contribution to Cognitive Deficits in Schizophrenia. *Schizophr. Bull.* **40**:952–957. [3, 16]

———. 2015. The Role of Endocannabinoid Signaling in Cortical Inhibitory Neuron Dysfunction in Schizophrenia. *Biol. Psychiatry* [16]

Volk, D. W., B. I. Siegel, C. D. Verrico, and D. A. Lewis. 2013. Endocannabinoid Metabolism in the Prefrontal Cortex in Schizophrenia. *Schizophr. Res.* **147**:53–57. [16]

Voon, V., K. Derbyshire, C. Rück, et al. 2015. Disorders of Compulsivity: A Common Bias Towards Learning Habits. *Mol. Psychiatry* **20**:345–352. [6, 9]

Voon, V., A. R. Mehta, and M. Hallett. 2011. Impulse Control Disorders in Parkinson's Disease: Recent Advances. *Curr. Opin. Neurol.* **24**:324. [12]

Vossel, S., M. Bauer, C. Mathys, et al. 2014. Cholinergic Stimulation Enhances Bayesian Belief Updating in the Deployment of Spatial Attention. *J. Neurosci.* **34**:15735–15742. [7]

Wagenaar, W. A. 1988. Paradoxes of Gambling Behavior. Hillsdale: Erlbaum. [2]

Wagstaff, A. J., S. M. Cheer, A. J. Matheson, D. Ormrod, and K. L. Goa. 2002. Spotlight on Paroxetine in Psychiatric Disorders in Adults. *CNS Drugs* **16**:425–434. [10]

Wakefield, J. C. 1992a. The Concept of Mental Disorder: on the Boundary between Biological Facts and Social Values. *Am. Psychol.* **47**:373. [2, 10]

———. 1992b. Disorder as Harmful Dysfunction: A Conceptual Critique of DSM-III-R's Definition of Mental Disorder. *Psychol. Rev.* **99**:232–247. [10]

———. 2007. The Concept of Mental Disorder: Diagnostic Implications of the Harmful Dysfunction Analysis. *World Psychiatry* **6**:149. [2, 10]

Walsh, T., J. M. McClellan, S. E. McCarthy, et al. 2008. Rare Structural Variants Disrupt Multiple Genes in Neurodevelopmental Pathways in Schizophrenia. *Science* **320**:539–543. [3, 16]

Walter, H., H. Kammerer, K. Frasch, M. Spitzer, and B. Abler. 2009. Altered Reward Functions in Patients on Atypical Antipsychotic Medication in Line with the revised Dopamine Hypothesis of Schizophrenia. *Psychopharmacology (Berl.)* **206**:121–132. [4]

Waltz, J. A., M. J. Frank, B. M. Robinson, and J. M. Gold. 2007. Selective Reinforcement Learning Deficits in Schizophrenia Support Predictions from Computational Models of Striatal-Cortical Dysfunction. *Biol. Psychiatry* **62**:756–764. [4]

Waltz, J. A., and J. M. Gold. 2007. Probabilistic Reversal Learning Impairments in Schizophrenia: Further Evidence of Orbitofrontal Dysfunction. *Schizophr. Res.* **93**:296–303. [4]

Waltz, J. A., J. B. Schweitzer, J. M. Gold, et al. 2009. Patients with Schizophrenia Have a Reduced Neural Response to Both Unpredictable and Predictable Primary Reinforcers. *Neuropsychopharmacology* **34**:1567–1577. [4]

Waltz, J. A., J. B. Schweitzer, T. J. Ross, et al. 2010. Abnormal Responses to Monetary Outcomes in Cortex, but Not in the Basal Ganglia, in Schizophrenia. *Neuropsychopharmacology* **35**:2427–2439. [4]

Wang, K., H. Zhang, D. Ma, et al. 2009. Common Genetic Variants on 5p14.1 Associate with Autism Spectrum Disorders. *Nature* **459**:528–533. [3]

Wang, M., S. Vijayraghavan, and P. S. Goldman-Rakic. 2004. Selective D2 Receptor Actions on the Functional Circuitry of Working Memory. *Science* **303**:853–856. [4]

Wang, M., Y. Yang, C. J. Wang, et al. 2013. NMDA Receptors Subserve Persistent Neuronal Firing During Working Memory in Dorsolateral Prefrontal Cortex. *Neuron* **77**:736–749. [16]

Wang, X. J. 1999. Synaptic Basis of Cortical Persistent Activity: The Importance of NMDA Receptors to Working Memory. *J. Neurosci.* **19**:9587–9603. [4]

———. 2001. Synaptic Reverberation Underlying Mnemonic Persistent Activity. *Trends Neurosci.* **24**:455–463. [6]

———. 2012. Neural Dynamics and Circuit Mechanisms of Decision-Making. *Curr. Opin. Neurobiol.* **22**:1039–1046. [6]

Wang, X. J., and J. H. Krystal. 2014. Computational Psychiatry. *Neuron* **84**:638–654. [6, 11, 16]

Ward, K. E., L. Friedman, A. Wise, and S. C. Schulz. 1996. Meta-Analysis of Brain and Cranial Size in Schizophrenia. *Schizophr. Res.* **22**:197–213. [3]

Watson, D. 2005. Rethinking the Mood and Anxiety Disorders: A Quantitative Hierarchical Model for DSM-V. *J. Abnorm. Psychol.* **114**:522–536. [8]

Weickert, T. W., T. E. Goldberg, J. H. Callicott, et al. 2009. Neural Correlates of Probabilistic Category Learning in Patients with Schizophrenia. *J. Neurosci.* **29**:1244–1254. [4]

Weickert, T. W., A. Terrazas, L. B. Bigelow, et al. 2002. Habit and Skill Learning in Schizophrenia: Evidence of Normal Striatal Processing with Abnormal Cortical Input. *Learn. Mem.* **9**:430–442. [4]

Weidenfeld, J., M. E. Newman, A. Itzik, and S. Feldman. 2005. Adrenocortical Axis Responses to Adrenergic and Glutamate Stimulation Are Regulated by the Amygdala. *Neuroreport* **16**:1245–1249. [12]

Weiler, J. A., C. Bellebaum, M. Brune, G. Juckel, and I. Daum. 2009. Impairment of Probabilistic Reward-Based Learning in Schizophrenia. *Neuropsychology* **23**:571–580. [4]

Weinberger, D. R. 1987. Implications of Normal Brain Development for the Pathogenesis of Schizophrenia. *Arch. Gen. Psychiatry* **44**:660–669. [3, 16]

Weinberger, D. R., and K. F. Berman. 1996. Prefrontal Function in Schizophrenia: Confounds and Controversies. *Phil. Trans. R. Soc. B* **351**:1495–1503. [3]

Wells, A., P. Fisher, S. Myers, et al. 2012. Metacognitive Therapy in Treatment-Resistant Depression: A Platform Trial. *Behav. Res. Ther.* **50**:367–373. [15]

Wells, T. T., E. M. Clerkin, A. J. Ellis, and C. G. Beevers. 2014. Effect of Antidepressant Medication Use on Emotional Information Processing in Major Depression. *Am. J. Psychiatry* **171**:195–200. [15]

Whelan, R., R. Watts, C. A. Orr, et al. 2014. Neuropsychosocial Profiles of Current and Future Adolescent Alcohol Misusers. *Nature* **512**:185–189. [3, 14]

Whiteford, H. A., L. Degenhardt, J. Rehm, et al. 2013. Global Burden of Disease Attributable to Mental and Substance Use Disorders: Findings from the Global Burden of Disease Study 2010. *Lancet* **382**:1575–1586. [15]

Whitton, A. E., M. T. Treadway, and D. A. Pizzagalli. 2015. Reward Processing Dysfunction in Major Depression, Bipolar Disorder and Schizophrenia. *Curr. Opin. Psychiatry* **28**:7–12. [3, 12]

WHO. 2012. Measuring Health and Disability: Manual for WHO Disability Assessment Schedule (WHODAS 2.0). Geneva: World Health Organization. [8]

Wichers, M., I. Myin-Germeys, N. Jacobs, et al. 2007. Genetic Risk of Depression and Stress-Induced Negative Affect in Daily Life. *Br. J. Psychiatry* **191**:218–223. [15]

Widiger, T. A., and D. B. Samuel. 2005. Diagnostic Categories or Dimensions: A Question for the Diagnostic and Statistical Manual of Mental Disorders, 5th edition. *J. Abnorm. Psychol.* **114**:494–504. [8]

Wiecki, T. V., J. Poland, and M. J. Frank. 2015. Model-Based Cognitive Neuroscience Approaches to Computational Psychiatry: Clustering and Classification. *Clin. Psychol. Sci.* **3**:378–399. [3, 5, 6, 12, 15, 16]

Wieland, S., S. Schindler, C. Huber, et al. 2015. Phasic Dopamine Modifies Sensory-Driven Output of Striatal Neurons through Synaptic Plasticity. *J. Neurosci.* **35**:9946–9956. [11]

Wilkinson, S. T., R. Radhakrishnan, and D. C. D'Souza. 2014. Impact of Cannabis Use on the Development of Psychotic Disorders. *Curr. Addict. Rep.* **1**:115–128. [16]

Williams, H. J., N. Craddock, G. Russo, et al. 2011. Most Genome-Wide Significant Susceptibility Loci for Schizophrenia and Bipolar Disorder Reported to Date Cross-Traditional Diagnostic Boundaries. *Hum. Mol. Genet.* **20**:387–391. [1]

Wilson, R. C., A. Geana, J. M. White, E. A. Ludvig, and J. D. Cohen. 2014. Humans Use Directed and Random Exploration to Solve the Explore-Exploit Dilemma. *J. Exp. Psychol. Gen.* **143**:2074–2081. [6]

Winterer, G., and D. R. Weinberger. 2004. Genes, Dopamine and Cortical Signal-to-Noise Ratio in Schizophrenia. *Trends Neurosci.* **27**:683–690. [12]

Wittenborn, J. R., J. D. Holzberg, and B. Simon. 1953. Symptom Correlates for Descriptive Diagnosis. *Genet. Psychol. Monogr.* **47**:237–302. [8]

Wobrock, T., B. Guse, J. Cordes, et al. 2015. Left Prefrontal High-Frequency Repetitive Transcranial Magnetic Stimulation for the Treatment of Schizophrenia with Predominant Negative Symptoms: A Sham-Controlled, Randomized Multicenter Trial. *Biol. Psychiatry* **77**:979–988. [16]

Wolf, D. H., T. D. Satterthwaite, J. J. Kantrowitz, et al. 2014. Amotivation in Schizophrenia: Integrated Assessment with Behavioral, Clinical, and Imaging Measures. *Schizophr. Bull.* **40**:1328–1337. [4]

Wolkowitz, O. M. 1993. Rational Polypharmacy in Schizophrenia. *Ann. Clin. Psychiatry* **5**:79–90. [16]

Wolwer, W., A. Lowe, J. Brinkmeyer, et al. 2014. Repetitive Transcranial Magnetic Stimulation (rTMS) Improves Facial Affect Recognition in Schizophrenia. *Brain Stimul.* **7**:559–563. [16]

Wong, K. F., and X. J. Wang. 2006. A Recurrent Network Mechanism of Time Integration in Perceptual Decisions. *J. Neurosci.* **26**:1314–1328. [16]

Woodruff, A. R., L. M. McGarry, T. P. Vogels, et al. 2011. State-Dependent Function of Neocortical Chandelier Cells. *J. Neurosci.* **31**:17872–17886. [4]

Woods, B. T., D. Yurgelun-Todd, F. M. Benes, et al. 1990. Progressive Ventricular Enlargement in Schizophrenia: Comparison to Bipolar Affective Disorder and Correlation with Clinical Course. *Biol. Psychiatry* **27**:341–352. [16]

Woods, S. W., B. C. Walsh, K. A. Hawkins, et al. 2013. Glycine Treatment of the Risk Syndrome for Psychosis: Report of Two Pilot Studies. *Eur. Neuropsychopharmacology* **23**:931–940. [16]

Woodward, N. D., H. Karbasforoushan, and S. Heckers. 2012. Thalamocortical Dysconnectivity in Schizophrenia. *Am. J. Psychiatry* **169**:1092–1099. [16]

Woolley, C. S., and B. S. McEwen. 1994. Estradiol Regulates Hippocampal Dendritic Spine Density via an N-Methyl-D-Aspartate Receptor-Dependent Mechanism. *J. Neurosci.* **14**:7680–7687. [16]

Wright, A. G., R. F. Krueger, M. J. Hobbs, et al. 2013. The Structure of Psychopathology: Toward an Expanded Quantitative Empirical Model. *J. Abnorm. Psychol.* **122**:281–294. [5]

Wright, A. G., K. M. Thomas, C. J. Hopwood, et al. 2012. The Hierarchical Structure of DSM-5 Pathological Personality Traits. *J. Abnorm. Psychol.* **121**:951–957. [5]

Wright, I. C., S. Rabe-Hesketh, P. W. Woodruff, et al. 2000. Meta-Analysis of Regional Brain Volumes in Schizophrenia. *Am. J. Psychiatry* **157**:16–25. [3]

Wu, K., G. L. Hanna, D. R. Rosenberg, and P. D. Arnold. 2012. The Role of Glutamate Signaling in the Pathogenesis and Treatment of Obsessive-Compulsive Disorder. *Pharmacol. Biochem. Behav.* **100**:726–735. [10]

Wunderlich, K., P. Dayan, and R. J. Dolan. 2012a. Mapping Value Based Planning and Extensively Trained Choice in the Human Brain. *Nat. Neurosci.* **15**:786–791. [10]

Wunderlich, K., P. Smittenaar, and R. J. Dolan. 2012b. Dopamine Enhances Model-Based over Model-Free Choice Behavior. *Neuron* **75**:418–424. [12]

Wylie, G. R., E. A. Clark, P. D. Butler, and D. C. Javitt. 2010. Schizophrenia Patients Show Task Switching Deficits Consistent with N-Methyl-D-Aspartate System Dysfunction but Not Global Executive Deficits: Implications for Pathophysiology of Executive Dysfunction in Schizophrenia. *Schizophr. Bull.* **36**:585–594. [4]

Xu, B., J. L. Roos, S. Levy, et al. 2008. Strong Association of *de Novo* Copy Number Mutations with Sporadic Schizophrenia. *Nat. Genet.* **40**:880–885. [3]

Yamins, D. L. K., H. Kong, C. F. Cadieu, et al. 2014. Performance-Optimized Hierarchical Models Predict Neural Responses in Higher Visual Cortex. *PNAS* **111**:8619–8624. [5]

Yang, G. J., J. D. Murray, G. Repovs, et al. 2014. Altered Global Brain Signal in Schizophrenia. *PNAS* **111**:7438–7443. [16]

Yasumoto, S., E. Tanaka, G. Hattori, H. Maeda, and H. Higashi. 2002. Direct and Indirect Actions of Dopamine on the Membrane Potential in Medium Spiny Neurons of the Mouse Neostriatum. *J. Neurophysiol.* **87**:1234–1243. [4]

Yilmaz, A., F. Simsek, and A. S. Gonul. 2012. Reduced Reward-Related Probability Learning in Schizophrenia Patients. *Neuropsychiatr. Dis. Treat.* **8**:27–34. [4]

Yizhar, O. 2012. Optogenetic Insights into Social Behavior Function. *Biol. Psychiatry* **71**:1075–1080. [1]

Young, P. C. 2002. Advances in Real-Time Flood Forecasting. *Phil. Trans. R. Soc. A* **360**: 1433–1450. [11]

Yousefi, S., A. Wein, K. C. Kowalski, A. G. Richardson, and L. Srinivasan. 2015. Smoothness as a Failure Mode of Bayesian Mixture Models in Brain-Machine Interfaces. *IEEE Trans. Neural Syst. Rehabil. Eng.* **23**:128–137. [9]

Yu, A. J., and J. D. Cohen. 2009. Sequential Effects: Superstition or Rational Behavior. *Adv. Neural Inform. Processing Syst.* **21**:1873–1880. [14]

Yu, A. J., and P. Dayan. 2005. Uncertainty, Neuromodulation, and Attention. *Neuron* **46**:681–692. [15]

Zemel, R. S., P. Dayan, and A. Pouget. 1998. Probabilistic Interpretation of Population Codes. *Neural Comput.* **10**:403–430. [5]

Zhang, F., L. Qiu, L. Yuan, et al. 2014. Evidence for Progressive Brain Abnormalities in Early Schizophrenia: A Cross-Sectional Structural and Functional Connectivity Study. *Schizophr. Res.* **159**:31–35. [16]

Zhu, X., A. C. Need, S. Petrovski, and D. B. Goldstein. 2014. One Gene, Many Neuro-psychiatric Disorders: Lessons from Mendelian Diseases. *Nature* **17**:773–781. [3]

Zwaigenbaum, L., S. Bryson, C. Lord, et al. 2009. Clinical Assessment and Management of Toddlers with Suspected Autism Spectrum Disorder: Insights from Studies of High-Risk Infants. *Pediatrics* **123**:1383–1391. [3]

Subject Index

Further Titles in the Strüngmann Forum Report Series[1]

---

[1]   available at https://mitpress.mit.edu/books/series/str%C3%BCngmann-forum-reports-0